Aspects of Microbial Metabolism and Ecology

Special Publications of the Society for General Microbiology

Publications Officer: Colin Ratledge, 62 London Road, Reading, UK.

1. Coryneform Bacteria,
    eds I. J. Bousfield and A. G. Callely
2. Adhesion of Microorganisms to Surfaces,
    eds D. C. Ellwood, J. Melling and P. Rutter
3. Microbial Polysaccharides and Polysaccharases,
    eds R. C. W. Berkeley, G. W. Gooday and D. C. Ellwood
4. The Aerobic Endospore-forming Bacteria: Classification and Identification,
    eds R. C. W. Berkeley and M. Goodfellow
5. Mixed Culture Fermentations,
    eds M. E. Bushell and J. H. Slater
6. Bioactive Microbial Products: Search and Discovery,
    eds J. D. Bu'Lock, L. J. Nisbet and D. J. Winstanley
7. Sediment Microbiology,
    eds D. B. Nedwell and C. M. Brown
8. Sourcebook of Experiments for the Teaching of Microbiology,
    eds S. B. Primrose and A. C. Wardlaw
9. Microbial Diseases of Fish,
    ed R. J. Roberts
10. Bioactive Microbial Products 2: Development and Production,
    eds L. J. Nisbet and D. J. Winstanley
11. Aspects of Microbial Metabolism and Ecology,
    ed G. A. Codd

This book is based on a Symposium of the S.G.M. held at the University of Dundee, Dundee, Scotland, 7–9 September 1982.

Aspects of Microbial Metabolism and Ecology

Edited by

G. A. Codd

Department of Biological Sciences
University of Dundee
Dundee, Scotland, United Kingdom

1984

Published for the
Society for General Microbiology
by
ACADEMIC PRESS
(Harcourt Brace Jovanovich, Publishers)
London   Orlando   San Diego   New York
Toronto   Montreal   Sydney   Tokyo

COPYRIGHT © 1984, BY ACADEMIC PRESS, INC. (LONDON) LTD.
ALL RIGHTS RESERVED.
NO PART OF THS PUBLICATION MAY BE REPRODUCED OR
TRANSMITTED IN ANY FORM OR BY ANY MEANS, ELECTRONIC
OR MECHANICAL, INCLUDING PHOTOCOPY, RECORDING, OR ANY
INFORMATION STORAGE AND RETRIEVAL SYSTEM, WITHOUT
PERMISSION IN WRITING FROM THE PUBLISHER.

ACADEMIC PRESS, INC. (LONDON) LTD.
24-28 Oval Road,
London NW1 7DX

United States Edition published by
ACADEMIC PRESS, INC.
Orlando, Florida 32887

British Library Cataloguing in Publication Data

Aspects of microbial metabolism and ecology.
----(Special publications of the Society for
General Microbiology)
1. Microbial ecology
I. Codd, G. A.   II. Society for General
Microbiology   III. Series
576'.15   QR100

ISBN 0-12-178050-3

Library of Congress Cataloging in Publication Data
Main entry under title:

Aspects of microbial metabolism and ecology.

  (Special publications of the Society for General
Microbiology ; 11)
    Includes index.
    1. Microbial metabolism. 2. Microbial ecology.
I. Codd, G. A. (Geoffrey A.) II. Society for General
Microbiology. III. Series.
QR88.A78 1984   576'.1'33   84-16734
ISBN 0-12-178050-3 (alk. paper)

PRINTED IN THE UNITED STATES OF AMERICA

84 85 86 87   9 8 7 6 5 4 3 2 1

Contributors

G. A. CODD *Department of Biological Sciences, University of Dundee, Dundee DD1 4HN, Scotland, United Kingdom*

A. FATTOM[1] *Division of Microbial and Molecular Ecology, Life Science Institute, Hebrew University, Jerusalem, Israel*

C. E. GIBSON *Department of Agriculture for Northern Ireland, Freshwater Biological Investigation Unit, Greenmount, Muckamore, Northern Ireland*

W. A. HAMILTON *Department of Microbiology, University of Aberdeen, Marischal College, Aberdeen AB9 1AS, Scotland, United Kingdom*

D. H. JEWSON *Limnology Laboratory, New University of Ulster, Ballyronan, Northern Ireland*

D. P. KELLY *Department of Environmental Sciences, University of Warwick, Coventry CV4 7AL, England, United Kingdom*

J. G. KUENEN *Laboratory of Microbiology, Delft University of Technology, Delft 8, The Netherlands*

H. J. LAANBROEK[2] *Department of Microbiology, University of Groningen, 9751 NN Haren, The Netherlands*

J. G. MORRIS *Department of Botany and Microbiology, University College of Wales, Aberystwyth, Dyfed SY23 3DA, Wales, United Kingdom*

V. A. SAUNDERS *Department of Biology, Liverpool Polytechnic, Liverpool L3 3AF, England, United Kingdom*

H. G. SCHLEGEL *Institut für Mikrobiologie der Universität Göttingen, D-3400 Göttingen, Federal Republic of Germany*

M. SHILO *Division of Microbial and Molecular Ecology, Life Science Institute, Hebrew University, Jerusalem, Israel*

F. B. VAN ES[3] *Department of Microbiology, University of Groningen, 9751 NN Haren, The Netherlands*

H. VELDKAMP *Department of Microbiology, University of Groningen, 9751 NN Haren, The Netherlands*

---

[1] Present address: Department of Biology and Biochemistry, Birzeit University, Birzeit, West Bank, Via Israel.

[2] Present address: Delta Institute for Hydrobiological Research, 4401 Ea Yerseke, The Netherlands.

[3] Present address: Provinciale Waterstaat Drente, 9400 AC Assen, The Netherlands.

Preface

This book is based on lectures presented at a symposium of the Society for General Microbiology on microbial metabolism and ecology held at the University of Dundee. The meeting reviewed advances in aspects of microbial metabolism and energetics and considered their implications in both cellular and ecological terms. In this volume adaptive strategies of microbes to changes in the nature and supply of nutrients and energy are elucidated, and responses to changes in the levels of oxygen, organic and inorganic carbon, light quantity and quality in aquatic environments, and reduced sulphur compounds are considered.

Appreciation of the diversity and roles of autotrophic prokaryotes in natural environments from soils and surface waters to the hydrothermal vents of the Pacific floor is increasing, and it is particularly appropriate that autotrophs of both the phototrophic and chemolithotrophic varieties are covered in these chapters. Metabolic versatility among autotrophs (from the specialists, able to grow only in the presence of specific inorganic compounds or light, plus inorganic carbon, through to the versatile members, which can also grow chemoheterotrophically) is considered by several authors. For example, the discovery of "new" versatile sulphur-oxidising bacteria has required extension of the conventional spectrum of physiological types to include chemolithoheterotrophs. Facultative chemolithoautotrophic hydrogen bacteria are also reviewed, and adaptive processes of these organisms and of photographic bacteria are considered at the genetic level.

This volume reflects research effort which has contributed to the understanding of particular groups of microbes and of microbial processes. In some cases, research has been largely at the physiological–biochemical level under defined laboratory conditions using pure cultures. In other cases, more direct field observations have been made. Attempts to seek ecological implications from the former approach, while trying to appreciate ecological findings in metabolic terms, can be synergistic to individual interests in microbial metabolism in the laboratory and in microbial ecology. It is hoped that this book will help to increase the combined attention of laboratory and field microbiologists in providing further understanding of microbial metabolism in natural environments. I am grateful to the authors for their cooperation and forebearance, and to Roger Berkeley and Colin Ratledge of the Society of General Microbiology for their help with the conference and with publication.

Contents

Contributors v

Preface vii

1. Microbial Ecology: An Overview
   F. B. van Es, H. J. Laanbroek, and H. Veldkamp   1

2. Energy Sources for Microbial Growth: An Overview
   W. A. Hamilton   35

3. Changes in Oxygen Tension and the Microbial Metabolism of Organic Carbon
   J. G. Morris   59

4. The Utilisation of Light by Microorganisms
   C. E. Gibson and D. H. Jewson   97

5. Aspects of Carbon Dioxide Assimilation by Autotrophic Prokaryotes
   G. A. Codd   129

6. The Ecology and Adaptive Strategies of Benthic Cyanobacteria
   M. Shilo and Ali Fattom   175

7. Studies on the Regulation and Genetics of Enzymes of *Alcaligenes eutrophus*
   H. G. Schlegel   187

8. Ecology of the Colourless Sulphur Bacteria
   D. P. Kelly and J. G. Kuenen   211

9. Genetics, Metabolic Versatility, and Differentiation in Photosynthetic Prokaryotes
   V. A. Saunders   241

Index   277

1

Microbial Ecology: An Overview

F. B. VAN ES,[1] H. J. LAANBROEK,[2] AND H. VELDKAMP

Department of Microbiology
University of Groningen
Haren, The Netherlands

Microbial ecology is a rapidly expanding area of research. Methods to measure metabolic activities of microbes in the field are being continually improved in order to quantify the role these organisms play in the recycling of chemical elements and in the energy flow through ecosystems. Autecological laboratory studies mimicking field conditions are also steadily increasing. Very illuminating overviews of the activities of microorganisms in a variety of habitats have been given in the textbooks of Lynch and Poole (1979) and Atlas and Bartha (1981). These show that at present, microbial ecological knowledge has become more than a collection of field data which cannot be interpreted adequately and of autecological laboratory studies carried out under conditions which no organism will ever encounter in soil or natural water environments. Promising as this may be, it is still difficult to measure microbial activities in their natural environment and to create in the laboratory conditions for studies which are simple enough to be controlled adequately and complex enough to make experiments ecologically relevant. However, the tools to tackle these problems are steadily improving, and therefore Beijerinck's adage "Lucky are they who start today" certainly holds for the beginning student of microbial ecology.

Microbial diversity is tremendous, and there are many different habitats in which microbes play an essential role. In this introductory chapter, special attention has been given to the aquatic environments and to studies of chemoorganotrophic bacteria which inhabit aerobic waters and their anaerobic sediments.

---

[1]*Present address:* Provinciale Waterstaat Drente, 9400 AC Assen, The Netherlands.
[2]*Present address:* Delta Institute for Hydrobiological Research, 4401 Ea Yerseke, The Netherlands.

The Role of Microorganisms in the Energy Flow in Ecosystems

Decomposition of Organic Material

The conversion of light energy into chemically bound energy by plants and algae, called "primary production," forms the basis of life in terrestrial and aquatic ecosystems. One other possible basis for an ecosystem is the synthesis of cell material by bacteria in the dark deep sea. The energy for this process is derived from the oxidation of hydrogen sulphide escaping from vents in the ocean floor (Kelly and Kuenen, Chapter 8, this volume).

Plant material forms the source of food for herbivores, which in turn are eaten by carnivores. Together, these form the grazing food chain. However, only part of living plants and algal populations is consumed by herbivores. Dead plant material and animal faeces form the source of food for the decomposers, which are mainly microbes (fungi and bacteria in terrestrial ecosystems and bacteria in aquatic ecosystems). Their cell material forms the basis of the detritus food chain. As will be shown below, a considerable part of organic matter originating in primary production ends up (in one form or another) in what is represented in flow diagrams as a box labeled "decomposers" or "microheterotrophs," both in terrestrial and in aquatic ecosystems.

For montane grasslands in Wales, Perkins (1978) calculated the contribution of herbivores and decomposers to the annual flow of organic carbon. Sheep (the predominant herbivore) and slugs together ingested no more than 17% of the net annual primary production (Fig. 1). Half of the ingested plant material became available to decomposers as faeces and urine, and 8% was released in the form of methane due to bacterial activities in the rumen of sheep. No less than 68% of the net annual primary production became available to the decomposers. Similar high percentages have been found for salt marshes (Teal, 1962), forests, and other terrestrial ecosystems (Odum, 1971; Phillipson, 1973; Heal and MacLean, 1975; Fenchel and Blackburn, 1979).

Until recently, it was thought that the major part of phytoplanktonic primary production in marine waters was consumed by zooplankton, and therefore that regeneration of nutrients occurred mainly in the grazing food chain (e.g., Steele, 1974). Only a small fraction of the primary production was considered to enter the detritus food chain, mainly through animal excretory products. There seems to be no doubt that bacteria, because of their high-affinity uptake systems as compared to eukaryotes (Sepers, 1977; Stephens, 1981) and their relatively high surface/volume ratio (Williams, 1981a), are indeed better adapted to uptake of the nanomolar concentrations of organic nutrients in the sea than eukaryotes. This has been shown in microautoradiographic studies (Munro and Brock, 1968; Hoppe, 1976) and in experiments in which labelled organic substrates were

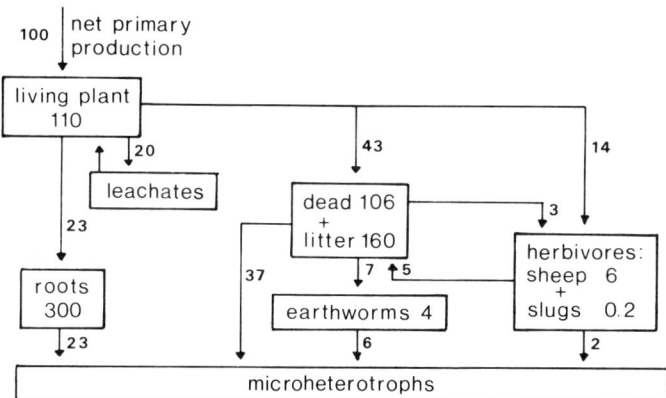

Fig. 1. Simplified diagram of the annual organic carbon flow in a montane grassland system (Wales). Boxes: average annual values of biomass or pool size of dissolved organic carbon (g C·m^{-2}). Arrows: transfer rates as a percentage of net primary production. Adapted from Perkins (1978).

added to water samples and after incubation the organisms present were size-fractionated (van Es and Meyer-Reil, 1982, Table 5).

It has been shown, however, that in the sea much less organic matter is channelled through the grazing food chain than was previously assumed (Chervin et al., 1981; Fransz and Gieskes, 1982; Fuhrman and Azam, 1982), Williams (1981b) reviewed the evidence that approximately one-half of the primary phytoplankton production enters the detritus food chain, as is shown in Fig. 2. A considerable part of this material may become available directly in the form of algal exudates (Wolter, 1982; Larsson and Hagström, 1982), and through lysis of decaying algae (together 30% in Fig. 2) or sloppy feeding of herbivores (10% in Fig. 2).

In estuaries the relative importance of the detritus food chain with respect to the energy flow through the ecosystem is even greater than that in the open sea due to the importance of allochthonous organic matter (van Es, 1982). Meyer-Reil et al. (1980) determined the *in situ* uptake rate of glucose in the upper centimetres of estuarine sediments from [^{14}C]glucose uptake and the *in situ* glucose concentration. Sixty percent of the label was incorporated in cell material. Assuming that microbial glucose uptake reflects one-quarter of the total uptake of organic substrates, the microbial biomass production was calculated to be 50% of sediment primary production. Hence, the total flow of organic matter through the microbial decomposer level must have been appreciably above 50% of primary production.

It should be realized that the term "mineralization" with respect to the breakdown of organic matter by microbes may be misleading. In decaying algae,

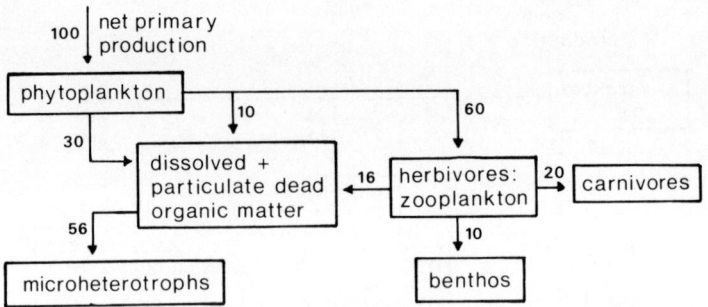

Fig. 2. Simplified diagram of the annual organic carbon flow for a planktonic system in the sea. Boxes: average annual values of biomass or pool size of dissolved organic carbon (g C·m^{-2}). Arrows: transfer rates as a percentage of net primary production. Adapted from Williams (1981b).

proteins and nucleic acids are rapidly hydrolysed, and during oxidation of amino acids, purines, and pyrimidines, ammonia, and phosphate are released. However, when later cell wall material with high carbon/nitrogen (C/N) and carbon/phosphorus (C/P) ratios is decomposed, ammonia and phosphate are taken up again, as occurs during cellulose degradation in soil. The ammonia and phosphate thus immobilised do not become available again until bacteria are digested by bacterivores or detritivores.

Microbes as a Food Source

Of prime importance for the energy flow through the detritus food chain is the amount of cell material produced by microbes during the decomposition of dead organic material. Laboratory studies of bacterial cell yields are numerous, and reviews have been given by Payne and Wiebe (1978) and Stouthamer (1979). As will be discussed in a following section, there does seem to be a discrepancy between laboratory and field studies. Many field observations indicate that more biomass is formed from organic substrates by chemoorganotrophic bacteria under conditions of energy limitation than would be expected on the basis of laboratory studies.

The field of predator–prey relations with microbes as a food source is still little explored. For eukaryotes, there are essentially two ways to feed on microorganisms. Either single cells, microcolonies, or fungus mycelium can be ingested, or entire detritus, soil, or sediment particles can be taken up, the microbes stripped off, and the remains excreted. With respect to terminology, the former group has been defined as "bacterivores" and the latter as "detritivores."

Because of their relatively low C/N and C/P ratios, bacteria form an excellent food source, which in addition can rapidly recover from grazing. Grazing may significantly increase the metabolic activity of a bacterial community. Reasons put forward include the quick regeneration of growth-limiting nutrients by the predator and mechanical breakdown and mixing of detritus particles during feeding and passage through the intestinal tract (Fenchel and Jørgensen, 1977).

It is difficult to capture bacterial cells because of their small size. This holds in particular for bacteria living in nutrient-poor environments such as the open sea, where bacteria with a diameter smaller than 0.4 μm are quite common (Zimmermann, 1977; Fuhrman, 1981). As will be discussed below, particle-bound bacteria may be larger and have a higher metabolic activity than free-floating ones, and may be relatively intensively grazed by detritivores.

Typical bacterivores are a variety of ciliates, amoebae, and nonphotosynthetic nanoflagellates (Fenchel, 1969, 1980a,b,c, 1982,a,b). In addition, higher organisms such as copepods (Rieper, 1982) and nematodes (Duncan *et al.*, 1974; L. A. Bouman and F. B. van Es, unpublished results) can be considered as typical bacterivores. The functional morphology and feeding mechanisms of bacterivorous ciliates and heterotrophic nanoflagellates, as well as the energy problems which filter-feeding invertebrates encounter in capturing even bacterial cells 0.2–0.5 μm in size, were considered by Fenchel. He concluded that in marine plankton, bacterial numbers are too low to support growth of ciliates, which require a bacterial density of 10^7 to 10^8 cells per milliliter for successful predation (Fenchel, 1980c). The smaller nanoflagellates, however, may well be important bacterial predators in the open sea (Fenchel, 1982b). In coastal, estuarine, and most freshwater environments, bacterial numbers seem to be high enough for ciliates to be of quantitative significance as bacterial predators.

Wright *et al.* (1982) were the first to show that even a bivalve (*Geukensia demissa*) is capable of effectively filtering single bacterial cells from coastal waters. Most other bivalves seem to be detritivores rather than bacterivores.

In soils and sediments, bacterial concentrations are up to 1000 times higher than those in natural waters, but these cells are normally surrounded by a vast majority of particles of similar size, both organic and inorganic. The question of how sediment-inhabiting bacterivores recognize bacterial cells in a way that enables feeding rates of hundreds to thousands of bacteria per hour, possibly selective even on the species level, is still largely unanswered. Perhaps microcolonies are actively searched for, which would make it advantageous for bacteria to avoid this situation by swimming away when possible.

Detritivores such as the earthworm and the lugworm have chosen an alternative strategy, and ingest particles in a much less selective way. This also holds for zooplankton organisms and bivalves which are able to live off bacteria associated with particles. There are many indications that larger detritivores feed exclusively on microorganisms associated with detritus particles, whereas the detritus itself remains unchanged in the digestive tract (Fenchel and Jørgensen, 1977).

Alexander (1981) reviewed mechanisms which prevent microbial predators and parasites from eliminating their preys and hosts. He concluded that the decline in the number of bacteria in water and soil is under predatory control, but that the equilibrium level of coexistence depends on several possible types of refuge, the distance between the individual prey or host cells, and the balance between the growth rate of the bacteria and the rate of predation.

Experimental studies on predation in a single-stage chemostat in which both predator and prey are growing have been reviewed by Frederickson (1977). Assuming Monod (1942, 1950) growth kinetics for both predator and prey, model equations have been given (Bungay and Bungay, 1968) which predict that in a wide range of operating conditions, a perpetual transient state of oscillation in both populations, as well as in the concentration of the growth-limiting bacterial nutrient, will occur. Although this has been confirmed experimentally, substantial discrepancies occur between the predictions of the model and experimental results, as was discussed by Frederickson (1977). In general, even in simple two- or three-membered cultures, a variety of unexpected extra variables not accounted for by the simple model do occur, and these variables are different for different predator–prey combinations. The approach as such, however, is very promising for revealing such variables under controlled conditions.

Jost et al. (1973) showed that selective feeding of the ciliate *Tetrahymena* influenced population dynamics of two bacterial species, *Escherichia coli* and *Azotobacter vinelandii*, that were competing for glucose. In the absence of *Tetrahymena*, *Azotobacter* was selectively eliminated at any dilution rate in a glucose-limited chemostat. However, in the presence of the ciliate, both bacterial populations could coexist when grown under glucose limitation.

Growth Under Natural Conditions

Measurements of Bacterial Growth Rates and Productivity

As stated by van Niel (1949), "Growth is the expression *par excellence* of the dynamic nature of living organisms. Among the general methods available for scientific investigation of dynamic phenomena, the most useful ones are those which deal with the kinetic aspects [p. 102]."

Because of the importance of bacteria as a basis for detritus food chains, many efforts have been made to measure their growth rates and biomass production in natural environments. The methods developed for this purpose were reviewed by van Es and Meyer-Reil (1982). It still is, however, quite difficult to measure reliably growth rates in nature. Promising developments in this field are the determination of growth rates by establishing the percentage of dividing cells of the total bacterial population by direct microscopic examination (Hagström *et al.*, 1979; Newell and Christian, 1981), as well as the rate of incorporation of labelled adenine and thymidine into RNA and DNA of bacterial cells (Karl, 1979, 1981; Fuhrman and Azam, 1980, 1982). The former method is almost ideal, since no incubation or other manipulations are required and environmental disturbance thus is minimal. However, the method is quite time-consuming, and the exact relation between the frequency of dividing cells and the growth rate has to be established in continuous culture, a system from which nongrowing cells are washed out. Under natural conditions, a considerable fraction of the bacterial community may not be growing at all, a factor which must be accounted for.

The second method, determining the incorporation rate of labelled adenine and thymidine, seems to be very useful since DNA synthesis is approximately proportional to total biomass production because the cellular DNA content is fairly constant and DNA does not turn over appreciably. According to Fuhrman and Azam (1980, 1982), the incorporation of methyl [^{3}H]thymidine into DNA reflects *in situ* bacterial biomass production when added at very low concentrations (less than 11 n*M*), and the incubation time is kept short (15 min). Although the use of precursor labelled in the 3-methyl position reduces incorporation in other cell components, Moriarty and Pollard (1981) stressed the importance of a DNA purification step. They also described a procedure to account for *de novo* synthesis of thymidine, which interferes with the rate of incorporation of labelled thymidine. Karl (1981) worked out a procedure to measure the rate of biosynthesis of both RNA and DNA from the incorporation of ^{3}H-labeled adenine.

Surprisingly, although the methods applied were quite different, published values of bacterial growth rates in comparable aquatic environments cover a relatively small range. Doubling times in coastal waters during the summer range from 3 to 40 hr, corresponding to specific growth rates of $\mu = 0.025 - 0.33$ hr^{-1} (van Es and Meyer-Reil, 1982). As will be described below, the range of growth rates measured in aquatic environments can be obtained in the laboratory with the aid of continuous culture techniques. Physiological and molecular biological studies under controlled laboratory conditions are desirable to evaluate growth measurements carried out in the field. This holds in particular for the measurement of heterotrophic productivity. The growth yield coefficient Y is defined by the equation

$$Y = \frac{\mu}{q} \qquad (1)$$

where μ is the specific growth rate (gram of biomass formed per gram of biomass per hour) and q is the specific rate of substrate uptake (gram of organic substrate consumed per gram of biomass per hour). As in the field, neither μ nor q can be measured directly, an apparent growth yield (Y_{app}) is derived from the incorporation and respiration of U-^{14}C-labelled organic substrates added to a sample, as follows (Williams, 1973):

$$Y_{app} = \frac{\text{net uptake (dpm)}}{\text{net uptake (dpm) + respiration (dpm)}} \qquad (2)$$

in which net uptake is the amount of ^{14}C present in the cells at the end of the incubation. Since natural chemoorganotrophic bacteria usually consume simultaneously many organic substrates, it is assumed that on average all organic substrates in the water sample are incorporated and respired in the same way as the trace amount of labelled substrate added. Since this assumption cannot be proven, the yield measurements necessarily give only apparent values. Reported field values for many organic substrates range from 0.6 to 0.8 (Williams, 1973; Hoppe, 1978).

As a rule, in natural environments, heterotrophic (= chemoorganotrophic) bacteria are carbon and energy limited. A still unsolved problem arose from laboratory studies of heterotrophic bacteria growing in the chemostat (described in a following section) under carbon and energy limitation. Yield values determined according to Eq. 1 appear to be consistently lower than the apparent growth yields observed in the field. This is the more surprising since it has been found that in the chemostat, growth yields decrease with decreasing growth rates under conditions of energy limitation (e.g., Tempest *et al.*, 1967). Further, most yield data obtained in the laboratory were collected at higher growth rates than those usually observed in nature. As yet, no explanation for the discrepancy is available.

Small-Scale and Short-Term Variations in Bacterial Metabolic Activity

The distribution of microorganisms in terrestrial environments is determined to a large extent by that of plants, which provide nutrients and which also greatly affect the microclimate in the top layer of soils. Within a small area, large temperature differences may occur in this layer (Fig. 3); therefore, it is not surprising that even from soils in temperate climates, thermophilic fungi and bacteria with optimum temperatures above 40°C can easily be isolated (Woldendorp, 1980). In most soils, microbes are not homogeneously distributed, and often more than 80% is associated with soil aggregates consisting of clay and silt particles mixed with organic material. Microscopic observation showed the local occurrence of single cells and microcolonies associated with particles, and indicated that only 0.1–0.2% of the solid phase in soil is covered with microbes (Woldendorp, 1980). The central part of the soil aggregates is often anaerobic (Greenwood and Goodman, 1967), and on their surface gram-positive bacteria resistant to desiccation (e.g., arthrobacters) are often dominant, whereas in the central part gram-negative rods are relatively numerous (Hattori, 1973).

In aquatic environments, which are much more homogeneous than soils and

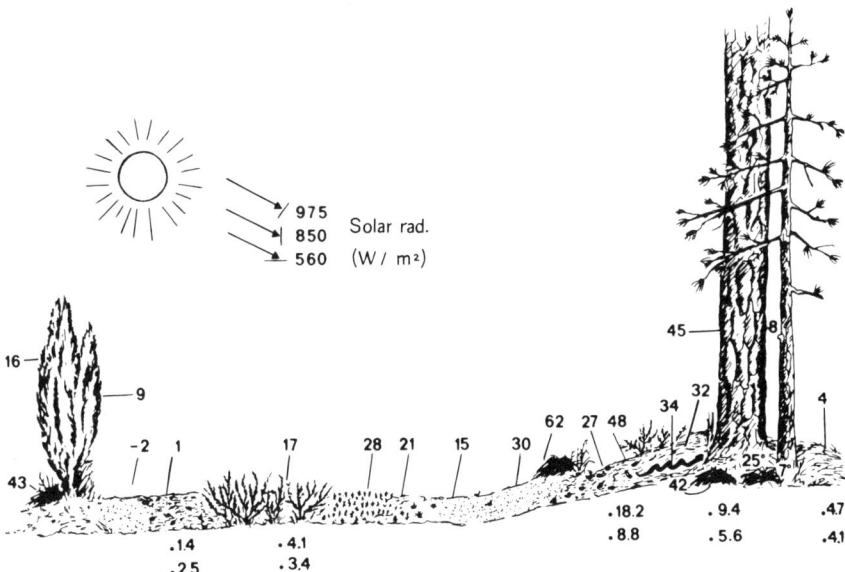

Fig. 3. Surface and soil temperatures at −4 and −9 cm (°C) measured in March at the south-facing edge of a pine wood in the Netherlands. From Stoutjesdijk (1977).

sediments, there also occur particulate microenvironments in which relatively high bacterial metabolic activity is observed, as compared to a bulk water phase with low activity (Harvey and Young, 1980). For the *in situ* study of microcommunities, the application of epifluorescence microscopy in combination with microautoradiography (Meyer-Reil, 1978) has proven to be very useful. Observations of aquatic environments (Paerl and Merkel, 1982; Kirchman and Mitchell, 1982) showed a 2–10 times higher metabolic activity of particle-associated bacteria as compared to free-floating cells. Newell and Christian (1981) occasionally encountered up to three times higher percentages of dividing cells among the former group in near-shore waters. Details of microbial growth at interfaces have been reviewed by Marshall (1976, 1980), and an overview of aspects of adhesion of microorganisms to surfaces was given by Ellwood *et al.* (1979). Despite the fact that particle-associated bacteria often show higher metabolic activities and growth rates, over 80% of bacterial cells in aquatic environments are generally either free-floating or only loosely associated with particles (van Es and Meyer-Reil, 1982). The greater part of bacterial uptake activity as determined by the addition of nanomolar quantities of organic substrates is observed in the sample fraction that passes through filters with a pore size of 1 to 3 μm. As indicated above, the advantage of growth on particles where nutrient concentrations are relatively high (Marshall, 1980) may be counterbalanced by a relatively high predation pressure.

Short-term variations in bacterial activities in aquatic environments are often directly coupled with diurnal variations in algal photosynthetic activity (Sieburth *et al.*, 1977), as illustrated in Fig. 4. The concentration of dissolved monosaccharides in the photic zone of aquatic environments (mainly glucose; Mopper *et al.*, 1980) has been shown to vary diurnally on the order of 10–50 n*M* (Burney *et al.*, 1982).

With respect to the concentrations of organic substrates in nature, one final remark should be made. A comparison of chemical and biological methods to determine the concentrations of dissolved monosaccharides and amino acids indicates that a large fraction of these nutrients is not directly available to bacterial cells. This was found by Gocke *et al.* (1981) for glucose in seawater and by Christensen and Blackburn (1980) for amino acids in sediment samples. The latter authors concluded that of the dissolved alanine in the pore water of a marine sediment, as determined chemically (800 n*M*), possibly only 10 n*M* was directly available. A model of amino acid turnover in a sediment is given in Fig. 5 (Christensen and Blackburn, 1980).

Generally speaking, aquatic and terrestrial environments both contain a vast

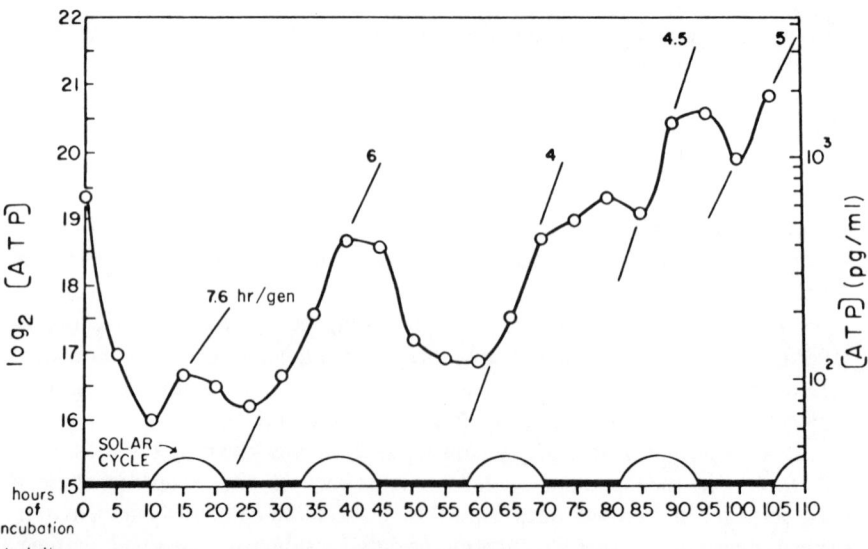

Fig. 4. Diurnal growth patterns of marine microplankton. An oceanic water sample from the photic zone was filtered through a 3-μm filter to remove larger organisms and was incubated in a diffusion chamber. A 0.1-μm Nucleopore filter separated the contents of the diffusion chamber from a continuous flow of fresh, untreated water from the same site. During the day, the microbial biomass (determined as cellular ATP content) increased, followed by a decrease during the night. From Sieburth *et al.* (1977).

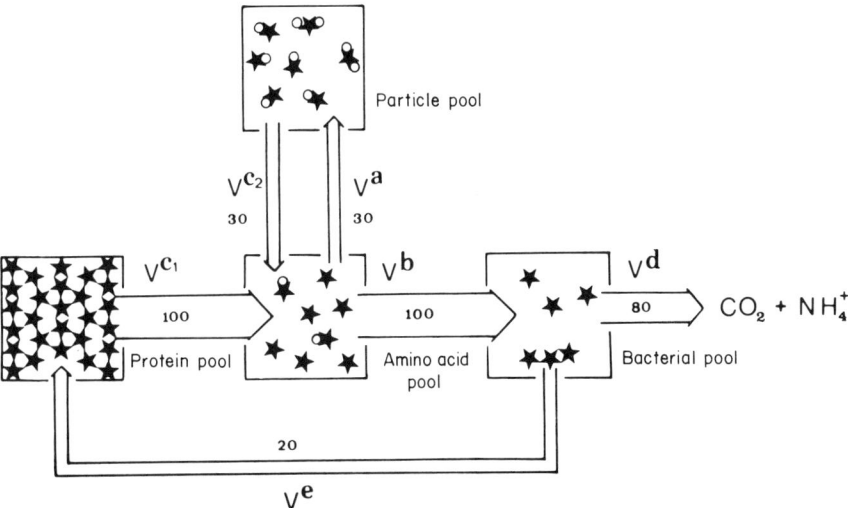

Fig. 5. Model of amino acid turnover in sediment. Relative rates are given. The following symbols are used: ★, an amino acid; ○, binding site for an amino acid. Protein is degraded at rate V^{c_1} to individual amino acids in a porewater pool. The amino acids leave this pool by biological uptake ($V^b = 100$) and by adsorption to particles ($V^a = 30$). Desorption from particles occurs at the same rate ($V^{c_2} = 30$). Some complexing of amino acids may occur in the free pool. The amino acids which are taken up by the bacterial pool are either oxidized ($V^d = 80$) or incorporated into cellular material ($V^e = 20$), which returns to the protein pool. From Christensen and Blackburn (1980).

reservoir of organic nutrients which is not directly available to bacterial cells and from which a continuous flow of available nutrients gradually becomes available in nanomolar concentrations. In fact, this flow has a stabilizing influence on the composition of the microbial flora in these ecosystems.

Autecological Studies

From Batch Culture to Continuous Culture

Ecologists working in the field are collecting data pertaining to the overall activities of mixed populations in their very complex natural environment. These field studies are comparatively recent, however. Until recently, the microbiologist's definition of ecology was that given by van Niel (1955), who stated that "Ecological studies aim at supplying the data on which to base an interpretation of the manner in which environmental factors determine the outcome of the struggle for existence [p. 202]" of individual species, which is in fact the definition of autecology.

In contrast to ecologists studying higher plants and animals, microbial ecologists can observe relatively little of the properties of individual species in the field. Therefore, a method for selective enrichment of particular species was developed by Winogradski which was later greatly extended by Beijerinck, who described this principle as the "ecological approach" to microbiology. The collected works of these microbiologists (Winogradski, 1949; Beijerinck, 1921–1940) show many examples of the successful application of this enrichment culture technique (Schlegel, 1965; Veldkamp, 1970). In his paper on urea bacteria, Beijerinck (1901; see Beijerinck 1921–1940) described the aim of the enrichment culture as being the creation in the laboratory of culture conditions which would select for the mass development of one particular organism, once the liquid had been inoculated with a sample taken from a natural environment. This organism then had to be isolated in pure culture on which its properties could be further studied. In this way, a study could be made of its ecological niche, which can be defined as the sum of the properties which enable the organism to succeed in the struggle for existence, thereby simultaneously defining its place and its function in the natural environment. Complementary to studies on the population level, then, should be ecophysiological work which reveals how the cellular machinery responds to varying environmental conditions. This early method of isolating bacteria in batch culture from their natural environment and the subsequent study of their physiology revealed the existence of many types of energy metabolism and initiated studies of metabolic pathways and their regulation, as elegantly described by Kluyver and van Niel (1956) in their overview of "The microbe's contribution to biology."

One important aspect of applying the enrichment culture technique, already noted by Beijerinck, is that decomposition of particular organic compounds is sometimes achieved by a microbial community rather than by one particular microbial species. Renewed attention to this phenomenon arose from studies on microbial breakdown of synthetic compounds (Slater and Bull, 1982), which will not be discussed here.

From an ecological point of view, one important restriction of the early studies described above was that during the first half of this century it was not yet possible to expose microbes to conditions similar to those encountered in nature in such a way that their metabolism could be studied adequately. Winogradski was the first to try to study microbes in the laboratory under the nutritionally poor conditions similar to those that occur in nature. Around 1924 (cf. Winogradski, 1949), he made a thorough study of soil microorganisms and observed a striking morphological similarity between the coccoid cells encountered by direct microscopic examination of soil samples and those which developed on silica gel plates enriched with a mineral base and humus. For this reason, he considered the autochthonous population of soil bacteria to be functional humivores. It was not until some 25 years later that the taxonomic position of these coccoid soil bacteria was elucidated by Conn (Conn and Dimmick, 1947; Conn, 1948) and Jensen (1952), which resulted in the creation of the genus *Arthrobacter* (Conn and Dimmick, 1947). At that time, however, it was still not possible to grow such organisms at extremely low concentrations of limiting nutrients in such a way that ecophysiologists could obtain dense populations of cells which could be analysed for metabolic properties. Studies of this kind became possible after the introduction of the continuous culture technique by Monod (1950) and Novick and Szilard (1950a,b), who named the single-stage, flow-controlled system the "chemostat." More recent descriptions of the possibilities of this technique were given by Tempest (1970a,b), Veldkamp (1976, 1977), Harder *et al.* (1977), and Tempest and Neijssel (1978).

One of the main features of the chemostat is that it allows microbial growth under conditions in which the growth rate is determined by the rate at which one or more nutrients are made available to the cells. It therefore allows bacterial growth at the submaximal rates which generally occur in nature. The specific growth rate (μ) of a bacterial population is given by the Monod (1942, 1950) equation:

$$\mu = \mu_{max} \frac{s}{k_s + s} \qquad (3)$$

where μ_{max} is the maximum specific growth rate (hr^{-1}), s is the concentration of the growth rate-limiting substrate, and k_s is a saturation constant numerically equal to the concentration of the growth rate-limiting nutrient at 0.5 μ_{max}.

When applying a chemostat, sterile culture medium is pumped from a reservoir at rate F (milliliters per hour) into the culture vessel. As culture liquid is removed at the same rate, the culture volume V (milliliters) is constant. The dilution rate is defined as $D = F/V$(hr^{-1}). The composition of the sterile medium is such that all components needed for growth are present in excess except one, the growth-limiting substrate. At a low dilution rate, D equals approximately μ (cf. Veldkamp, 1976).

In their natural environment, microbes are often exposed to nanomolar concentrations of essential nutrients. Such concentrations can be obtained in the chemostat. If, for instance, a bacterium is grown in a mineral medium with growth rate-limiting glucose as the carbon and energy source, and if $\mu_{max} = 0.5$ hr^{-1} and k_s for glucose $= 1$ μM, at a dilution rate of 0.01 hr^{-1} the glucose concentration in the culture vessel is 20 nM, as can be seen from Eq. 3.

The bacterial concentration under steady-state conditions ($\bar{x}$) is

$$\bar{x} = Y(S_R - \bar{s}) \qquad (4)$$

where $\bar{x}$ is the steady-state concentration of biomass (milligrams of dry weight per milliliter), $\bar{s}$ is the steady-state concentration of the growth-limiting substrate (milligrams per milliliter), S_R is the concentration of the growth-limiting substrate in the reservoir (milligrams per milliliter), and Y is the growth yield coefficient (weight of biomass formed per weight of substrate used).

It is therefore clear that high population densities can be obtained at extremely low concentrations of the growth-limiting substrate s at relatively high values of S_R. This means that ecophysiologists now have a tool with which to obtain enough cell material for their metabolic studies of cells grown at the extremely low nutrient concentrations that occur in nature. An additional important development with respect to techniques for autecological studies in the laboratory is the creation of gradients of physicochemical factors which affect microbial growth (Wimpenny, 1981). One of the possibilities of studies of microbial behaviour in gradients is to explore relatively rapidly optimal growth conditions to be applied in the chemostat.

One objection to the application of the chemostat is that bacteria are grown under steady-state conditions with one particular growth-limiting substrate. Due to the continually changing conditions in natural microenvironments, the metabolic machinery of bacteria living there is generally in a transient state. Moreover, it seems likely that frequently more than one energy substrate is consumed simultaneously. However, a mixture of growth-limiting substrates can easily be applied in the chemostat (Harder and Dijkhuizen, 1982) and fluctuations of any kind can be introduced, so that selective advantages due to such conditions can be revealed as well. Typical examples are studies on the ecological niches of specialistic and versatile colourless sulphur bacteria (Beudeker *et al.*, 1982) and on the effect of changing light intensities on phototrophic microbes (van Gemerden, 1974, 1980; Mur *et al.*, 1977). One disadvantage of the chemostat is that it cannot be used for studies of nongrowing cells.

Up to now, chemostat studies have mainly been applied to bacteria occurring in homogeneous aqueous suspension. However, the occurrence of bacteria which show a selective advantage in a chemostat in the presence of a suspension of clay particles was revealed by Jannasch and Pritchard (1972). Further studies of this kind should be rewarding.

In summary, it can be said that after the discovery in the early decades of this century of microbes with different types of energy metabolism using the batch culture enrichment technique, followed by the description of metabolic pathways and the discovery of basic principles of metabolic regulation, it is now possible for ecophysiologists to analyse the metabolism of bacteria growing at submaximal rates in media with nutrient concentrations as low as those found in natural environments.

Natural environments can be classified in various ways. With respect to bacterial energy metabolism, extremes are aerobic waters and their anaerobic sediments. In aerobic waters, the large majority of heterotrophic bacteria have an energy metabolism which allows complete oxidation of organic substrates. These bacteria are generally quite versatile in their utilisation of commonly occurring monomers of proteins and polysaccharides for which they have to compete. In the anaerobic sediments, however, microbial interactions are more complicated, as anaerobic mineralization is the result of a network of processes in which metabolic products of one species form the substrates for others. Further, the range of energy substrates which can be used by anaerobes is generally much smaller than that of aerobes. For this reason, aerobes and anaerobes will be discussed in separate sections.

The Aerobic Aquatic Environment

All natural environments are heterogeneous, but some are more so than others. Soil conditions are extremely complex and difficult to assess. Natural waters, however, are less complex and heterogeneous, and therefore more amenable to ecological studies. As discussed above, in natural environments the bulk of organic nutrients, as well as some of the inorganic ones such as phosphate and iron, are generally present in a form which is not directly available to bacterial cells. These cells are confronted with a flow of available nutrients, often present in nanomolar concentrations, which continuously changes both quantitatively and qualitatively. Their environment is characterized by concentration gradients, and therefore studies of spatial order in microbial ecosystems (Wimpenny, 1981) are particularly relevant to the discovery of optimal conditions for growth of the members of naturally occurring microbial populations.

As yet, little is known of species composition and the properties of individual species occurring in aquatic environments. One issue is the diversity of the mixed population of, for instance, bacteria which can use amino acids as a carbon and energy source. The stability of such a community increases with increasing numbers of species which show few differences in metabolic properties. Small perturbances of environmental conditions may result in decreased activities of some members of the community, which are counterbalanced by an increase in the activity of other metabolically related species. Such conditions lead to a situation in which the chemical composition of the environment is rather constant. One example is the decomposition of proteins in natural waters and the subsequent mineralization of amino acids. The composition of the fraction of dissolved amino acids in marine environments is remarkably constant, as shown in Table 1. This indicates that the microbial population involved is highly diverse. Although organic compounds in aquatic environments can be taken up by phytoplankton and invertebrates, their quantitative importance in this respect generally is negligibly small as compared to that of the bacterial population (Sepers, 1977).

The diversity of the bacterial population which can use amino acids as a carbon and energy source was studied by Sepers (1981). Ammonifying bacteria

Table 1. *Composition of the fraction of dissolved amino acids in natural aquatic environments (in mol %)[a]*

Amino acid	Baltic Sea[b]	Pacific Ocean[c]	North Sea[d]
Glycine	21.1	16.4	18.7
Alanine	9.6	10.7	7.9
Valine	2.0	2.3	2.6
Isoleucine	1.0	3.8	2.7
Leucine	1.8	3.8	2.0
Lysine	7.9	2.1	12.1
Arginine	2.3	—	0.4
Histidine	2.2	2.2	3.3
Serine	18.2	29.5	31.3
Threonine	4.2	7.6	5.4
Aspartic acid	5.3	3.8	6.8
Glutamic acid	6.0	2.7	2.3
Phenylalanine	1.9	2.2	1.3
Ornithine	5.8	14.5	n.d.[e]
Tyrosine	0.9	2.2	1.8
Others	8.3	—	3.2
Total amino acid concentration (μg · liter^{-1})	23.7–38.6	16.3–56.8	120–600

[a] These data refer to surface samples.
[b] Data from Dawson and Gocke (1978).
[c] Data from Degens et al. (1964).
[d] Data from Bohling 1972).
[e] n.d., Not determined.

occurring in the Hollands Diep-Haringvliet basin (the Netherlands), through which waters of the rivers Rhine and Meuse flow to the North Sea, were isolated in a variety of ways. Altogether, 169 strains were selected, 68 of which were subsequently tested for the ability to utilise 19 amino acids and 23 other organic compounds. Only two couples of strains appeared to be identical with respect to the spectrum of organic compounds that could be used as an energy, carbon, and nitrogen source. All other 66 strains grew on more than five substrates, as shown in Fig. 6. This study thus confirmed the hypothesis that the diversity among ammonifying bacteria in natural waters is quite large. Indications were also that most members of the total population of chemoorganotrophic bacteria in the environment studied could use several amino acids as an energy substrate, as well as a carbon and nitrogen source for growth. It seems very likely that in their natural environment these bacteria can take up a variety of organic substrates simultaneously. This means that rather complex regulatory mechanisms must be present to channel the substrates taken up into flows of energy metabolism and biosynthesis. The way in which this is done still has to be elucidated.

One additional possibility, not considered by Sepers (1981), is that a particular amino acid which cannot be used by a bacterial species as a single carbon, nitrogen, and energy source may well be cometabolized in the presence of others, as was observed by van Es (1982; unpublished results). Generally speaking, cometabolism is probably a more general phenomenon than is realized at present. Its importance has been reviewed by Dalton and Stirling (1982).

Observations of the kind described above lead to the conclusion that it is the ever-changing microenvironment which allows the coexistence of so many different, though metabolically related, species. Short periods in which optimal conditions for growth of one organism occur are followed by larger periods in which it still can grow, though more slowly than its competitors. In addition, strategies should have developed during evolution to allow the organism to survive during periods in which no growth is possible. To survive during periods in which no energy source is available, the presence of cellular reserve material is of great importance. This is due to the fact that for the uptake of nutrients when these again become available, the electrochemical potential ($\Delta\bar{\mu}_{H^+}$) across the cell membrane must be above zero. Therefore, either energy has to be generated in the breakdown of reserve material, or the cell should have other mechanisms to prevent $\Delta\bar{\mu}_{H^+}$ from decreasing below a certain level. If these are not available, pool constituents leak out of the cells, inevitably followed by death (Konings and Veldkamp, 1980).

The cultivation of heterotrophic bacteria in the laboratory under a limitation

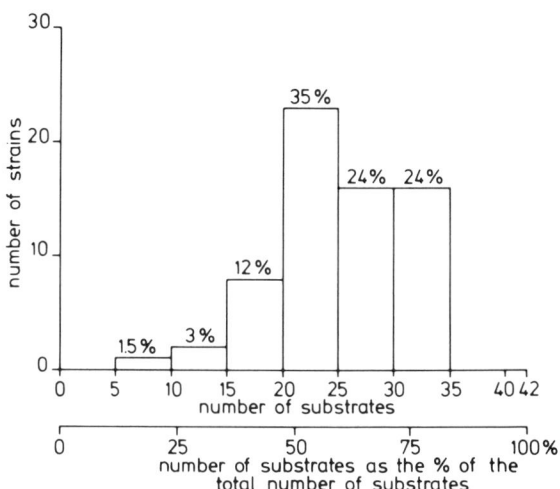

Fig. 6. Substrate specificity of ammonifying bacteria. Percentages on top of the bars indicate the number of bacterial strains as the percentage of the total number of strains (Sepers, 1981).

other than the carbon and energy source commonly leads to the synthesis of reserve material (polyglucose or poly-β-hydroxybutyric acid). It has been found, however, that some bacteria also produce reserve material at low growth rates under carbon and energy limitation. One example is a *Hyphomicrobium* species which produced a considerable amount of polyhydroxybutyrate when grown in a methylamine (carbon, nitrogen, and energy source)-limited chemostat at a rate of 0.02 hr^{-1} (Meiberg, 1979).

Many studies have been done on microbial starvation (Strange, 1976). One example is the study of *Staphylococcus epidermidis* by Horan *et al.* (1981), who showed that in washed suspensions starved under anaerobic conditions, the viability declined to <10% within 12 hr. During that time, $\Delta\bar{\mu}_{H^+}$ decreased considerably and the intracellular amino acid pool became negligibly small.

Other organisms, however, do survive much longer periods of starvation. Boylen and Ensign (1970) determined the half-life of a population of the soil bacterium *Arthrobacter crystallopoites* (100 days) by extrapolation. Compared to other bacteria, the marine psychrophilic *Vibrio* sp. studied by Novitsky and Morita (1978) exceeded the longest reported starvation survival by at least 2.5 times. When suspended in a starvation menstruum the rod-shaped cells divided to form small spheres, 0.4 μm in diameter, without increasing the cultural biomass. After 70 weeks, over 15 times the original number of cells were still viable. On addition of low-level nutrients, cell numbers increased, and the cells increased in size and changed in shape to resemble a normally growing laboratory culture. The physiological background of this behaviour still has to be elucidated. It is not yet clear how such organisms can prevent the electrochemical gradient across their membrane from being dissipated entirely during prolonged periods of starvation.

In nature, short periods of relatively rapid growth are followed by periods of slower growth or dormancy, and only a few variables are needed to explain the enormous diversity among chemoorganotrophic bacteria occurring in natural waters. These factors include the variety of substrates which can be used by a single species, responses to such factors as pO_2, pH, and temperature, and nutrient concentration. The importance of the last factor was revealed by Jannasch (1967) and has subsequently been confirmed by other authors. One of them (Kuenen *et al.*, 1977) showed that in phosphate-limited chemostats with lactate as the carbon and energy source, which were inoculated with samples from the same source, different bacteria became predominant at relatively high and low phosphate concentrations. The organism which came to the fore at low phosphate concentrations was a *Spirillum* species, whereas an unidentified rod-shaped bacterium became predominant at higher phosphate concentrations. After these bacteria had been obtained in pure culture, it was found that mixtures of them showed the same behaviour with respect to concentrations of other growth rate-limiting nutrients, such as succinate, NH_4^+, and K^+ (Fig. 7).

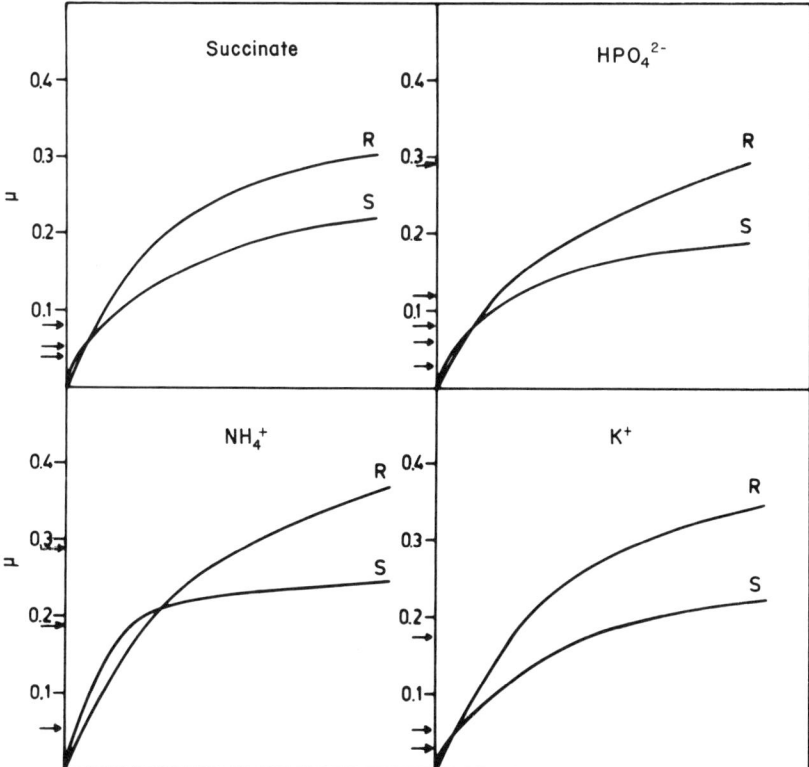

Fig. 7. Specific growth rate (μ) of a rod-shaped bacterium (R) and a *Spirillum* species (S) as a function of the concentration of growth-limiting substrates. Lactate was the carbon and energy source in the cultures limited by inorganic ions. Temperature, 25°C. From Kuenen *et al.* (1977).

Studies of this kind (Harder *et al.*, 1977) have firmly established the fact that in natural waters, bacteria are common which have a selective advantage at extremely low nutrient concentrations. Little is known, however, of the physiological background of this property. It seems that all bacteria which exhibit it have a relatively high surface/volume ratio (Table 2) which affects the maximal rate of growth, not the substrate affinity. A comparative physiological study of a couple of organisms growing relatively rapidly at low and higher lactate concentrations was made by Matin and Veldkamp (1978). This study showed that not only k_s for growth but also k_m for lactate transport was lower in the *Spirillum* sp. which grew faster at relatively low lactate concentrations. Furthermore the *Spirillum* sp. possessed a higher V_{max} for lactate transport; it had a greater surface/volume ratio and a fourfold lower maintenance requirement than its competitor. This study should be considered, however, as a mere beginning in

Table 2. Surface/volume ratios of four couples of chemostat-grown spiral and rod-shaped bacteria competing for a growth-limiting substrate[a,b]

Organism	Limiting substrate	Specific growth rate (hr^{-1})	Average length (μm)	Average diameter (μm)	Surface/volume ratio (μm^{-1})
1. *Pseudomonas* sp.[c]	Lactate	0.1	2.9	1.1	4.3
Spirillum sp.	Lactate	0.1	3.5	0.55	8.0
2. *Pseudomonas* sp.[d]	Lactate	0.15	2.3	1.0	4.9
Spirillum sp.	Lactate	0.15	2.6	0.66	6.8
3. Unidentified rod[e]	Phosphate	0.2	2.8	1.1	4.3
Spirillum sp.	Phosphate	0.2	3.8	0.6	7.2
4. *Thiobacillus thioparus*[e]	Thiosulfate	0.1	2.0	0.4	11
Thiomicrospira pelophila	Thiosulfate	0.1	2.5	0.2	21

[a] From Kuenen et al. (1977).
[b] Surface/volume ratios were calculated from the dimensions of the bacterial cells, treating them as small cylinders. Dimensions of bacterial cells were estimated from photomicrographs and electron micrographs.
[c] W. Harder and H. Veldkamp (unpublished observations).
[d] Matin and Veldkamp (1978).
[e] Kuenen et al. (1977).

elucidating the physiological aspects of selective advantages developed by chemoorganotrophs with respect to life in nutrient-poor environments. Discussions of the properties of oligotrophs appear in Hirsch (1979) and Poindexter (1981).

In both aquatic and terrestrial environments, it has been found that there is a large discrepancy between total and viable bacterial counts. At most, only a few percent of the bacteria present are able to grow on a variety of growth media tested. Van Es and Meyer-Reil (1982), who discussed this discrepancy, concluded that this cannot be due to the fact that most naturally occurring bacteria are dead. Many microautoradiographic and other microscopic studies (Zimmermann et al., 1978) have shown that on average 10–60% of naturally occurring bacterial cells take up and utilise organic and inorganic nutrients. As these techniques yield a conservative estimate of the fraction of active cells, it may be concluded that we are not yet able to create in the laboratory the conditions for growth for a majority of naturally occurring bacteria. This may well be due, at least in part, to their oligotrophic nature.

Naturally occurring heterotrophic bacteria have a very high affinity for many organic substrates. Maximum values, ranging from 1 to several hundred nanomoles, have been reported by Sepers (1977), Hoppe (1978), and Bell and Albright (1981).

One point which should be kept in mind is that maximal specific growth rate values (μ_{max}) and substrate affinity (k_s) are not growth constants. They pertain only to a particular combination of environmental factors which should always be defined. One example is the observation of Law and Button (1977), who showed that the k_s value for glucose of a marine *Corynebacterium* was relatively low in a growth medium which contained, in addition to growth-limiting glucose, a mixture of other growth-limiting organic nutrients which could be used as a carbon and energy source.

Of particular interest is the metabolism of aerobic bacteria which possess different types of energy metabolism. One example is the occurrence of mixotrophic colourless sulphur bacteria. These can grow heterotrophically, autotrophically, and mixotrophically, which means that they can use an inorganic and an organic energy source simultaneously. The ecological niche of such bacteria was described by Beudeker *et al.* (1982). They are losers when exposed to competition for organic or inorganic energy sources with specialists in these fields. The selective advantage of the mixotrophs is expressed only under the limitation of both organic and inorganic energy sources. A similar phenomenon was observed in anaerobic bacteria, as shown by Laanbroek *et al.* (1979) in a study on specialized and versatile *Clostridium* species. The specialistic *Cl. cochlearium,* which could grow only on L-glutamate, L-glutamine, and L-histidine, grew faster at all concentrations of glutamate than the more versatile *Cl. tetanomorphum,* which could also use glucose as a carbon and energy source. In a chemostat limited by glucose as well as by L-glutamate, coexistence occurred. However, *Cl. tetanomorphum* then used not only all of the glucose but also some of the glutamate. Therefore, the ratio between the populations of these organisms should depend on the ratio in which these substrates become available in their natural environment, the anaerobic sediment of natural waters. Here microbial interactions are much more frequent than in the aerobic overlying waters, determined by incomplete oxidation of energy substrates, which gives rise to products on which other bacteria are entirely dependent. In the next section, interrelations of this type will be described.

As was shown in Eq. 1, the specific growth rate is the product of the specific substrate uptake rate and the yield coefficient. This means that theoretically a bacterium with a relatively low growth yield may still be able to grow relatively quickly. An example of this is *Cl. cochlearium,* which always grows faster on L-glutamate than *Cl. tetanomorphum,* whereas its growth yield on L-glutamate is smaller (Laanbroek *et al.,* 1979).

The Anaerobic Sediment of Natural Waters

Except for the surface layer, most sediments are permanently anaerobic, and in shallow waters such as estuaries, anaerobic mineralization is of considerable quantitative importance. In the sediment of such environments, about 50% of the total degradation of organic matter may occur in the anaerobic part of the ecosystem (Jørgensen, 1977). Although in the surfaces of sediments denitrification is generally observed, the main terminal processes of mineralization in anoxic sediments are sulphate reduction and methanogenesis (Fenchel and Blackburn, 1979). Further, since the bacteria involved cannot hydrolyse polymers such as proteins and polysaccharides or use their monomers as substrates for energy metabolism (Pfenning et al., 1981; Mah and Smith, 1981), these organisms are dependent on fermentative bacteria which initiate the mineralization process (Bryant, 1976, 1979; Laanbroek and Veldkamp, 1982). Acetate, H_2, and CO_2 are considered to be major products of the first stages of anaerobic mineralization. Acetate and H_2 are formed not only by bacteria which ferment substrates such as glucose but also by several sulphate-reducing as well as acetogenic bacteria. The last type of organism was discovered by Bryant et al. (1967). These bacteria can produce acetate and molecular hydrogen from ethanol (Bryant et al., 1967), propionate (Boone and Bryant, 1980), or butyrate, caproate, and caprylate (McInerney et al., 1979). However, for thermodynamic reasons, such conversions are feasible only when other bacteria consume the molecular hydrogen produced, thus keeping its concentration very low. The activities of acetogenic bacteria are probably of importance only in freshwater sediments, in anaerobic digesters, and in the rumen, in which methanogens are major consumers of acetate and molecular hydrogen. In addition to these substrates, methanogens can use only a few others, such as formate, methanol, and methylamines (Mah and Smith, 1981). It has been shown that sulphate-reducing bacteria have a higher affinity than methanogens for acetate as well as for hydrogen (Kristjansson et al., 1982; Lovley et al., 1982; Schönheit et al., 1982). Therefore, methanogens can dominate only in anaerobic environments in which the activities of sulphate reducers are limited by low sulphate concentrations, which is the case in fresh waters. The concentration of sulphate in freshwater is around 0.15 mM, whereas that of seawater is 20–30 mM. It is therefore not surprising that the main terminal process in the mineralization of organic matter in estuarine and marine sediments is sulphate reduction (Jørgensen, 1977, 1980; Oremland and Taylor, 1978; Sørensen et al., 1979, 1981; Mountfort et al., 1980).

Until now, acetogenic bacteria have never been shown to occur in such environments, and are probably not very common there. This can be explained by the fact that in the marine environment, sulphate becomes depleted only in the lower (e.g., 20–30-cm-deep) layers of the sediments. Since sulphate-reducing bacteria as a group can use a whole range of fermentation products, among which are ethanol, propionate, and butyrate (Pfennig et al., 1981; Laanbroek and Veldkamp, 1982), it seems unlikely that acetogens can successfully compete for such substrates in the marine environment. In conclusion, it seems that the properties of sulphate-reducing bacteria allow them to dominate both methanogens and acetogens to a large extent in any anaerobic environment that is sulphate sufficient.

Ecological Niches of Sulphate-Reducing Bacteria

In sediments of the Ems-Dollard estuary (The Netherlands), a lactate-consuming, sulphate-reducing bacterium, *Desulfovibrio baculatus* (Rozanova and Nazina, 1976), has been shown to occur in approximately equal numbers as the lactate-fermenting *Veillonella alcalescens* and an *Acetobacterium* species (Laanbroek et al., 1983). Competition experiments in the chemostat have shown, however, that under lactate limitation with excess sulphate, the competition for lactate is won by the sulphate reducer (Laanbroek et al., 1983).

The vertical distribution of L-lactate-, propionate-, and acetate-consuming, sulphate-reducing bacteria in sediments of the Ems-Dollard estuary was studied by Laanbroek and Pfennig (1981). The result is shown in Fig. 8.

Acetate added to estuarine mud disappeared rapidly with concomitant sulphide production. From the mud, the acetate-utilising *Desulfobacter postgatei* (Widdel and Pfennig, 1981) could easily be isolated. Addition of propionate to a similar sample resulted in acetate and sulphide production. The propionate-utilising *Desulfobulbus propionicus* (Widdel and Pfennig, 1982) appeared to be common in these mud samples. The L-lactate-utilizing *Desulfovibrio* species appeared to be the most abundant lactate-utilising sulphate reducer. Thus, from the same habitat, three sulphate-reducing bacteria were isolated, with acetate, propionate, and L-lactate as the substrates (Laanbroek and Pfennig, 1981).

Further study of the ecological niches of these bacteria showed that the only electron donors which could be used by *Desulfobacter* were ethanol and acetate. In an ethanol-limited chemostat inoculated with pure cultures of *Desulfovibrio baculatus* and *Desulfobacter postgatei* in the presence of excess of sulphate, coexistence occurred, with the latter bacterium growing mainly on the acetate produced by *Desulfovibrio*. However, when the mixture was exposed to competition for sulphate as well in an ethanol + sulphate-limited chemostat, *Desulfovibrio* won (Laanbroek et al., 1984).

Of particular interest was a comparison of the ecological niches of *Desulfovibrio baculatus* and *Desulfobulbus propionicus*, both of which can use ethanol and L-lactate as electron donors for sulphate reduction. Competition experiments in the chemostat under the following limitations—L-lactate, ethanol, L-lactate + SO_4^{2-}, and ethanol + SO_4^{2-}—led all rapidly to dominance of *Desulfovibrio baculatus*. However, under L-lactate + propionate limitation, coexistence was observed in which *Desulfobulbus* consumed all of the propionate and perhaps also some of the L-lactate. More or less accidentally, it was subsequently discovered, however, that it was not only the presence of propionate + sulphate which caused the survival of *Desulfobulbus*. An ethanol-limited chemostat inoculated with mud from the Ems-Dollard estuary rapidly resulted in the dominance of *Desulfovibrio*, as expected. However, due to sulphide production, the culture became iron-limited, which gave rise to the development of

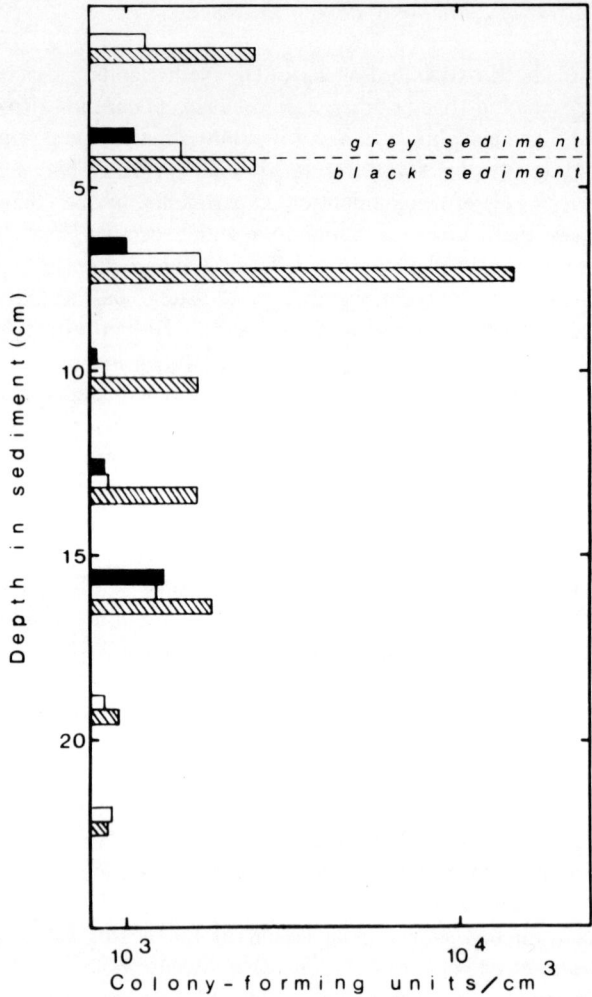

Fig. 8. Vertical distribution of anaerobic bacteria growing on sulphate plus acetate (black bars), propionate (open bars), or L-lactate (striped bars) in an estuarine sediment. From Laanbroek and Pfennig (1981).

Desulfobulbus with concomitant propionate production. It appeared that this bacterium was fermenting ethanol in the presence of CO_2. A further study of its energy metabolism showed that it was even more versatile than indicated by the studies of Widdel and Pfennig (1982), who showed that propionate, lactate, and alcohols could be used as electron donors with sulphate, propionate, and nitrate, and that lactate could also be fermented. Table 3 and Fig. 9 show the known reactions which *Desulfobulbus* can achieve to provide energy for growth. Of all of the 12 reactions (Table 3), *Desulfovibrio* can perform only 4 (numbers 1, 2, 4, and 5). As *Desulfobulbus* it can also ferment pyruvate (not shown in Table 3), but this seems to be of little ecological importance since pyruvate is not a normal fermentation product.

Despite the fact that *Desulfobulbus* is much more versatile than *Desulfovibrio*, the latter organism is more numerous at all depths of the estuarine mud (Fig. 8). As *Desulfovibrio* shows a competitive advantage under the conditions described above, it must be concluded that the frequency with which these conditions occur dominates those which *Desulfobulbus* encounters, to give it a selective advantage. Propionate appears to be a central compound in its energy metabolism. It is produced in the absence of sulphate and again consumed in its presence. In addition, it has developed an ability to use ethanol and lactate as an energy source in both the presence and the absence of sulphate. Further, although its specific growth rate on these compounds in the presence of sulphate is lower than that of *Desulfovibrio* when these are the only limiting substrates, it can probably grow mixotrophically on propionate + lactate as well as on ethanol. It has been shown that it can grow mixotrophically on lactate + H_2, as well as on ethanol + H_2 (Laanbroek *et al.*, 1982).

These autecological studies again show that versatility in itself is not so advantageous that the most versatile organism becomes dominant in its natural en-

Table 3. *Reactions achieved in energy metabolism of* Desulfobulbus propionicus[a]

Reaction	$\Delta G^{o\prime}$ value
1. $H_2 + \frac{1}{4}H^+ + \frac{1}{4}SO_4^{2-} \rightarrow \frac{1}{4}HS^- + H_2O$	−38.0
2. Formate$^-$ + $\frac{1}{4}H^+$ + $\frac{1}{4}SO_4^{2-}$ → HCO_3^- + $\frac{1}{4}HS^-$	−36.7
3. Propionate$^-$ + $\frac{3}{4}SO_4^{2-}$ → acetate$^-$ + HCO_3^- + $\frac{3}{4}HS^-$ + $\frac{1}{4}H^+$	−37.9
4. Lactate$^-$ + $\frac{1}{2}SO_4^{2-}$ → acetate$^-$ + HCO_3^- + $\frac{1}{2}HS^-$ + $\frac{1}{2}H^+$	−80.2
5. Ethanol + $\frac{1}{2}SO_4^{2-}$ → acetate$^-$ + $\frac{1}{2}HS^-$ + $\frac{1}{2}H^-$ + H_2O	−66.4
6. 1-Propanol + $\frac{5}{4}SO_4^{2-}$ → acetate$^-$ + HCO_3^- + $\frac{5}{4}HS^-$ + $\frac{3}{4}H^+$ + H_2O	−101.8
7. Lactate$^-$ → $\frac{1}{3}$acetate$^-$ + $\frac{2}{3}$propionate$^-$ + $\frac{1}{3}HCO_3^-$ + $\frac{1}{3}H^+$	−54.9
8. Ethanol + $\frac{2}{3}HCO_3^-$ → $\frac{1}{3}$acetate$^-$ + $\frac{2}{3}$propionate$^-$ + $\frac{1}{3}H^+$ + H_2O	−41.4
9. 1-Propanol + $\frac{2}{3}$acetate$^-$ + $\frac{2}{3}HCO_3^-$ → $\frac{5}{3}$propionate$^-$ + $\frac{1}{3}H^+$ + H_2O	−38.6
10. Ethanol + HCO_3^- + H_2 → propionate$^-$	−66.5
11. Acetate$^-$ + HCO_3^- + H^+ + $3H_2$ → propionate$^-$ + $3H_2O$	−76.1
12. Propionate$^-$ + $\frac{3}{4}NO_3^-$ + $\frac{1}{2}H^+$ + $\frac{3}{4}H_2O$ → acetate$^-$ + HCO_3^- + $\frac{3}{4}NH_4^+$	−373.5

[a] The $\Delta G^{o\prime}$ values are expressed as kilojoules per reaction (pH 7.0) and were calculated from data of Thauer *et al.* (1977).

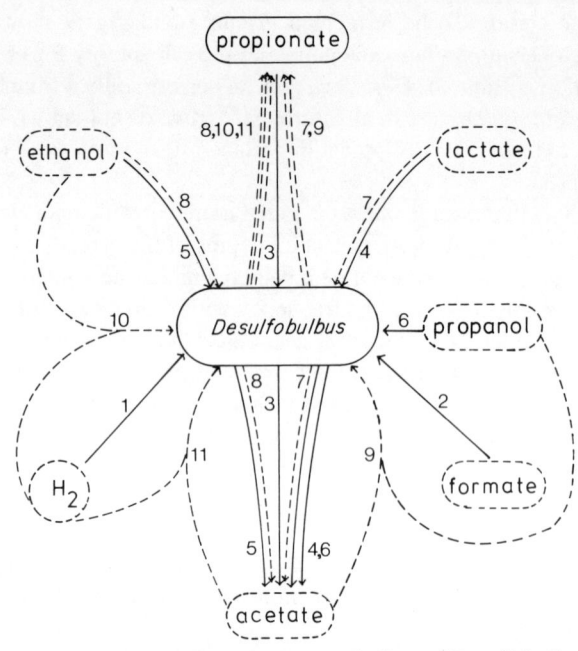

Fig. 9. Schematic presentation of the energy metabolism of *Desulfobulbus propionicus*. Solid lines in the presence of sulphate and dotted lines in its absence. The numbers correspond to the reactions presented in Table 3. From Laanbroek *et al.* (1982b).

vironment. Specialists with limited abilities always become dominant under conditions in which the turnover rate of their energy substrates is relatively high. In this particular case, turnover rates of energy substrates still have to be determined in the field. The observations were mentioned only to indicate that in order to learn why a particular bacterial species occurs in a particular environment and to explain its population density, a combination of fieldwork and laboratory studies is essential. Neither fieldwork nor laboratory studies (on the population, cellular, and molecular levels) alone will allow us to understand why a bacterial species can survive in the wilderness in which it is found. Generally speaking, there seems to be little doubt that exciting new developments in microbial ecology are expected at the interface between field and laboratory.

References

Alexander, M. (1981). Why microbial predators and parasites do not eliminate their prey and hosts. *Annu. Rev. Microbiol.* **35**, 113–133.
Atlas, M. and Bartha, R. (1981). "Microbial Ecology: Fundamentals and Applications." Addison-Wesley, Reading, Massachusetts.

Bell, C. R. and Albright, L. J. (1981). Attached and free-floating bacteria in the Fraser River Estuary, British Columbia, Canada. *Mar. Ecol.: Prog. Ser.* **6**, 317–327.
Beijerinck, M. W. (1921–1940). "Verzamelde Geschriften," Vols. 1–6. Nijhoff, The Hague.
Beudeker, R. F., Gottschal, J. C. and Kuenen, J. G. (1982). Reactivity versus flexibility in thiobacilli. *Antonie van Leeuwenhoek* **48**, 39–51.
Bohling, H. (1972). Gelöste Aminosäuren in Oberflachenwasser der Nordsee bei Helgoland: Konzentrationsveränderungen im Sommer 1970. *Mar. Biol. (Berlin)*, **16**, 281–289.
Boone, D. R. and Bryant, M. P. (1980). Propionate-degrading bacterium, *Syntrophobacter wolinii* sp.nov. gen.nov., from methanogenic ecosystems. *Appl. Environ. Microbiol.* **40**, 626–632.
Boylen, C. W. and Ensign, J. C. (1970). Long-term starvation survival of rod and spherical cells of *Arthrobacter crystallopoietes*. *J. Bacteriol.* **103**, 569–577.
Bryant, M. P. (1976). The microbiology of anaerobic degradation and methanogenesis with special reference to sewage. *In* "Microbial Energy Conversion" (Eds. H. G. Schlegel and J. Barnea), pp. 107–117. E. Goltze KG, Göttingen.
Bryant, M. P. (1979). Microbial methane production—theoretical aspects. *J. Anim. Sci.* **48**, 193–201.
Bryant, M. P., Wolin, E. A., Wolin, M. J. and Wolfe, R. S. (1967). *Methanobacillus omelianskii*, a symbiotic association of two species of bacteria. *Arch. Mikrobiol.* **59**, 20–31.
Bungay, H. R., III and Bungay, M. L. (1968). Microbial interactions in continuous culture. *Adv. Appl. Microbiol.* **10**, 269–290.
Burney, C. M., Davis, P. G., Johnson, K. M. and Sieburth J. Mc.N. (1982). Diel relationships of microbial trophic groups and in situ dissolved carbohydrate dynamics in the Caribbean Sea. *Mar. Biol. (Berlin)* **67**, 311–322.
Chervin, M. B., Malone, T. C. and Neale, P. J. (1981). Interactions between suspended organic matter and copepod grazing in the plume of the Hudson River. *Estuarine Coastal Shelf Sci.* **13**, 169–183.
Christensen, D. and Blackburn, T. H. (1980). Turnover of tracer (^{14}C, ^{3}H labelled) alanine in inshore marine sediments. *Mar. Biol. (Berlin)* **58**, 97–103.
Conn, H. J. (1948). The most abundant groups of bacteria in soil. *Bacteriol. Rev.* **12**, 257–273.
Conn, H. J. and Dimmick, I. (1947). Soil bacteria similar in morphology to *Mycobacterium* and *Corynebacterium*. *J. Bacteriol.* **54**, 291–303.
Dalton, H. and Stirling, D. J. (1982). Co-metabolism. *Philos. Trans. R. Soc. London, Ser. B* **297**, 481–496.
Dawson, R. and Gocke, K. (1978). Heterotrophic activity in comparison to the free amino acid concentrations in Baltic Sea water samples. *Oceanol. Acta* **1**, 45–54.
Degens, E. T., Reuter, J. H. and Shaw, K. N. F. (1964). Biochemical compounds in offshore California sediments and sea waters. *Geochim. Cosmochim. Acta* **28**, 45–66.
Duncan, A., Schiemer, F. and Klekowski, R. Z. (1974). A preliminary study of feeding rates on bacterial food by adult females of a benthic nematode, *Plectus palustris* De Man 1880. *Pol. Arch. Hydrobiol.* **21**, 249–258.
Ellwood, D. C., Melling, J. and Rutter, P. (Eds.) (1979). "Adhesion of Microorganisms to Surfaces." Academic Press, London.
Fenchel, T. (1969). The ecology of marine microbenthos. IV. *Ophelia* **6**, 1–182.
Fenchel, T. (1980a). Suspension feeding in ciliated protozoa: Functional response and particle size selection. *Microb. Ecol.* **6**, 1–11.

Fenchel, T. (1980b). Suspension feeding in ciliated protozoa: Feeding rates and their ecological significance. *Microb. Ecol.* **6,** 13–25.
Fenchel, T. (1980c). Relation between particle size selection and clearance in suspension-feeding ciliates. *Limnol. Oceanogr.* **25,** 733–738.
Fenchel, T. (1982a). Ecology of heterotrophic microflagellates. I. Some important forms and their functional morphology. *Mar. Ecol.: Prog. Ser.* **8,** 211–223.
Fenchel, T. (1982b). Ecology of heterotrophic microflagellates. II. Bioenergetics and growth. *Mar. Ecol.: Prog. Ser.* **8,** 225–231.
Fenchel, T. and Blackburn, T. H. (1979). "Bacteria and Mineral Cycling." Academic Press, London.
Fenchel, T. and Jørgensen, B. B. (1977). Detritus food chains of aquatic ecosystems: The role of bacteria. *Adv. Microb. Ecol.* **1,** 3–37.
Franz, H. G. and Gieskes, W. W. C. (1984). The unbalance of phytoplankton production and copepod production in the North Sea. Symposium of Biological Productivity of the North Atlantic Shelf Areas. *Rapp. P. V. Réun. Cons. Int. Explor. Mer* **183,** 218–225.
Frederickson, A. G. (1977). Behavior of mixed cultures of microorganisms. *Annu. Rev. Microbiol.* **31,** 63–89.
Fuhrman, J. A. (1981). Influence of method on the apparent size distribution of bacterioplankton cells: epifluorescence microscopy compared to scanning electron microscopy. *Mar. Ecol.: Prog. Ser.* **5,** 103–106.
Fuhrman, J. A. and Azam, F. (1980). Bakterioplankton secondary production estimates for coastal waters of British Columbia, Antarctica, and California. *Appl. Environ. Microbiol.* **39,** 1085–1095.
Fuhrman, J. A. and Azam, F. (1982). Thymidine incorporation as a measure of heterotrophic bacterioplankton production in marine surface waters: evaluation and field results. *Mar. Biol. (Berlin)* **66,** 109–120.
Gocke, K., Dawson, R. and Liebezeit, G. (1981). Availability of dissolved free glucose to heterotrophic microorganisms. *Mar. Biol. (Berlin)* **62,** 209–216.
Greenwood, D. J. and Goodman, D. (1967). Direct measurement of the distribution of oxygen in soil aggregates and in columns of fine soil crumbs. *J. Soil Sci.* **18,** 182–196.
Hagström, Å., Larsson, U. Hörstedt, P. and Normark, S. (1979). Frequency of dividing cells, a new approach to the determination of bacterial growth rates in aquatic environments. *Appl. Environ. Microbiol.* **37,** 805–812.
Harder, W. and Dijkhuizen, L. (1982). Strategies of mixed substrate utilization in microorganisms. *Philos. Trans. R. Soc. London, Ser. B* **297,** 459–480.
Harder, W., Kuenen, J. G. and Matin, A. (1977). Microbial selection in continuous culture—a review. *J. Appl. Bacteriol.* **43,** 1–24.
Harvey, R. W. and Young, L. Y. (1980). Enumeration of particle-bound and unattached respiring bacteria in the salt marsh environment. *Appl. Environ. Microbiol.* **40,** 156–160.
Hattori, T. (1973). "Microbial Life in the Soil." Dekker, New York.
Heal, O. W. and MacLean, S. F., Jr. (1975). Comparative productivity in ecosystems—secondary production. *In* "Unifying Concepts in Ecology" (Eds. W. H. van Dobben and R. H. Lowe-McConnell), pp. 89–109. Junk Publ., The Hague.
Hirsch, P. (1979). Life under conditions of low nutrient concentrations. *In* "Strategies of Microbial Life in Extreme Environments" (Ed. M. Shilo), Dahlem Conf. Life Sci. Res. Rep. 13, pp. 357–372. Verlag Chemie, Weinheim.
Hoppe, H.-G. (1976). Determination and properties of actively metabolizing bacteria in the sea, investigated by means of microautoradiography. *Mar. Biol. (Berlin)* **36,** 291–302.

Hoppe, H.-G. (1978). Relations between active bacteria and heterotrophic potential in the sea. *Neth. J. Sea Res.* **12**, 78–98.

Horan, N. J., Midgley, M. and Dawes, E. A. (1981). Effect of starvation on transport, membrane potential and survival of *Staphylococcus epidermidis* under anaerobic conditions. *J. Gen. Microbiol.* **127**, 223–230.

Jannasch, H. W. (1967). Enrichments of aquatic bacteria in continuous culture. *Arch. Mikrobiol.* **59**, 165–173.

Jannasch, H. W. and Pritchard, P. H. (1972). The role of inert particulate matter in the activity of aquatic microorganisms. *Mem. Ist. Ital. Idrobiol. Dott, Marco de Marchi* **29**, Suppl., 289–308.

Jensen, H. L. (1952). The coryneform bacteria. *Annu. Rev. Microbiol.* **6**, 77–90.

Jørgensen, B. B. (1977). The sulfur cycle of a coastal marine sediment (Limfjorden, Denmark). *Limnol. Oceanogr.* **22**, 814–832.

Jørgensen, B. B. (1980). Mineralization and the bacterial cycling of carbon, nitrogen and sulphur in marine sediments. *In* "Contemporary Microbial Ecology" (Eds. D. C. Ellwood, J. N. Hedger, M. J. Latham, J. H. Slater and L. M. Lynch), pp. 239–251. Academic Press, London.

Jost, J., Drake, J. F., Frederickson, A. G. and Tsuchiya, H. M. (1973). Interactions of *Tetrahymena pyriformis, Escherichia coli, Azotobacter*, and glucose in a minimal medium. *J. Bacteriol.* **113**, 834–840.

Karl, D. M. (1979). Measurement of microbial activity and growth in the ocean by rates of stable ribonucleic acid synthesis. *Appl. Environ. Microbiol.* **38**, 850–860.

Karl, D. M. (1981). Simultaneous rates of ribonucleic acid and deoxyribonucleic acid syntheses for estimating growth and cell division of aquatic microbial communities. *Appl. Environ. Microbiol.* **42**, 802–810.

Kirchman, D. and Mitchell, R. (1982). Contribution of particle-bound bacteria to total microheterotrophic activity in five ponds and two marshes. *Appl. Environ. Microbiol.* **43**, 200–209.

Kluyver, A. J. and van Niel, C. B. (1956). "The Microbe's Contribution to Biology." Harvard Univ. Press, Cambridge Massachusetts.

Konings, W. N. and Veldkamp, H. (1980). Phenotypic responses to environmental change. *In* "Contemporary Microbial Ecology" (Eds. D. C. Ellwood, J. N. Hedger, M. J. Latham, J. H. Slater, and J. M. Lynch), pp. 161–193. Academic Press, London.

Kristjansson, J. K., Schönheit, P. and Thauer, R. K. (1982). Different K_s values for hydrogen of methanogenic bacteria and sulfate-reducing bacteria: an explanation for the apparent inhibition of methanogenesis by sulfate. *Arch. Microbiol.* **131**, 278–282.

Kuenen, J. G., Boonstra, J., Schröder, H. G. J. and Veldkamp, H. (1977). Competition for inorganic substrates among chemoorganotrophic and chemolithotrophic bacteria. *Microb. Ecol.* **3**, 119–130.

Laanbroek, H. J. and Pfennig, N. (1981). Oxidation of short-chain fatty acids by sulfate-reducing bacteria in freshwater and in marine sediments. *Arch. Microbiol.* **128**, 330–335.

Laanbroek, H. J. and Veldkamp, H. (1982). Microbial interactions in sediment communities. *Philos. Trans. R. Soc. London, Ser. B* **297**, 533–550.

Laanbroek, H. J., Smit, A. J., Klein-Nulend, G. and Veldkamp, H. (1979). Competition for L-glutamate between specialised and versatile *Clostridium* species. *Arch. Microbiol.* **120**, 61–66.

Laanbroek, H. J., Abee, T. and Voogd, I. L. (1982). Alcohol conversions by *Desulfobulbus propionicus* Lindhorst in the presence and absence of sulfate and hydrogen. *Arch. Microbiol.* **133**, 178–184.

Laanbroek, H. J., Geerligs, H. J., Peijnenburg, A. A. C. M. and Siesling, J. (1983). Competition for L-lactate between *Desulfovibrio baculatus*, *Veillonella alcalescens* and an *Acetobacterium* species, isolated from intertidal sediments. *Microb. Ecol.* **9**, 341-354.

Laanbroek, H. J., Geerligs, H. J., Sijtsma, L. and Veldkamp, H. (1984). Competition for sulfate and ethanol among *Desulfobacter*, *Desulfabulbus*, and *Desulfovibrio* species isolated from intertidal sediments. *Appl. Environ. Microbiol.* **47**, 329-334.

Larssen, U. and Hagström, A. (1982). Fractionated phytoplankton primary production, exudate release and bacterial production in a balthic eutrophication gradient. *Mar. Biol. (Berlin)*, **67**, 57-70.

Law, A. T. and Button, D. K. (1977). Multiple-carbon-source-limited growth kinetics of a marine coryneform bacterium. *J. Bacteriol.* **129**, 115-123.

Lovley, D. R., Dwyer, D. F. and Klug, M. J. (1982). Kinetic analysis of competition between sulfate reducers and methanogens for hydrogen in sediments. *Appl. Environ. Microbiol.* **43**, 1373-1379.

Lynch, J. M. and Poole, N. J. (1979). "Microbial Ecology." Blackwell, Oxford.

McInerney, M. J., Bryant, M. P. and Pfennig, N. (1979). Anaerobic bacterium that degrades fatty acids in syntrophic association with methanogens. *Arch. Microbiol.* **122**, 129-135.

Mah, R. A. and Smith, M. R. (1981). The methanogenic bacteria. *In* "The Prokaryotes: A Handbook on Habitats, Isolation, and Identification of Bacteria" (Eds. M. P. Starr, H. Stolp, H. G. Trüper, A. Balows and H. G. Schlegel), pp. 948-977. Springer-Verlag, Berlin and New York.

Marshall, K. C. (1976). "Interfaces in Microbial Ecology." Harvard Univ. Press, Cambridge Massachusetts.

Marshall, K. C. (1980). Reactions of microorganisms, ions and macromolecules at interfaces. *In* "Contemporary Microbial Ecology" (Eds. D. C. Ellwood, J. N. Hedger, M. J. Latham, J. H. Slater and J. M. Lynch), pp. 93-107. Academic Press, London.

Matin, A. and Veldkamp, H. (1978). Physiological basis of the selective advantage of a *Spirillum* sp. in a carbon-limited environment. *J. Gen. Microbiol.* **105**, 187-197.

Meiberg, J. B. M. (1979). Metabolism of Methylated Amines in *Hyphomicrobium* spp. Ph.D. Thesis, University of Groningen, The Netherlands.

Meyer-Reil, L.-A. (1978). Autoradiography and epifluorescence microscopy combined for the determination of number and spectrum of actively metabolizing bacteria in natural waters. *Appl. Environ. Microbiol.* **36**, 506-512.

Meyer-Reil, L.-A., Bölter, M., Dawson, R., Liebezeit, G., Szwerinski, H. and Wolter, K. (1980). Interrelationships between microbiological and chemical parameters of sandy beach sediments, a summer aspect. *Appl. Environ. Microbiol.* **39**, 797-802.

Monod, J. (1942). "Recherches sur la croissance des cultures bactériennes." Hermann, Paris.

Monod, J. (1950). La technique de culture continu; théorie et applications. *Ann. Inst. Pasteur, Paris* **79**, 390-410.

Mopper, K., Dawson, R., Liebezeit, G. and Ittekkot, V. (1980). The monosaccharide spectra of natural waters. *Mar. Chem.* **10**, 55-66.

Moriarty, D. J. W. and Pollard, P. C. (1981). DNA synthesis as a measure of bacterial productivity in seagrass sediments. *Mar. Ecol.: Prog. Ser.* **5**, 151-156.

Mountfort, D. O., Asher, R. A., Mays, E. L. and Tiedje, J. M. (1980). Carbon and electron flow in mud and sandflat intertidal sediments at Delaware Inlet, Nelson, New Zealand. *Appl. Environ. Microbiol.* **39**, 686-694.

Munro, A. L. S. and Brock, T. D. (1968). Distinction between bacterial and algal utilization of soluble substances in the sea. *J. Gen. Microbiol.* **51**, 35–42.
Mur, L. R., Gons, H. J. and van Liere, L. (1977). Some experiments on the competition between green algae and blue-green bacteria in light-limited environments. *FEMS Microbiol. Lett.* **1**, 335–338.
Newell, S. Y. and Christian, R. R. (1981). Frequency of dividing cells as an estimator of bacterial productivity. *Appl. Environ. Microbiol.* **42**, 23–31.
Novick, A. and Szilard, L. (1950a). Description of the chemostat. *Science* **112**, 715–716.
Novick, A. and Szilard, L. (1950b). Experiments with the chemostat on spontaneous mutation of bacteria. *Proc. Natl. Acad. Sci. U.S.A.* **36**, 708–719.
Novitsky, J. A. and Morita, R. Y. (1978). Possible strategy for the survival of marine bacteria under starvation conditions. *Mar. Biol. (Berlin)* **48**, 289–295.
Odum, E.P. (1971). "Fundamentals of Ecology," 3rd ed. Saunders, Philadelphia.
Oremland, R. S. and Taylor, B. F. (1978). Sulphate reduction and methanogenesis in marine sediments. *Geochim. Cosmochim. Acta* **42**, 209–214.
Paerl, H. W. and Merkel, S. M. (1982). Differential phosphorus assimilation in attached vs. unattached microorganisms. *Arch. Hydrobiol.* **93**, 125–134.
Payne, W. J. and Wiebe, W. J. (1978). Growth yield and efficiency in chemosynthetic microorganisms. *Annu. Rev. Microbiol.* **32**, 155–185.
Perkins, D. F. (1978). The distribution and transfer of energy and nutrients in the *Agrostis-Festuca* grassland ecosystem. *In* "Production Ecology of British Moors and in Montane Grasslands" (Eds. O. W. Heal and D. F. Perkins), pp. 375–395. Springer-Verlag, Berlin.
Pfennig, N., Widdel, F. and Trüper, H. G. (1981). The dissimilatory sulfate-reducing bacteria. *In* "The Prokaryotes: A Handbook on Habitats, Isolation and Identification of Bacteria" (Eds. M. P. Starr, H. Stolp, H. G. Trüper, A. Balows and H. G. Schlegel), pp. 928–940. Springer-Verlag, Berlin and New York.
Phillipson, J. (1973). The biological efficiency of protein production by grazing and other land-based systems. *In* "The Biological Efficiency of Protein Production" (Ed. J. G. W. Jones), pp. 217–235. Cambridge Univ. Press, London and New York.
Poindexter, J. S. (1981). Oligotrophy. *Adv. Microb. Ecol.* **5**, 63–91.
Rieper, M. (1982). Feeding preferences of marine harpacticoid copepods for various species of bacteria. *Mar. Ecol.: Prog. Ser.* **7**, 303–307.
Rozanova, E. P. and Nazina, T. N. (1976) A mesophilic sulfate-reducing rod-shaped, nonspore-forming bacterium. *Microbiology* **45**, 711–716.
Schlegel, H. G. (Ed.) (1965). "Anreicherungskultur und Mutantenauslese." Fischer, Stuttgart.
Schönheit, P., Kristjansson, J. K. and Thauer, R. K. (1982). Kinetic mechanism for the ability of sulfate reducers to outcompete methanogens for hydrogen and acetate. *In* "Abstracts of FEMS Symposium on Physiology, Ecology and Taxonomy of Sulfate-Reducing Bacteria." Freiburg, Federal Republic of Germany.
Sepers, A. B. J. (1977). The utilization of dissolved organic compounds in aquatic environments. *Hydrobiologia* **52**, 39–54.
Sepers, A. B. J. (1981). Diversity of ammonifying bacteria. *Hydrobiologia* **83**, 343–350.
Sieburth, J. McN., Johnson, K. M., Burney, C. M. and Lavoie, D. M. (1977). Estimation of in situ rates of heterotrophy using diurnal changes in dissolved organic matter and growth rates of picoplankton in diffusion culture. *Helgol. Wiss. Meeresunters.* **30**, 565–574.
Slater, J. H. and Bull, A. T. (1982). Environmental microbiology: Biodegradation. *Philos. Trans. R. Soc. London, Ser. B* **297**, 575–597.

Sørensen, J., Jørgensen, B. B. and Revsbech, N. R.(1979). A comparison of oxygen, nitrate and sulfate respiration in coastal marine sediments. *Microb. Ecol.* **5**, 105–115.

Sørensen, J., Christensen, D. and Jørgensen, B. B. (1981). Volatile fatty acids and hydrogen as substrates for sulfate-reducing bacteria in anaerobic marine sediments. *Appl. Environ. Microbiol.* **42**, 5–11.

Steele, J. H. (1974). "The Structure of Marine Ecosystems." Harvard Univ. Press, Cambridge Massachusetts.

Stephens, G. C. (1981). The trophic role of dissolved organic material. *In* "Analysis of Marine Ecosystems" (Ed. A. R. Longhurst), pp. 271–291. Academic Press, London.

Stouthamer, A. H. (1979). The search for correlation between theoretical and experimental growth yields. *Int. Rev. Biochem.* **21**, 1–48.

Stoutjesdijk, R. (1977). High surface temperatures of trees and pine litter in the winter and their biological importance. *Int. J. Biometeorol.* **21**, 325–331.

Strange, R. E. (1976). "Microbial Response to Mild Stress." Meadowfield Press, Shildon.

Teal, J. M. (1962). Energy flow in the salt marsh ecosystem in Georgia. *Ecology* **43**, 614–624.

Tempest, D. W. (1970a). The place of continuous culture in microbiological research. *Adv. Microb. Physiol.* **4**, 223–251.

Tempest, D. W. (1970b). The continuous cultivation of micro-organisms. I. Theory of the chemostat. *In* "Methods in Microbiology" (Eds. J. R. Norris and D. W. Ribbons), Vol. 2, pp. 259–276. Academic Press, London.

Tempest, D. W. and Neijssel, O. M. (1978). Eco-physiological aspects of microbial growth in aerobic nutrient-limited environments. *Adv. Microb. Ecol.* **2**, 105–154.

Tempest, D. W., Herbert, D. and Phipps, P. J. (1967). Studies of the growth of *Aerobacter aerogenes* at low dilution rates in a chemostat. *In* "Microbial Physiology and Continuous Culture, Proceedings of the 3rd International Symposium" (Eds. E. D. Powell, C. G. T. Evans, R. E. Strange and D. W. Tempest), pp. 240–254. H. M. Stationery Office, London.

Thauer, R. K., Jungermann, K. and Decker, K. (1977). Energy conservation in chemotrophic anaerobic bacteria. *Bacteriol. Rev.* **41**, 100–180.

van Es, F. B. (1982). Community metabolism of intertidal flats in the Ems-Dollard Estuary. *Mar. Biol. (Berlin)* **66**, 95–102.

van Es, F. B. and Meyer-Reil, L.-A. (1982). Biomass and metabolic activity of heterotrophic marine bacteria. *Adv. Microb. Ecol.* **6**, 111–170.

van Gemerden, H. (1974). Coexistence of organisms competing for the same substrate: an example among the purple sulfur bacteria. *Microb. Ecol.* **1**, 104–109.

van Gemerden, (1980). Survival of *Chromatium vinosum* at low light intensities. *Arch. Microbiol.* **125**, 115–121.

van Niel, C. B. (1949). The kinetics of growth of microorganisms. *In* "The Chemistry and Physiology of Growth" (Ed. A. K. Parpart), pp. 91–106. Princeton Univ. Press, Princeton, New Jersey.

van Niel, C. B. (1955). Natural selection in the microbial world. *J. Gen. Microbiol.* **13**, 201–217.

Veldkamp, H. (1970). Enrichment cultures of prokaryotic organisms. *In* "Methods in Microbiology" (Eds. J. R. Norris and D. W. Ribbons), Vol. 3A, pp. 305–362. Academic Press London.

Veldkamp, H. (1976). "Continuous Culture in Microbial Physiology and Ecology." Meadowfield Press, Durham.

Veldkamp, H. (1977). Ecological studies with the chemostat. *Adv. Microb. Ecol.* **1,** 59–95.
Widdel, F. and Pfennig, N. (1981). Studies on dissimilatory sulfate-reducing bacteria that decompose fatty acids. I. Isolation of new sulfate-reducing bacteria enriched with acetate from saline environments. Description of *Desulfobacter postgatei* gen.nov., sp.nov. *Arch. Microbiol.* **129,** 395–400.
Widdel, F. and Pfennig, N. (1982). Studies on the dissimilatory sulfate-reducing bacteria that decompose fatty acids. II. Incomplete oxidation of propionate by *Desulfobulbus propionicus* gen.nov., sp. nov. *Arch. Microbiol.* **131,** 360–365.
Williams, P. J. LeB. (1973). On the question of growth yields of natural heterotrophic populations. *In* "Modern Methods in the Study of Microbial Ecology" (Ed. T. Rosswall), pp. 400–401. Bull. Ecol. Res. Comm., Stockholm.
Williams, P. J. LeB. (1981a). Microbial contribution to overall marine plankton metabolism: Direct measurements of respiration. *Oceanol. Acta* **4,** 359–364.
Williams, P. J. LeB. (1981b). Incorporation of microheterotrophic processes into the classical paradigm of the planktonic food web. *Kiel. Meeresforsch, Sonderh.* **5,** 1–28.
Wimpenny, J. W. T. (1981). Spatial order in microbial ecosystems. *Biol. Rev. Cambridge Philos. Soc.,* **56,** 295–342.
Winogradski, S. (1949). "Microbiologie du sol. Oeuvres complètes." Masson, Paris.
Woldendorp, J. W. (1980). Micro-organismen in de bodem en hun relatie tot planten. *In* "Oecologie van micro-organismen" (Ed. H. Veldkamp), pp. 107–137. Pudoc, Wageningen.
Wolter, K. (1982). Bacterial incorporation of organic substances released by natural phytoplankton populations. *Mar. Ecol.: Prog. Ser.* **7,** 287–295.
Wright, R. T., Coffin, R. B., Ersing, C. P. and Pearson, D. (1982). Field and laboratory measurements of bivalve filtration of natural marine bacterioplankton. *Limnol. Oceanog.* **27,** 91–98.
Zimmermann, R. (1977). Estimation of bacterial number and biomass by epifluorescence microscopy and scanning electron microscopy. *In* "Microbial Ecology of a Brackish Water Environment" (Ed. G. Rheinheimer), pp. 103–120. Springer-verlag, Berlin.
Zimmermann, R., Iturriaga, R. and Becker-Birck, J. (1978). Simultaneous determination of the total number of aquatic bacteria and the number thereof involved in respiration. *Appl. Environ. Microbiol.* **36,** 926–935.

2

Energy Sources for Microbial Growth: An Overview

W. A. HAMILTON

Department of Microbiology
University of Aberdeen
Marischal College
Aberdeen, Scotland, United Kingdom

Introduction

For the microbial physiologist, particularly one used to thinking in terms of pure cultures growing in batch, or closed, systems, the first impression of microbial ecology is almost certainly one of complexity and confusion. Similarly, the protein chemist or molecular biologist might cast a rather jaundiced eye over the myriad variable but interdependent metabolic activities that constitute the essence of cellular physiology. Both of these views are, of course, quite wrong, or at least incomplete. In moving from one level of biological organisation to the next, it is necessary first to establish new frames of reference and then to seek out the underlying themes of unity that can both direct our research efforts and constitute a framework upon which to build knowledge and develop understanding. In cellular physiology and microbial ecology, one such common underlying theme is bioenergetics. By examining each system in terms of the mechanistic, kinetic, and thermodynamic parameters of the processes of energy transduction, the basic structure and biological activity of that system are made manifest, however overtly complex they may otherwise appear to be.

In a stimulating and thought-provoking article, Wimpenny (1981) has put forward an argument similar to that presented here, but with a particular emphasis on the significance of spatial heterogeneity within ecosystems.

Ecology

Odum (1971) has defined an ecosystem as a biological community interacting with its abiotic surroundings in a manner that results in energy flow, trophic structure, and materials cycling. Valid studies in microbial ecology therefore properly include analyses of the individual organisms active in and isolatable from a given ecosystem, characterisation of the physical and chemical parameters of the environment, and descriptions, both qualitative and quantitative, of the various biological activities of the ecosystem. The most meaningful of such studies are those which seek to unravel the interactions among these components, and so give a fuller description of the ecosystem as an integrated dynamic functional whole. It is my thesis that the standpoint of energetics offers the best experimental and conceptual framework for the study of the individual organisms and the ecosystem as a whole. As with nutrition and product formation in cellular metabolism, trophic structure and materials cycling are essentially secondary expressions of energy flow.

Physiology

The parallels that I have drawn between cellular physiology and microbial ecology are quite deliberate. Bioenergetics has proved to be central both to the metabolic functioning of the cell itself and to our understanding of these processes, in particular their integration and control. Indeed, it was the purpose of the Symposium of the Society for General Microbiology on "Microbial Energetics" in 1977 to "focus attention on the cross-fertilisation that is currently enriching the fields of bioenergetics and microbial physiology." Remarkably, however, bioenergetics has often been associated more with confusion and controversy than with clarity and consensus. That this is no longer the case stems largely from the power of the chemiosmotic theory to illuminate not only oxidative and photosynthetic phosphorylation but also such otherwise apparently diverse functions as reversed electron flow, nutrient transport, regulation of intracellular pH and ionic and osmotic balance, and motility. Although it would be unwise to extrapolate detailed molecular mechanisms to specific problems in microbial ecology, it is salutary to examine the basic concepts on which chemiosmosis is founded, and to consider whether these also constitute ideas and approaches which might generate a unifying view of the manifold interactions of microorganisms with their biotic and abiotic surroundings.

Chemiosmosis

The fundamental principle at the heart of bioenergetics is the coupling of reactions by group transfer. For example, in dehydrogenation reactions, the group transferred from the donor AH_2 to the acceptor B

$$AH_2 \rightarrow B$$
$$A \leftarrow BH_2$$

2. ENERGY SOURCES FOR MICROBIAL GROWTH: AN OVERVIEW

is a pair of hydrogen atoms. The fact that B is most commonly NAD allows for further hydrogen transfer from NADH, either in substrate reduction (most commonly via the NADPH/NADP couple) or by way of the electron transport chain to oxygen or an alternative terminal electron acceptor. Similarly, in a phosphoryl group transfer reaction, ATP may serve as the donor and a sugar such as glucose as the acceptor, being thus converted to glucose-6-phosphate (G-6-P).

$$\text{ATP} \searrow \nearrow \text{Glucose} \\ \text{ADP} \nearrow \searrow \text{G-6-P}$$

Such coupled group transfer reactions fulfill two general functions: They facilitate the conduction of groups between individual reactants and between the metabolic pathways of which these reactants are intermediates, and they equilibrate the group potential in that, for example, a more reduced compound can donate hydrogens and so reduce a more oxidised compound. Group transfer reactions thus respond to the free energy parameters of the reaction or system in question (the thermodynamic or energetic function), while at the same time constituting the necessary reaction pathway for this response (the mechanistic and kinetic function).

In chemiosmosis it is further recognised that reactions are vectorial. That is, they have a directional component in which the path of diffusion of the reacting species to and from the enzyme active site is a legitimate part of the molecular mechanism. Such a vectorial character has supramolecular dimensions and can be readily demonstrated experimentally when the enzyme, or transport carrier, is sited within a membrane and is functionally anisotropic. Group transfer can thus give rise to group translocation. As a striking example of this, one can consider the phosphotransferase (PTS) system for sugar uptake, which is widely distributed among anaerobic and facultative bacteria. The cytoplasmic Enzyme I catalyses the phosphoryl group transfer from phosphoenol pyruvate (PEP) to the protein HP_r. The Enzyme II complex then catalyses the further phosphoryl group transfer to the sugar, such as glucose.

$$\text{Pyruvate} \swarrow \nearrow HP_r\sim P \searrow \nearrow \text{Glucose} \\ \text{PEP} \nearrow \searrow HP_r \nearrow \searrow \text{G-6-P}$$

Enzyme II is, however, an integral membrane protein and functionally anisotropic such that glucose reacts from the periplasm, with G-6-P being released into the cytoplasm. The PTS is thus a molecular enzymatic and transport mechanism that catalyses glucosyl group translocation and for which the energetic driving force is in the form of phosphoryl group transfer from a "high-energy" to a "low energy" compound.

Other group translocating mechanisms are, of course, the H^+-translocating ATPase and electron transport systems in which, associated with the drop in free energy resulting from phosphoryl group transfer from ATP to water, or from hydrogen group transfer from a low to a higher redox potential, is the conservation of energy in the form of a transmembrane proton gradient, the proton motive force. Thus, we have the basis of the primary mode of biological energy transduction.

Equally important, however, at least in the present context, is the stress that is laid by chemiosmosis on the functional importance of the structural organisation of the cell, particularly, of course, in the case of eukaryotic organisms. Not only do membranes assume a cardinal importance, but membrane-bounded organelles allow for specialisation of function with consequent increases in efficiency and versatility. This degree of complexity further draws to our attention the need for mechanisms of communication and energy flow among the various cellular compartments. Thus, the ubiquitous proton current between two aqueous phases separated by an ion-impermeable membrane is the primary energy conservation mechanism: ATP, in addition to being the main energy currency directly involved in metabolic transformations, is also the form in which energy generated in the mitochondria is translocated to, for example, the H^+- or Na^+/K^+-translocating ATPase of the cytoplasmic membrane: It is pyruvate, first produced by glycolysis in the cytoplasm, which then enters the mitochondria for further energy-generating metabolism to acetyl CoA and the reactions of the tricarboxylic acid (TCA) cycle.

Finally, this brief consideration of the essence of chemiosmosis highlights the total dependence of bioenergetics on chemical transformations (largely carbon flux) involving nutrients, metabolic intermediates, and products, either cellular material or excretion products. In particular, redox reactions are seen to be of central importance in regard to both the mechanisms of oxidative and photosynthetic phosphorylation and the exergonic reactions of metabolism itself.

An interesting and much fuller account of the development of these fundamental concepts of chemiosmosis is contained in a retrospective view by Mitchell (1981).

In turning our attention to microbial ecology and the interactions between microbial communities and their immediate environment, it is clear, therefore, that we must first consider the energetic capabilities and potentials of the individual organisms. In terms of their concerted action, however, and the activity of the ecosystem as a whole, is there in addition anything to be gained from attempting to extend and develop the ideas that have proved to be so powerful in clarifying the complexities of cellular physiology? Can we identify a similar interdependence between energetics and carbon flux? Are redox considerations of the same determinative importance? Does the flow of energy through the ecosystem and the coordination of the individual cellular activities depend upon mechanisms suggestive of coupling, group transfer, and group translocation?

What are the identities of any ionic or molecular species serving such a communicative role between microbial species? In combination, do such factors offer a unifying rationale for an integrated view of the structural and functional organisation which is an evident characteristic of microbial ecosystems?

In support of this approach to energy transformations within complex communities of organisms, I cannot do better than quote from Mitchell (1981): "I can hardly overemphasize that I never intended the concepts of specific ligand conduction and chemical group translocation, on which the chemiosmotic theory is based, to be confined to systems coupled by proticity."

Energy-Generating Mechanisms

Although microorganisms play such a crucial role in the various elemental cycles within the biosphere, it is evident that the individual organisms, although satisfying their own particular requirements, have no concept of any part they may be playing in some grand design. In the same vein, microorganisms have no knowledge of the Second Law of Thermodynamics, although it is the dictates of this universal law of nature which determine their particular requirements. The considerable individuality among the microorganisms stems from the various mechanisms that they have developed in response to environmental pressures to trap, conserve, and transduce the energy required for their biosynthesis and growth. Many of these energy-transducing mechanisms will be dealt with in depth by other contributors to this symposium (see also Hamilton, 1979). Here, only a comparatively brief reference is possible to this most striking characteristic, in particular of prokaryotes.

Phototrophy

Ultimately, all biological reactions are driven by the radiant energy of the sun, and phototrophy is duly found among both eukaryotic and prokaryotic microorganisms. Whereas the algae and cyanobacteria demonstrate oxygenic photosynthesis in which water is the primary reductant or source of reducing equivalents required for the conversion CO_2 to cell carbon, the eubacterial phototrophs are nonoxygenic and are found in anaerobic environments where the primary reductant is either inorganic (generally a reduced sulphur compound) or organic. It is interesting to note in passing that the predominant method of obtaining organic material in anaerobic environments is by diffusion from aerobic environments rather than from primary productivity within the anaerobic environment itself. The interface between aerobic and anaerobic (O_2/AnO_2) regions is therefore of considerable importance, and we shall have occasion to return to this point later.

Lithotrophy

Chemolithotrophy is a mode of nutrition uniquely found within a limited range of bacterial species. Both the energy and the reducing equivalents required for autotrophic growth by CO_2 fixation are supplied by the oxidation of reduced inorganic compounds such as ammonia, nitrite, hydrogen sulphide, ferrous iron, or hydrogen, although the last substrate can also be oxidised by many heterotrophic species. It is probably also acceptable to include here the methylotrophs for which the energy-generating substrates are strictly inorganic (lacking a carbon:carbon covalent bond), although the routes of carbon assimilation are from formaldehyde rather than CO_2.

Organotrophy

All other microbial and higher life forms obtain their energy and reducing equivalents, as well as their cell carbon, from the oxidation of organic compounds. Two particular features of chemoorganotrophy are evident only among microorganisms, however, and are of crucial importance to their dominant role in the biological activity of ecosystems. Firstly, the range of materials that can be degraded by different microbial species is virtually limitless; it extends from simple sugars and amino acids to hydrocarbons and even hitherto nonexistent products of human synthetic chemical activities. Secondly, oxygen is not necessarily the obligatory terminal electron acceptor. This point is the keystone of the archway that this chapter is attempting to build, and it must therefore be discussed and elaborated on at greater length.

Apart from the direct involvement of molecular oxygen in such reactions as the initial step in the breakdown of hydrocarbons, biological oxidations are carried out by the removal from the substrate of hydrogen atoms or electrons. The normal hydrogen acceptors are the pyridine nucleotides NAD and NADP, although with certain substrates other components of the terminal respiratory system, such as flavoproteins or cytochromes, can serve this function. As these components are present only in catalytic amounts, it is necessary for the continuing oxidative catabolism of the cell that they also serve as hydrogen donors and are themselves reoxidised.

$$AH_2 \rightarrow NAD^+ \rightarrow CH_2$$
$$A \leftarrow NADH + H^+ \leftarrow C$$

Since the anabolic reactions of cell synthesis are largely reductive, the catabolic and anabolic processes are in this manner directly coupled through the pyridine nucleotide-dependent hydrogen group transfer. But in addition to supplying the carbon intermediates and reducing equivalents required for biosynthesis, the oxidative reaction sequences of catabolism are the mechanism of energy generation in chemoorganotrophy. Thus, more reducing equivalents are produced from the necessary excess catabolic activity than are required for anabolism, and consequently some other mechanism must exist for their reoxidation.

Fermentation

In fermentation a relatively simple expedient is adopted. The NADH produced during the oxidative breakdown of a sugar residue to pyruvate is itself reoxidised to NAD by donating its hydrogens to convert the pyruvate to a more reduced fermentation product. Ethanol is the principal fermentation product from yeast, and lactic acid from mammalian muscle. Once again, the metabolic activity and versatility of the bacteria are strikingly demonstrated by the far wider range of fermentation substrates and products that are characteristic of the prokaryotes, both as a group and within individual species. The essential features of all fermentations are as follows: (1) they occur in the absence of, and indeed are generally inhibited by, oxygen; (2) the determining influence is the need to achieve an intracellular redox balance; (3) substrate-level amounts of reduced fermentation products are produced extracellularly; (4) energetically, the process is inefficient, as much of the free energy of the primary substrate is retained in the fermentation product(s), and is therefore at least potentially still available to the ecosystem in which the fermentation is taking place.

Respiration

Much more widespread is the process of respiration. Through the agency of the so-called electron transport or terminal respiratory system, the hydrogens and electrons removed in substrate dehydrogenation are passed from NADH through a series of carriers to oxygen, which as the terminal (hydrogen and) electron acceptor is reduced to water. The characteristic features of respiration are as follows: (1) it requires the presence of oxygen; (2) the cells retain intracellular redox balance; (3) normally, respiration is associated with a functional TCA cycle and the consequent catabolism of the substrate to CO_2, with the absence of any partial breakdown products; (4) the process is coupled to ATP synthesis by oxidative phosphorylation and is energetically efficient.

Anaerobic Respiration

The use of alternative terminal electron acceptors in anaerobic respiration is, however, exclusive to the bacteria. A significant number of genera have the capacity to reduce nitrate to nitrite, nitrogen, or even ammonia, or fumarate to succinate, rather than reducing oxygen to water. Nitrate and fumarate are genuine alternative electron acceptors in that the organisms in question can generally also use oxygen, and will do so preferentially. The sulphate-reducing bacteria and methanogens, on the other hand, are obligate anaerobes whose biological activity and growth are dependent on the dissimilatory reduction of sulphate to sulphide and of CO_2 to methane, respectively. Whereas the reductions of nitrate, fumarate, and sulphate involve relatively minor modifications to the terminal respiratory system as employed in aerobic respiration, and are likewise coupled to oxidative (more correctly, electron transport-linked) phosphorylation, the methanogens operate by a fundamentally different mechanism which is only incompletely understood. A particularly full account of the various bacterial anaerobic energy conservation mechanisms has been given by Thauer *et al.* (1977). The characteristic features of anaerobic respiration are these: (1) it is found only in the absence of oxygen; (2) the cells again maintain intracellular redox balance; (3) carbon flux is generally considerably restricted in terms of substrates that can be oxidised and by the absence of a functional TCA cycle; (4) the reduced electron acceptors represent a significant reservoir of free energy which can then be utilised by other members of the microbial community, either within the anaerobic environment or after diffusion across the O_2/AnO_2 interface.

Environmental Implications

What emerges from these very general statements regarding microbial energy metabolism is an intriguing picture of the different modes of structural and functional organisation displayed by naturally occurring ecosystems. Within aerobic environments there is a plethora of organic materials, formed initially by biological primary productivity, with possible modification thereafter by various biotic and abiotic processes. These organics form the nutrient sources for the growth and activity of chemoorganotrophic organisms. By virtue of the respiratory mode of energy transduction characteristic of these species, each organism is capable of extracting the maximum benefit from its chosen nutrient(s) for cell synthesis and the accompanying generation of the necessary energy and reducing equivalents. To a large extent, each cellular species acts independently, with an economy firmly based on the conversion of organic compounds as extracellular energy sources into ATP as the intracellular energy currency.

In anaerobic environments, on the other hand, although the principal energy source is again the complement of organic compounds, their utilisation by bacteria capable only of fermentation or anaerobic respiration exhibits major qualitative differences. Individual organisms have restricted energy-conserving capabilities, and a significant proportion of the free energy available from the nutrients is returned to the environment in the form of reduced and therefore energy-rich compounds, organic in the case of fermentation, inorganic with anaerobic respiration. These reduced products of anaerobic metabolism then become sources of energy and of reducing equivalents for other organisms. Thus, the turnover of organic material and the flow of energy through an anaerobic ecosystem are dependent upon the concerted action of organisms with differing but complementary activities, with functional integration being achieved by the intercellular diffusion of materials such as hydrogen, acetate, hydrogen sulphide, and methane. In many cases, this communication extends into an adjacent aerobic region such that maximal biological activity within an anaerobic ecosystem is generally close to the O_2/AnO_2 interface. This mechanism of energy transduction within a community of organisms is analogous, at least in general concept, to the coupling and group translocation which similarly direct and control the equivalent process within an individual cell.

Molecular Mechanisms

What is necessary now is to explain in more detail the molecular mechanisms of these effects, and to illustrate the general principle with specific examples.

When a reduced substrate AH_2 is oxidised to A by dehydrogenation, we refer to the two forms, AH_2 and A, as a "redox couple." The so-called redox potential is a measure of the tendency of the couple to act as hydrogen or electron donor (being itself oxidised to A) or to function as an acceptor (being consequently reduced to AH_2). The redox potential is measured by comparison with the standard redox couple, which is the hydrogen electrode, where hydrogen gas is in contact with protons in solution in the presence of platinum as a catalyst. The reaction is

$$H_2 \rightleftharpoons 2H^+ + 2e$$

and the tendency to donate reducing equivalents, in this case as electrons, is the measured potential when the electrode is electrically coupled with a reference redox couple electrode. Standard conditions are said to be 25°C at pH 7 with, in the case of the hydrogen electrode, hydrogen at a partial pressure of 1 atm. The symbol $E^{0'}$ is used for the standard redox potential, which for hydrogen has a value of -420 mV. For comparison, the half reaction at the oxygen electrode is

$$O^{2-} \rightleftharpoons \tfrac{1}{2}O_2 + 2e$$

with $E^{0'}$ equal to 820 mV. In combination with the hydrogen electrode, the complete redox reaction is

$$H_2 + \tfrac{1}{2}O_2 = H_2O$$

in which hydrogen donates electrons and is oxidised, whereas oxygen accepts electrons and is consequently reduced. Virtually all biologically important redox reactions have standard potentials between these two extreme values, and they are coupled according to the same principle evident in the reaction of hydrogen and oxygen to form water: A redox couple of low potential will always donate its reducing equivalents to a couple of higher potential. Where the flow of reducing equivalents is in the direction of decreasing redox potential, the reaction must be driven or coupled to an energy-yielding reaction in some manner, as in ATP-driven reversed electron transport. The magnitude of the free energy change is given by

$$\Delta G^{0'} = -nF\Delta E^{0'}$$

where $\Delta G^{0'}$ is the standard free energy change, n is the number of electrons transferred, F is the Faraday constant (96.649 kJ V^{-1} mol^{-1}) and $\Delta E^{0'}$ is the difference between the two redox potentials expressed in volts. For the oxidation of hydrogen, therefore, $\Delta G^{0'} = -240$ kJ mol^{-1}.

Thus, we have a measure of the potential free energy of any given oxidisable substrate, together with a basis for quantifying the redox-dependent energy transformations that occur in individual cells and in microbial communities where the basic mode of nutrition is chemoorganotrophy. In the terminal respiratory system, for example, a series of hydrogen and electron carriers of increasing redox potential mediate the passage of reducing equivalents from the NADH/NAD couple ($E^{0'} = -320$ mV) to oxygen or to an alternative electron acceptor. This system is so organised, of course, that the free energy released in the redox reactions is transduced into phosphoryl group energy in the form of ATP. Since the redox potentials of the alternative electron acceptors used in anaerobic respiration are all lower than that of oxygen (for example, $E^{0'}$ values for the various redox couples involved in sulphate reduction to sulphide span the range +225 to −402 mV; for CO_2/CH_4, $E^{0'} \rightleftharpoons -244$ mV), it is evident that correspondingly less redox, and ultimately less phosphoryl group energy, is available to the organisms using this mode of metabolism.

The redox potential of any given environment can also be determined in the same manner as for a particular redox couple. The measured values are recorded in terms of E_h. (Note that it is the $E^{0'}$, or standard potential, that is given in tables for individual redox couples; the actual E_h values will differ, depending on the pH and the concentrations of the reacting species.) An environment from which oxygen has been removed will have an E_h in the region of +200 mV. Only in the presence of reducing agents, such as sulphide, for example, will the E_h be lowered to −100 mV or less. It is the environmental E_h, rather than the presence or absence of oxygen, that determines what respiratory or fermentation processes are operating in any particular ecosystem. Sulphate reduction is normally associated with an E_h of −100 to −150 mV and methanogenesis with an E_h of about −300 mV.

Nutritional Interrelationships

Interspecies Hydrogen Transfer

Perhaps the first major statement on the detailed nature of the intimate metabolic relationships between populations of different microbial species along the lines that I am proposing here was presented in the symposium on "Syntrophism and Other Microbial Interactions" held at the International Congress of Microbiology in Munich in 1978, and since published in book form (Gottschalk et al., 1980). In discussing hexose fermentation by *Clostridium thermocellum, Selenomonas ruminantium,* and in particular by *Ruminococcus albus,* Tewes and Thauer (1980) showed that the products of fermentation, and consequently ATP yield, were radically affected by the hydrogen partial pressure. When this was kept low, the fermentation products were 2 acetate, 2 CO_2, and 4 hydrogen per molecule of hexose fermented. The ATP yield was 4, with 2 of these coming from the conversion of acetyl CoA to acetate. When, on the other hand, the hydrogen partial pressure was allowed to remain close to 1 atm. *R. albus* produced less hydrogen and instead converted some of the acetyl CoA to the reduced product ethanol, with a consequent decrease in ATP yield. The fermentation balance altered to 1.3 acetate, 0.7 ethanol, 2 CO_2, 2.6 hydrogen, and 3.3 ATP. As discussed more fully by Thauer et al. (1977), a particularly effective way of maintaining a low partial pressure of hydrogen is to grow *R. albus* in a chemostat in co-culture with a hydrogen-oxidising anaerobe such as *Vibrio succinogenes.* Through the process of interspecies hydrogen transfer, the fermentative organism gains in that it achieves a higher energy yield from its catabolism, whereas the hydrogen-utilising species is provided with a source of both energy and reducing equivalents. In the community as a whole, the carbon turnover is greater in that there is increased synthesis of cell material and the fermentation products remaining in the medium are more oxidised. Such a metabolic interdependence, or coupling, between species has been named "syntrophism."

Syntrophy based on interspecies hydrogen transfer is by no means restricted to fermentative reactions. Probably the most celebrated example is *Methanobacillus omelianskii,* which was shown by Bryant et al. (1967) to be an association of two separate bacterial species. The so-called S organism grows on ethanol, with the production of acetate and hydrogen, and the second organism, a methanogen, grows on hydrogen and CO_2, which is reduced to methane. The remarkable feature of this consortium is that the conversion of ethanol to acetate and hydrogen is endergonic with $\Delta G^{0'} = +9.62$ kJ mol^{-1}, and is therefore normally incapable of supporting growth. Only the removal of the hydrogen by the methanogen allows the reaction to proceed significantly in the direction of product formation, and ethanol to serve as an energy source for the S organism.

Bryant and his colleagues (McInerney et al., 1979, 1981; Boone and Bryant, 1980) have reported other examples of such obligate syntrophy in which the growth of newly identified genera on saturated fatty acids up to octonoate can be demonstrated, but only in the presence of a second hydrogen-utilising species. Growth rates of the consortia are extremely low, with doubling times being quoted in the range 54–84 hr, and it must be stressed that there is a close physical association between the species; the fatty acid oxidisers *Syntrophobacter wolinii* and *Syntrophomonas wolfei* have not been isolated and grown in pure culture. Since the fatty acid oxidation produces hydrogen, acetate, and, in the case of odd-numbered fatty acids, propionate, these organisms are termed *hydrogen-producing acetogenic bacteria*.

In the Munich symposium, McInerney and Bryant (1980) suggested that the complete degradation of complex substrates such as carbohydrate, protein, or lipid materials in anaerobic environments is dependent on three broad groups of microorganisms: the fermenters, the hydrogen-producing acetogenic bacteria, and those organisms capable of utilising hydrogen. In marine environments, this last group is likely to be the sulphate reducers, and otherwise the methanogens. Where the acetogenic species are growing on fatty acids, only the removal of the hydrogen allows the reaction to proceed. Therefore, the interspecies hydrogen transfer fulfills a mechanistic and a thermodynamic function, both of which are central to the trophic structure and materials cycling of the complete ecosystem.

Acetate Food Chain Dependence

Winter and Wolfe (1980a,b) have studied the acetogenic bacterium *Acetobacterium woodii*. This organism is fundamentally different from the acetogens discussed above in that it is perfectly capable of growing in pure culture on a fermentable substrate, such as fructose, which is converted to 3 molecules of acetate with a doubling time on the order of 6 hr. *Acetobacterium woodii* is also capable of growth driven by acetate formation from hydrogen and CO_2, and from the distribution of radioactive labelling it appears that carbon atoms 1 and 2, and 5 and 6 of hexoses are fermented directly to acetate, with the third molecule arising by the reduction of CO_2 produced from carbon atoms 3 and 4. Moreover, when *A. woodii* is grown in coculture with any one of a number of methanogens, the fermentation balance is altered to 2 acetate, 1 CO_2, and 1 methane. Thus, the transfer of hydrogen (and of 1-carbon units) from *A. woodii* allows methane formation and growth of the methanogenic species. One particular species of methanogen, *Methanosarcina barkeri*, can be trained to grow on acetate, although methanogenesis is maintained at significant rates only at acetate concentrations of 10 mM or more. Satisfactory cocultures of *A. woodii* and acetate-trained *M. barkeri* were obtained only when the acetogen plus fructose were inoculated into a grown culture of the methanogen. Under these circumstances, fructose was first converted to 3 molecules of acetate, which were subsequently degraded completely to CO_2 and methane. To quote from Winter and Wolfe (1980a), "while the interaction of hydrogen:CO_2-grown methanogens with *Acetobacterium woodii* in mixed cultures represents a real syntrophic association, the combination of acetate-grown *Methanosarcina barkeri* and *Acetobacterium woodii* is more of the type of a food chain dependence."

Thus, we see that microbial fermentative and acetogenic activities, individually and in combination, can account for the anaerobic degradation of the majority of organic materials to two principal products, hydrogen and acetate, and that mechanisms exist for the further metabolism of these products. The mechanisms are, however, of quite different character. In interspecies hydrogen transfer, the associations, both physical and metabolic, are tight, with each partner in the syntrophy deriving direct benefit. Indeed, where ethanol or a fatty acid is the primary substrate, it is only through the continuous utilisation of hydrogen that the growth of the acetogenic organism is thermodynamically feasible. It is therefore legitimate to consider interspecies hydrogen transfer as an example at the cellular level of energetic coupling by group translocation.

Where acetate produced by one organism is used as a carbon and energy source by a second species, any comparisons with intracellular mechanisms of energy transduction must be far less direct, although a parallel may be drawn with the flux of pyruvate from cytosol to mitochondrion. The passage of acetate within an ecosystem will be diffusion limited and will allow for significant spatial and/or temporal separation between the reacting species.

In both cases, however, it is clear that those anaerobic bacteria capable of growth on hydrogen and/or acetate must be of considerable importance in terms of the overall biological activity displayed by anoxic environments. As already indicated, two groups of organisms are responsible, the methanogens and the sulphate-reducing bacteria.

Methanogenic and Sulphate-Reducing Bacteria

Both laboratory and field studies have shown the methanogens to be capable of growth on hydrogen: CO_2 and on acetate (see Taylor, 1982, for a review). Evidence has been found that methanogenesis also occurs in saltwater environments, but rather with formaldehyde, methanol, and methylamines as growth substrates (Oremland *et al.*, 1982a,b; Nedwell and Banat, 1981; Banat *et al.*, 1983). The situation with the sulphate reducers is more complicated, but probably only in the sense that more factual information is available concerning these organisms. Many species of *Desulfovibrio* are capable of growth on hydrogen as the source of energy and reducing equivalents (Brandis and Thauer 1981). Until relatively recently, however (Postgate, 1979), it was considered that growth with acetate as energy source was not possible. (During growth on hydrogen, 70% of the assimilated carbon comes from acetate and 30% from CO_2.) In general, the nutritional spectrum of the sulphate reducers was believed to be comparatively restrictive, with growth on substrates such as lactate being associated with the production of acetate as a metabolic end product. This would be consistent with the lack of a complete TCA cycle functioning in an energy-generating sense, although the two linear pathways leading to α-oxoglutarate (oxidative) and succinate (reductive) would be required for biosynthesis. This last point accords with the presence of menaquinone ($E^{0'} = -74$ mV) in sulphate reducers rather than ubiquinone ($E^{0'} = +113$ mV). Menaquinone is more suited to act as a donor of reducing equivalents for the reduction of fumarate to succinate ($E^{0'} = +33$ mV).

This picture has now been radically altered by the findings of Pfennig and his co-workers, and in particular by Widdel (Pfennig and Biebl, 1976; Laanbroek and Pfennig, 1981; Widdel and Pfennig, 1981a,b; Pfennig et al., 1981). Sulphate-reducing bacteria are an extensive group, far more varied in form and function than was previously realised, including species that reduce sulphur rather than sulphate. They can be divided into two broad nutritional groups: (1) those which can grow on lactate, but not on acetate, propionate, or butyrate, and (2) those which can grow on fatty acids up to C_{18} (including acetate), benzoate, and aromatic compounds, and autotrophically on CO_2. Organisms in the first group will normally produce acetate as an end product, whereas the second group can be subdivided into (1) those which give incomplete oxidation of fatty acids (again with acetate as a product), and (2) those which can oxidise fatty acids and aromatics completely to CO_2. Brandis-Heep et al. (1982) examined the mechanism of acetate oxidation by *Desulfobacter postgatei*. There appears to be a fully functional TCA cycle, the only unusual feature of which is the absence of succinate thiokinase, with the alternative succinyl CoA:acetate CoA transferase also being the mechanism of acetate activation. Despite the reservations expressed above, succinate dehydrogenase and malate dehydrogenase are both functional, with menaquinone as an electron acceptor.

Thus, it is clear that both methanogens and sulphate reducers have the capacity to act as terminal oxidisers in anaerobic microbial ecosystems. Under certain conditions, it is possible, however, that the sulphate reducers might have a more complex role to play in the metabolism of hydrogen and acetate. As discussed above, many species of sulphate reducers not only cannot oxidise acetate but do in fact excrete this metabolite as the end product of their incomplete catabolism. Similarly, hydrogen can be both oxidised and produced by sulphate reducers, depending on the species and conditions (Tsuji and Yagi, 1980; Traore et al., 1981). Odom and Peck (1981) have even proposed a model for hydrogen cycling as a mode of energy transduction in hydrogenase-positive strains of *Desulfovibrio*. More important, however, in the present context are the studies of syntrophic growth on lactate or ethanol and interspecies hydrogen transfer between sulphate reducers and methanogens (Bryant *et al.*, 1977; McInerney and Bryant, 1981), and studies of Laube and Martin (1981) on the conversion of cellulose to methane and CO_2 by triculture of *Acetivibrio cellulolyticus, Desulfovibrio,* sp., and *Methanosarcina barkeri.* Whether such activities are significant in naturally occurring ecosystems, however, must remain an open question at the present time (see, however, Jones *et al.*, 1982).

Methane and Hydrogen Sulphide as Extracellular Energy Currencies

In discussing anaerobic respiration earlier, I pointed out that, at least in thermodynamic terms, the reduced alternative terminal electron acceptors are as valid a source of energy and reducing equivalents for microbial growth as are hydrogen and the various organic products of fermentation and acetogenic activity. The only difference lies in the organisms capable of reaping this rich harvest; they must be lithotrophs rather than organotrophs.

A striking feature of the chemolithotrophic organisms capable of oxidising reduced sulphur compounds (*Thiobacillus, Beggiatoa, Thiothrix*) or methane (methylotrophs) is that they are aerobes. Therefore, before hydrogen sulphide or methane can be reoxidised, or serve an energetic coupling role in terms of the holistic view of ecosystems being presented in this chapter, they must diffuse across the O_2/AnO_2 interface into an aerobic environment. This would therefore be a diffusion-limited food chain dependence comparable with that displayed by acetate, but over a longer range and crossing environmental boundaries. These latter qualifications will be eased by virtue of the gaseous nature of both hydrogen sulphide and methane. Additionally, the diffusion of sulphide out of and sulphate back into an anaerobic environment may be of particular importance in terms of sulphide toxicity and sulphate limitation.

The existence of the phototrophic green and purple sulphur bacteria allows for the cycling of sulphur totally within the confines of an anaerobic environment (Biebl and Pfennig, 1978). It is important to note that in these cases the reduced sulphur is supplying only reducing equivalents, the energy for growth coming directly from sunlight. That it is unwise to be too pedantic, however, about whether a given nutrient is supplying energy, reducing equivalents, or carbon is reinforced by the demonstration by these authors that the syntrophy between *Chlorobium* and *Desulfovibrio* species, in addition to a sulphide/sulphate cycle, involves the assimilation by the phototroph of acetate produced by the sulphate reducer. Such a relationship has the effect of increasing the physiological

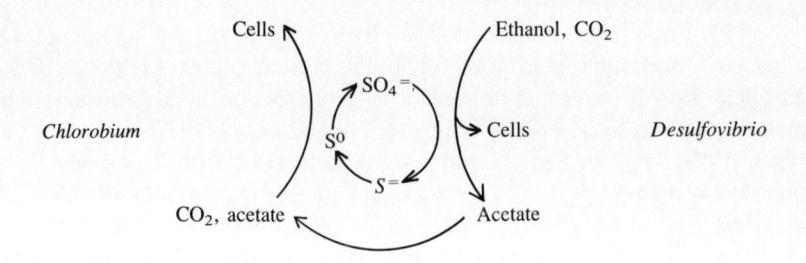

```
                    Cells ↖           ↗ Ethanol, CO₂
                          ⟩    ↗SO₄⁼↘(
 Chlorobium               ⟩  S⁰       ⟩→ Cells          Desulfovibrio
                          ⟩    ↖S⁼↙ 
                    CO₂, acetate ↖    ↘ Acetate
```

range of the green sulphur bacteria in terms of the organic compounds that can serve as sources of assimilatable carbon. This complexity was commented on further by Pfennig at the Munich symposium (Pfennig, 1980), where he drew attention to the as-yet unresolved nature of the interrelationships determining the form of certain syntrophic consortia between phototrophic green sulphur bacteria and unknown chemotrophic bacteria.

It seems clear, therefore, that bioenergetics offers a valid conceptual framework for the analysis of microbial ecosystems. The necessary factual information and supporting theory are available to allow definition of the thermodynamic potential of any given reaction or metabolic sequence and the limitations imposed on it by particular environmental parameters. Through the chemiosmotic theory, we can translate such rather abstract ideas into the recognisable reality of the metabolic activities of individual microbial species. The principles of coupling and group translocation, which are the embodiment of chemiosmosis, also have relevance in describing the thermodynamic and kinetic characteristics of microbial communities. Here, rather than metabolic sequences and organelles within a cellular matrix, it is discrete populations of organisms within an ecosystem which must be functionally coordinated. The necessary microbial species have all been identified, and many of the likely associations studied in some detail.

Now it is only necessary to question whether naturally occurring microbial ecosystems do in fact conform to the pattern suggested here, and to consider what experimental methods are required to grapple with such a problem.

Energy Flow in Natural Ecosystems

Organisms Present and Their Putative Activities

The most obvious method of establishing the workings of any given ecosystem involves first describing the biological activities of the system, and then isolating and characterising the organisms present that can fully account for these activities.

The estuarine sediments of the River Don in Aberdeenshire are rich in cellulosic material derived from paper mill effluent, and from sewage fungus which grows in the river immediately below the mills and then sloughs off and is carried down into the estuary. Associated with this nutrient load are anoxic conditions within the sediment and the production of hydrogen sulphide as an environmental pollutant (Poole et al., 1977; Madden et al., 1980). A new species of anaerobic cellulolytic bacterium, *Clostridium papyrosolvens*, has been isolated from these sediments (Madden et al., 1982). The fermentation products from cellulose include hydrogen, CO_2, ethanol, acetate, and lactate. In a companion study (Bryder, 1981), evidence has been obtained for sulphate-reducing bacteria of both lactate- and acetate-oxidising types in the estuarine sediment. Laboratory experiments were set up in which cellulose (filter paper strip) degradation was examined after inoculation with *Cl. papyrosolvens*, either alone or in combination with a sulphate reducer. The pure culture demonstrated limited cellulose breakdown, with a buildup of the fermentation products hydrogen and acetate. In coculture with hydrogenase-positive *Desulfovibrio salexigens*, there was increased cellulose degradation, no evidence of hydrogen accumulation, but an increased buildup of acetate. Coculture with *Desulfotomaculum acetoxidans* (acetate-oxidising and hydrogenase-negative) again showed greater breakdown of cellulose, but in the absence of any accumulation of acetate. It is therefore concluded that the anoxic sediments of the estuary develop as a result of increased oxygen demand from the nutrient overloading of the system, and that within such an environment, cellulose degradation and sulphide production are interdependent through the utilisation by the sulphate-reducing bacteria of the fermentation products from the cellulolytic clostridia.

A somewhat analogous situation occurs in seawater-displacement oil storage systems on North Sea oil production platforms, where hydrocarbons are the primary nutrient source for microbial activity leading to anoxic conditions and the production of dangerous levels of hydrogen sulphide. In a preliminary study, a number of hydrocarbon-degrading strains have been isolated and assigned to the genera *Pseudomonas* and *Flavobacterium*. Their ability to degrade hexadecane, with the production of fatty acids as partial breakdown products, is being examined, along with an analysis of the dependence on time and oxygen tension (Ross, 1982; Ross and Hamilton, 1982). A number of sulphate-reducing bacteria have also been isolated from a range of lactate, succinate, acetate, fatty acid, and benzoate enrichments. The nutrient spectra of these isolates are being established, and attempts will be made to reestablish a hydrocarbon-degrading, sulphide-producing microbial community and to determine the details of the nutritional interrelationships.

Such studies suffer from a number of limitations, however. First, one can hope to isolate only those organisms for which one enriches, and that in turn is strongly dependent on one's knowledge about what types of organisms might be present. The significance of this statement is strikingly borne out by consideration of the previous standard use of lactate media to isolate sulphate reducers. Since the findings of Pfennig and his co-workers regarding the great range of nutritional types, such a limited approach is clearly untenable. Even with due care and attention to the isolation of species from the environment under study, it is virtually impossible to know with any certainty the relative abundance of the various organisms. This uncertainty is rooted in such problems as the relative fastidiousness of different organisms, the comparative effectiveness of particular media, the laboratory growth of organisms that are dormant in the ecosystem under study, the physical heterogeneity of the system, and the difficulties associated with culturing organisms from surfaces or clumps of material. In short, the type of results discussed above describe the putative activities of the given ecosystem rather than necessarily identifying what is actually going on at any particular time or under any particular set of conditions.

Activity Assays and Turnover in Situ

Information of this type can be obtained most readily by direct chemical analysis of the ecosystem with regard to possible primary nutrients, intermediates, and final products. Such analysis can be carried out either *in situ* or in laboratory simulations and model systems, and can be extended by the use of specific inhibitors such as molybdate, which inhibits sulphate reduction at concentrations below 20 mM. The most telling studies have also relied on radioactive tracers, ^{14}C- and ^{35}S-labelled compounds, to give detailed information on the turnover of particular components and on the activity of the ecosystem as a whole (see Taylor *et al.*, 1981, for a discussion of the methodology). Summarising the findings of a number of such studies, it is evident that sulphate reduction and methanogenesis do play major roles in organic turnover in anaerobic environments, and a clearer picture is emerging regarding the substrates used and the interactions between the two terminal respiratory processes.

It is apparent that in the sediments of eutrophic lakes, both sulphate reduction and methanogenesis may occur. The addition of sulphate inhibits methanogenesis as a result of stimulation of the sulphate-reducing bacteria, and it is considered that the mechanism of this effect is competition for substrates, principally hydrogen and acetate (Jones *et al.*, 1982; Winfrey and Zeikus, 1977; Lovley *et al.*, 1982). Kristjansson *et al.* (1982) have confirmed this with their demonstration that K_S values for hydrogen are 1 μM for *Desulfovibrio vulgaris* and 6 μM for *Methanobrevibacter arboriphilus*. The earlier idea that sulphide toxicity is responsible for the loss of methanogenic activity (Cappenberg, 1975) does not appear to hold true. Smith and Klug (1981) claim that even with the low concentrations of sulphate in eutrophic sediments, dissimilatory sulphate reduction is important and can account for up to 30% of the mineralisation of organic carbon entering the sediment.

The situation in estuarine and salt marsh sediments is rather different in that sulphate reduction is by far the major process, and although methanogenesis can occur, there is less evidence of competition for substrates. Although the earlier papers of Nedwell and his colleagues (Abram and Nedwell, 1978a,b; Nedwell and Banat, 1981) suggested that hydrogen and acetate might serve as substrates for both sulphate reduction and methanogenesis, their more recent findings (Banat et al., 1983) are consistent with the idea that formaldehyde, methanol, and methionine are the principal methanogenic substrates. A similar conclusion was reached by Oremland et al. (1982a,b), who found methanol and trimethylamine to account for the bulk of methane production in salt marsh sediments. These substrates do not stimulate sulphate reduction, and Balba and Nedwell (1982) have found acetate, propionate, and butyrate to be substrates for this process in their salt marsh sediments.

A particularly extensive study of sulphate reduction in coastal marine sediments has been pursued by Jørgensen and his colleagues (Jørgensen, 1977, 1980, 1981; Sørensen et al., 1979, 1981. *In situ* rates of [^{35}S]sulphate turnover have been recorded over a 2-year period, and this has allowed calculation of a complete budget for the sulphur cycle. Although very low rates of sulphate reduction are still measurable at depths of 1.5 m, 65% of the activity down to 2 m is found in the top 10 cm of sediment. Rates range from 10 to 50 (winter) to 50 to 400 nmol cm^{-3} day^{-1} (summer). The main forms of acid-soluble sulphide in the upper 5 to 10 cm of sediment are mackinawite (FeS) and greigite (Fe$_3$S$_4$), whereas below 10 cm conversion takes place to the insoluble pyrites (FeS$_2$). Approximately 10% of the sulphide formed by dissimilatory sulphate reduction is precipitated as insoluble metal salts, with the remaining 90% being recycled by oxidation, biotic and abiotic, at the O$_2$/AnO$_2$ interface. The number of sulphate-reducing bacteria cultured from a given sampling site is not necessarily related directly to the [^{35}S]sulphate turnover measured at the same site. Sulphate reduction can account for about 50% of the total mineralisation of organic material in the sediments. The principal substrates for sulphate reduction are hydrogen (~10%), acetate (~50%), propionate (~10%), and butyrate (~10%). Since most of the energy available to the sulphate reducers remains in the form of hydrogen sulphide, the reduced alternative electron acceptor, it can be calculated that 30–40% of the total energy flow in the sediment ecosystem is carried by hydrogen sulphide. For completeness, one can add that Higgins et al. (1981) estimate that approximately 50% of the organic carbon degraded by anaerobic microflora in the biosphere as a whole is converted to methane.

Conclusion

Thus, it is clear that the qualitative and quantitative features of energy flow, trophic structure, and materials cycling (Odum, 1971) exhibited by naturally occurring anaerobic ecosystems are exactly those predicted from first principles and from the judicious application of our knowledge of the metabolic potential of the various microbial species constituting the biological community. Moreover, it can be stressed again that the "first principles" and "metabolic potential" are those of bioenergetics, particularly as developed within the tenet of chemiosmosis. The application of the powerful concept of coupling through group translocation as a possible general mechanism for the coordination of the individual cellular activities, each severely constrained by thermodynamic limitations, has provided the framework for the development of our understanding of the functioning of ecosystems, where the whole is greater than the sum of the parts.

References

Abram, J. W. and Nedwell, D. B. (1978a). Inhibition of methanogenesis by sulphate reducing bacteria competing for transferred hydrogen. *Arch. Microbiol.* **117**, 89–92.

Abram, J. W. and Nedwell, D. B. (1978b). Hydrogen as a substrate for methanogenesis and sulphate reduction in anaerobic saltmarsh sediment. *Arch. Microbiol.* **117**, 93–97.

Balba, M. T. and Nedwell, D. B. (1982). Microbial metabolism of acetate, propionate and butyrate in anoxic sediment from a UK saltmarsh. *J. Gen. Microbiol.* **128**, 1415–1422.

Banat, I. M., Nedwell, D. B. and Balba, M. T. (1983). Stimulation of methanogenesis by slurries of saltmarsh sediment after addition of molybdate to inhibit sulphate-reducing bacteria. *J. Gen. Microbiol.* **129**, 123–129.

Biebl, H. and Pfennig, N. (1978). Growth yields of green sulfur bacteria in mixed cultures and sulfur and sulfate reducing bacteria. *Arch. Microbiol.* **117**, 9–16.

Boone, D. R. and Bryant, M. P. (1980). Propionate-degrading bacterium, *Syntrophobacter wolinii* sp. nov. gen. nov., from methanogenic ecosystems. *Appl. Environ. Microbiol.* **40**, 626–632.

Brandis, A. and Thauer, R. K. (1981). Growth of Desulfovibrio species on hydrogen and sulphate as sole energy source. *J. Gen. Microbiol.* **126**, 249–252.

Brandis-Heep, A., Gebhardt, N. and Thauer, R. K. (1982). Dissimilatory sulfate reduction with acetate as electron donor. *In* "Proceedings of FEMS Symposium on "Physiology, Ecology and Taxonomy of Sulfate-Reducing Bacteria." Freiburg, Federal Republic of Germany.

Bryant, M. P., Wolin, E. A., Wolin, M. J. and Wolfe, R. S. (1967). *Methanobacillus omelianskii*, a symbiotic association of two species of bacteria. *Arch. Mikrobiol.* **59**, 20–31.

Bryant, M. P., Campbell, L. L., Reddy, C. A. and Crabill, M. R. (1977). Growth of Desulfovibrio in lactate or ethanol media low in sulfate in association with hydrogen-utilising methanogenic bacteria. *Appl. Environ. Microbiol.* **33**, 1162–1169.

Bryder, M. J. (1981). Interrelationships between Sulphate-reducing Bacteria and Other Bacterial Groups in an Anoxic Estuarine Sediment. Ph.D. Thesis, University of Aberdeen.

Cappenberg, T. E. (1975). A study of mixed continuous cultures of sulfate-reducing and methane-producing bacteria. *Microb. Ecol.* **2**, 60–72.

Gottschalk, G., Pfennig, N. and Werner, H. (1980). "Anaerobes and Anaerobic Infections." Fischer, Stuttgart.
Haddock, B. A. and Hamilton, W. A. (1977). "Microbial Energetics," *Symp. Soc. Gen. Microbiol.* Vol. 27. Cambridge Univ. Press, London and New York.
Hamilton, W. A. (1979). Microbial energetics and metabolism. *In* "Microbial Ecology: A Conceptual Approach" (Eds. J. H. Lynch and N. J. Poole), pp. 22–44. Blackwell, Oxford.
Higgins, I. J., Best, D. J., Hammond, R. C. and Scott, D. (1981). Methane-oxidising microorganisms. *Microbiol. Rev.* **45**, 556–590.
Jones, J. G., Simon, B. M. and Gardener, S. (1982). Factors affecting methanogenesis and associated anaerobic processes in the sediments of a stratified eutrophic lake. *J. Gen. Microbiol.* **128**, 1–11.
Jørgensen, B. B. (1977). The sulfur cycle of a coastal marine sediment. *Limnol. Oceanogr.* **22**, 814–832.
Jørgensen, B. B. (1980). Mineralization and the bacterial cycling of carbon, nitrogen and sulphur in marine sediments. *In* "Contemporary Microbial Ecology" (Eds. D. C. Ellwood, J. N. Hedger, M. J. Latham, J. H. Slater and J. M. Lynch), pp. 239–251. Academic Press, London.
Jørgensen, B. B. (1982). Mineralization of organic matter in the sea bed—the role of sulphate reduction. *Nature (London)* **296**, 643–645.
Kristjansson, J. K., Schönheit, P. and Thauer, R. K. (1982). Different K_S values for hydrogen of methanogenic bacteria and sulfate reducing bacteria: an explanation for the apparent inhibition of methanogenesis by sulfate. *Arch. Microbiol.* **131**, 278–282.
Laanbroek, H. J. and Pfennig, N. (1981). Oxidation of short-chain fatty acids by sulfate-reducing bacteria in freshwater and in marine sediments. *Arch. Microbiol.* **128**, 330–335.
Laube, V. M. and Martin, S. M. (1981). Conversion of cellulose to methane and carbon dioxide by triculture of *Acetivibrio cellulolyticus*, *Desulfovibrio* sp., and *Methanosarcina barkeri*. *Appl. Environ. Microbiol.* **42**, 413–420.
Lovley, D. R. Dwyer, D. F. and Klug, M. J. (1982). Kinetic analysis of competition between sulfate reducers and methanogens for hydrogen in sediments. *Appl. Environ. Microbiol.* **43**, 1373–1379.
McInerney, M. J. and Bryant, M. P. (1980). Syntrophic associations of H_2-utilizing methanogenic bacteria and H_2-producing alcohol and fatty acid—degrading bacteria in anaerobic degradation of organic matter. *In* "Anaerobes and Anaerobic Infections" (Eds. G. Gottschalk, N. Pfennig and H. Werner), pp. 117–126.
McInerney, M. J. and Bryant, M. P. (1981). Anaerobic degradation of lactate by syntrophic associations of *Methanosarcina barkeri* and *Desulfovibrio* species and effect of H_2 on acetate degradation. *Appl. Environ. Microbiol.* **41**, 346–354.
McInerney, M. J., Bryant, M. P. and Pfennig N. (1979). Anaerobic bacterium that degrades fatty acids in syntrophic association with methanogens. *Arch. Microbiol.* **122**, 129–135.
McInerney, M. J., Bryant, M. P., Hespell, R. B. and Costerton, J. W. (1981). *Syntrophomonas wolfei* gen. nov. sp. nov, an anaerobic syntrophic fatty acid oxidising bacterium. *Appl. Environ. Microbiol.* **41**, 1029–1039.
Madden, R. H., Bryder, M. J. and Poole, N. J. (1980). The cellulolytic community of an anaerobic estuarine sediment. *In* "Proceedings of the International Conference on Energy from Biomass" (Eds. P. Chartier and D. O. Hall), pp. 366–371. Applied Science, London.
Madden, R. H., Bryder, M. J. and Poole, N. J. (1982). Isolation and characterization of

ananaerobic, cellulolytic bacterium, *Clostridium papyrosolvens* sp. nov. *Int. J. Syst. Bacteriol.* **32**, 87–91.

Mitchell, P. (1981). Bioenergetic aspects of unity in biochemistry: evolution of the concept of ligand conduction in chemical, osmotic, and chemiosmotic reaction mechanisms. *In* "Of Oxygen, Fuels and Living Matter" (Ed. G. Semenza), Part 1, pp. 1–160. Wiley, New York.

Nedwell, D. B. and Banat, I. M. (1981). Hydrogen as electron donor for sulphate reducing bacteria in slurries of salt marsh sediment. *Microb. Ecol.* **7**, 305–313.

Odom, J. M. and Peck, H. D. (1981). Hydrogen cycling as a general mechanism for energy coupling in the sulfate-reducing bacteria, *Desulfovibrio* sp. *FEMS Microbiol. Lett.* **12**, 47–50.

Odum, E. P. (1971). "Fundamentals of Ecology." Saunders, Philadelphia, Pennsylvania.

Oremland, R. S., Marsh, L. M. and Des Marais, D. J. (1982a). Methanogenesis in Big Soda Lake, Nevada: an alkaline, moderately hypersaline desert lake. *Appl. Environ. Microbiol.* **43**, 462–468.

Oremland, R. S., Marsh, L. M. and Polcin, S. (1982b). Methane production and simultaneous sulphate reduction in anoxic salt marsh sediments. *Nature (London)* **296**, 143–145.

Pfennig, N. (1980). Syntrophic mixed cultures and symbiotic consortia with phototrophic bacteria: a review. *In* "Anaerobes and Anaerobic Infections" (Eds. G. Gottschalk, N. Pfennig and H. Werner), pp. 127–131. Fischer, Stuttgart.

Pfennig, N. and Biebl, H. (1976). *Desulfuromonas acetoxidans* gen. nov. and sp. nov., a new anaerobic, sulfur-reducing, acetate-oxidising bacterium. *Arch. Microbiol.* **110**, 3–12.

Pfennig, N., Widdel, F. and Trüper, H. G. (1981). The dissimilartory sulfate-reducing bacteria. *In* "The Prokaryotes: A Handbook on Habitats, Isolation, and Identification of Bacteria" (Eds. M. P. Starr, H. Stolp, H. G. Trüper, A. Balows and H. G. Schlegel), pp. 926–940. Springer-Verlag, Berlin and New York.

Poole, N. J., Parkes, R. J. and Wildish, D. J. (1977). Reaction of estuarine ecosystems to effluent from pulp and paper industry. *Helgol. Wiss. Meeresunters.* **30**, 622–632.

Postgate, J. R. (1979). "The Sulphate-Reducing Bacteria." Cambridge Univ. Press, London and New York.

Ross, D. (1982). Ecological Studies on Sulphate Reducing Bacteria in Offshore Oil Storage Systems. Ph.D. Thesis, University of Aberdeen.

Ross, D. and Hamilton, W. A. (1982). Ecological studies on sulphate-reducing bacteria in offshore oil storage systems. *In* "Proceedings of FEMS Symposium on Physiology, Ecology and Taxonomy of Sulfate-Reducing Bacteria." Freiburg, Federal Republic of Germany.

Smith, R. L. and Klug, M. J. (1981). Reduction of sulfur compounds in the sediments of a eutrophic lake basin. *Appl. Environ. Microbiol.* **41**, 1230–1237.

Sørensen, J., Jørgensen, B. B. and Revsbeck, N. P. (1979). A comparison of oxygen, nitrate and sulphate respiration in coastal marine sediments. *Microb. Ecol.* **5**, 105–115.

Sørensen, J., Christensen, D. and Jørgensen, B. B. (1981). Volatile fatty acids and hydrogen as substrates for sulfate-reducing bacteria in anaerobic marine sediment. *Appl. Environ. Microbiol.* **42**, 5–11.

Taylor, C. D., Ljungdahl, P. O. and Molongoski, J. J. (1981). Technique for simultaneous determination of [^{35}S] sulphide and [^{14}C] carbon dioxide in anaerobic aqueous samples. *Appl. Environ. Microbiol.* **41**, 822–825.

Taylor, G. T. (1982). The methanogenic bacteria. *Prog. Ind. Microbiol.* **16**, 231–329.

Tewes, F. J. and Thauer, R. K. (1980). Regulation of ATP-synthesis in glucose fermenting bacteria involved in interspecies hydrogen transfer. In "Anaerobes and Anaerobic Infections" (Eds. G. Gottschalk, N. Pfennig and H. Werner), pp. 97–104. Fischer, Stuttgart.

Thauer, R. K., Jungermann, K. and Decker, K. (1977). Energy conservation in chemotrophic anaerobic bacteria. *Bacteriol. Rev.* **41,** 100–180.

Traore, A. S., Hatchikian, C. E., Belaich, J.-P. and LeGall, J. (1981). Microcalorimetric studies of the growth of sulfate-reducing bacteria: energetics of *Desulfovibrio vulgaris* growth. *J. Bacteriol.* **145,** 191–199.

Tsuji, K. and Yagi, T. (1980). Significance of hydrogen burst from growing cultures of *Desulfovibrio vulgaris* Miyazaki, and the role of hydrogenase and cyt c_3 in energy production system. *Arch. Microbiol.* **125,** 35–42.

Widdel, F. and Pfennig, N. (1981a). Studies on dissimilatory sulfate-reducing bacteria that decompose fatty acids. 1. Isolation of new sulfate-reducing bacteria enriched with acetate from saline environments. Description of *Desulfobacter postgatei* gen. nov. sp. nov. *Arch. Microbiol.* **129,** 395–400.

Widdel, F. and Pfennig, N. (1981b). Sporulation and further nutritional characteristics of *Desulfotomaculum acetoxidans. Arch. Microbiol.* **129,** 401–402.

Wimpenny, J. W. T. (1981). Spatial order in microbial ecosystems. *Biol. Rev. Cambridge Philos. Soc.* **56,** 295–342.

Winfrey, M. R. and Zeikus, J. G. (1977). Effect of sulfate on carbon and electron flow during microbial methanogenesis in freshwater sediments. *Appl. Environ. Microbiol.* **33,** 275–281.

Winter, J. V. and Wolfe, R. S. (1980a). Syntrophism in methane formation and utilisation. In "Anaerobes and Anaerobic Infections" (Eds. G. Gottschalk, N. Pfennig and H. Werner), pp. 105–115. Fischer, Stuttgart.

Winter, J. V. and Wolfe, R. S. (1980b). Methane formation from fructose by syntrophic associations of *Acetobacterium woodii* and different strains of methanogens. *Arch. Microbiol.* **124,** 73–79.

3

Changes in Oxygen Tension and the Microbial Metabolism of Organic Carbon

J. G. MORRIS

Department of Botany and Microbiology
University College of Wales
Aberystwyth, Dyfed, Wales, United Kingdom

Molecular oxygen is utilised by living organisms in two ways: (1) as an avid electron acceptor, and (2) as a source of the oxygen atoms incorporated into various substrates in reactions catalysed by the specialist enzymes termed "oxygenases." The first of these processes—oxygen acting as a potent oxidant—is by far the more widespread and quantitatively important of its functions. This propensity is exploited for the purpose of free energy conservation by all organisms capable of aerobic respiration, but it is also the basis of the intolerance of oxygen that is displayed by obligate anaerobes.*

Whether or not an organism is aerotolerant is evidently of considerable significance, since this will determine both its natural habitat and the manner in which it must be cultivated and studied in the laboratory. ["Aerotolerance" is defined as the capacity to grow and multiply in air at 1 atm (containing about 20.9% O_2 v/v).] Obligate anaerobes are aerointolerant and grow optimally in the total absence of oxygen. Obligate aerobes, on the other hand, are both aerotolerant and wholly reliant on an adequate supply of oxygen. The more versatile facultative anaerobes can thrive in either the presence or the absence of air. A further group—microaerophiles—may be distinguished as aerobes or facultative anaerobes which, despite utilising oxygen in a respiratory mode, are aerointolerant, growing best at low dissolved oxygen tensions.

Here, it may be helpful to remind ourselves how the oxygen contents of culture media are measured and expressed, for this can be a source of unnecessary confusion. The dissolved oxygen tension in a microbial culture medium is the partial pressure of O_2 in that medium, which in turn equals the partial pressure of O_2 in the atmosphere with which the medium is in steady-state equilibrium. Thus, a medium equilibrated with air supplied at 1 atm pressure

---

*In this chapter, "oxygen," unless otherwise qualified, refers to dioxygen (O_2).

(101.3 kPa) will have a dissolved oxygen tension of 0.209 atm (159 mm Hg, 21.2 kPa). It is this tension that is measured by the usual type (voltammetric or galvanic) of submerged oxygen electrode, and its value represents the activity of O_2 in the medium. However, the concentration (C) of dissolved oxygen is related to the dissolved oxygen tension (P) by the expression $C = P/f$, where f is an activity coefficient whose magnitude depends upon the constitution of the medium. Thus, water equilibrated at 30°C with air at 101.3 kPa will have a dissolved oxygen tension of 21.2 kPa but a dissolved oxygen concentration of 240 μM (about 7.6 mg liter^{-1}). A nutrient medium containing 5% glucose w/v similarly equilibrated with air to yield a dissolved oxygen tension of 21.2 kPa will have a dissolved oxygen concentration of only 184 μM (Brown, 1970). The dissolved oxygen tension is a determinant of the force driving the diffusion of O_2 from the medium into the microbial cell, but such characteristics as the microbe's affinity for oxygen and the specific rate of oxygen consumption will be expressed in concentration terms. It is therefore generally wise to report both the tension and the concentration of dissolved oxygen that exist in a microbial culture. Unfortunately, the normal oxygen electrodes (of the Mackereth or Clark types) are of limited sensitivity, generally being unable to measure accurately oxygen tensions corresponding to dissolved oxygen concentrations smaller than about 1 μM. However, much lower concentrations of oxygen *can* be assayed, for example, by using a membrane-covered *Photobacterium* probe whose photoemission varies linearly with oxygen concentration over the range 35 nM to 8.4 μM (Lloyd et al., 1981).

Dioxygen as a Source of Incorporated Oxygen Atoms

Since many microorganisms can thrive in the absence of dioxygen, it is evident that this is not an indispensable source of the oxygen atoms incorporated into biomass. Indeed, biosynthesis is mainly an anaerobic process even in obligate aerobes, with the oxygen atoms of cellular materials having their origin in supplied nutrients, carbon dioxide, and water. Thus, in aerobically grown *Escherichia coli,* less than 0.1% of cell material is derived from dioxygen. The exception arises when an oxygenase-dependent reaction initiates the metabolism of a substance provided as the sole source of carbon and energy for growth, as in methane utilisation when the aerobic methanotroph may derive up to 40% of its cellular content of oxygen from dioxygen. In many more instances, we find that oxygenases and other dioxygen-dependent enzymes serve in metabolic routes that are alternatives to equivalent anaerobic pathways which either have been displaced in the course of evolution or have been retained alongside the ''novel'' aerobic process. Aerobic routes for the synthesis of nicotinic acid and unsaturated fatty acids are representative of such alternative pathways (Morris, 1975).

An interesting example of an oxygen-dependent biosynthetic pathway serving a supplementary role is provided by the synthesis of ubiquinone-8 (Q-8) in *E. coli* K-12. Biosynthesis of Q-8 is accomplished by the alternating introduction of three hydroxyl and three methyl groups into the aromatic ring of 2-octaprenylphenol. During anaerobic growth of *E. coli* K-12 on glycerol plus fumarate, the hydroxyl groups are introduced by the action of hydroxylase enzymes, but under aerobic conditions they are generated by a dioxygen-utilising monooxygenase system. This system, charged with a pool of substrate 2-octaprenylphenol, is kept in a "standby" condition even in anaerobically growing cells. When these cells are exposed to air, the system is immediately made operational, and the higher levels of Q-8 characteristic of the aerobically growing organism are attained remarkably quickly (Knoell, 1981).

There remain certain products of biosynthesis that can be produced only by oxygen-dependent routes. For example, the cyclisation reaction whereby squalene is converted into lanosterol cannot be accomplished in the absence of dioxygen (Bloch, 1976), with the result that although anaerobically growing microbes may produce squalene, they cannot synthesise sterols. Consequently, should they require sterols, these must be supplied in their growth media, as is the case with *Saccharomyces cerevisiae* and *S. carlsbergensis,* which display a requirement for ergosterol, and for unsaturated fatty acids and nicotinic acid, only when growing anaerobically (Schatzmann, 1975). In such circumstances, oxygen serves as a growth nutrient, and its insufficient provision may limit the growth even of microorganisms with fully functional fermentative capacities (Harrison, 1976). Even more strikingly, the presence or absence of oxygen can determine the gross cellular structure of a microorganism. This is so in yeasts wherein dioxygen is required for the synthesis of haem (Mattoon *et al.,* 1979), for the production of sterols and multiply unsaturated fatty acids (Coccucci *et al.,* 1975), and for the generation of functional mitochondria from the precursor promitochondria produced during anaerobic growth (Roodyn and Wilkie, 1968). In turn, a change in the constitution of some cellular component imposed by the concentration of oxygen which prevailed during growth may, on occasion, have unexpected secondary effects on the metabolism of a microorganism. Thus, oxygen limitation caused cessation of excretion of L-lysine by *Corynebacterium glutamicum,* with an attendant several-fold increase in lysine content of the cells (to 1.6% w/w of the biomass). These events were apparently caused by alteration in the structure of the cytoplasmic membrane, particularly its lipid composition (Hänel *et al.,* 1981).

Besides participating in several aerobic biosynthetic processes, oxygenases may play a crucial role in some aerobic catabolic pathways. They are, for example, implicated in the aerobic degradation of aromatic hydrocarbons by species of *Pseudomonas* (Dagley, 1978), although alternative oxygen-independent means of accomplishing hydroxylative "ring cleavage" reactions are available to various anaerobes (Evans, 1977). Utilisation of paraffinic hydrocarbons is generally a dioxygen-dependent process involving oxygenases of relatively broad specificity as, for example, the methane oxygenases of methanotrophic bacteria (Dalton, 1981). In certain of these species, such as *Methylosinus trichosporium* OB36, whether or not the oxygenase is associated with a system of intracytoplasmic membranes may also be determined by the oxygen tension prevailing during growth (Scott *et al.*, 1981). The microbial (aerobic) cooxidation with hydrocarbons of organic compounds which are themselves unable to support bacterial growth is not uncommon (Perry, 1979). On the other hand, hydrocarbons which enter anaerobic environments, such as anoxic sediments, are well preserved and may persist indefinitely (Atlas, 1981).

Finally, not only oxygen itself but also immediate products of its microbial utilisation may be essential reactants in primary catabolic routes. There is evidence that lignin degradation by white rot, wood-destroying fungi such as *Phanerochaete chrysosporium*, involves "oxygen radicals," possibly the hydroxyl free radical (Faison and Kirk, 1982; Forney and Reddy, 1982). If so, this would represent a nice combination of the roles of oxygen as electron acceptor and nutrient, with the reactant hydroxyl radicals being generated by its partial reduction.

Dioxygen as an Avid Electron Acceptor

Oxygen and the Metabolism of Obligate Anaerobes

Although all obligate anaerobes are hypersensitive to oxygen toxicity, they vary greatly in their tolerance of oxygen. At one extreme are those exceptionally oxygen-sensitive (EOS) anaerobes for which exposure to oxygen is rapidly fatal, whereas at the other end of the spectrum of oxygen tolerance are a number of relatively aeroduric, moderate anaerobes which may be able to grow suboptimally in the presence of some oxygen. Between these extremes are other moderate anaerobes which can survive brief exposure to air and thereafter renew their growth and multiplication when anaerobic conditions are restored (Morris, 1979).

It is now generally accepted that the causes of oxygen toxicity to microbes are several and various, and that although inactivation of key metabolic components (including enzymes and cofactors) as a consequence of direct interaction with dioxygen must not be discounted (Fee, 1981), a more general threat is posed by products of oxygen consumption by the organism. These hazardous by-products of oxygen utilisation, such as superoxide anion and hydrogen peroxide, result from partial reduction of the dioxygen molecule. Though the debate continues as to which of these is the more life-threatening and whether the actual lethal agent

3. OXYGEN TENSION AND MICROBIAL METABOLISM OF ORGANIC CARBON 63

is more likely to be the secondarily produced hydroxyl free radical (Bannister and Hill, 1982), it is evident that obligate anaerobes may be rendered more vulnerable to oxygen toxicity by their generally smaller content of those protective enzymes which scavenge and dispose of such agents (Fridovich, 1981). The distribution amongst anaerobes of enzymes such as catalase and superoxide dismutase is somewhat capricious, with superoxide dismutase activity even having been reported in *Methanobacterium bryantii* (Kirby et al., 1981). Less obvious mechanisms of self-protection may play significant roles in certain bacteria, including enzymes which efficiently and harmlessly reduce O_2 to water with minimal formation of O_2^- or H_2O_2 (Hoshino et al., 1978) and DNA repair systems which help to minimise the actual damage that may be done by a reagent such as H_2O_2 (Carlsson and Carpenter, 1980). In the case of the particular threat (of photooxidative death) that is posed by the conjunction of exposure to both dioxygen and high visible light intensities, singlet oxygen generation may be countered by the quenching action of carotenoids (Krinsky, 1979).

A few examples must suffice to illustrate the variety of possible interactions between oxygen and obligate anaerobes. The methanogens are generally regarded to be amongst the most oxygen-sensitive of anaerobes. In part this reflects their requirement for a highly reducing growth environment (of low E_h), but it also appears that the unique anaerobic respiratory pathway of ATP generation which produces methane involves several key components that are particularly oxygen labile. This is true not only of their methyl reductase (Ellefson and Wolfe, 1981) but also of Factor 420, a 8-hydroxy 5-deazaflavin which is rapidly degraded when cells of *Methanobacterium thermoautotrophicum* are exposed to air (Schönheit et al., 1981).

Amongst the moderate anaerobes are organisms such as *Clostridium acetobutylicum* (O'Brien and Morris, 1971), *Peptostreptococcus anaerobius* (Hoshino et al., 1978), and *Selenomonas ruminantium* (Wimpenny and Samah, 1978), which can direct part of their fermentatively derived reducing power to the task of harmlessly scavenging oxygen via a soluble NADH oxidase whose synthesis may be derepressed in response to exposure of the organism to oxygen. At low dissolved oxygen tensions this may be sufficient to allow growth of the culture to continue, albeit at a diminished rate. As a consequence of the presence of oxygen and its action as an alien electron acceptor, there is a shift in fermentation products towards more oxidised compounds. Thus, when *Cl. acetobutylicum* was grown in the presence of sublethal dissolved oxygen tensions, the yield of butyrate from glucose was markedly diminished and that of acetate correspondingly enhanced, though without concomitant extra production of gaseous hydrogen (O'Brien and Morris, 1971). Similarly, as the dissolved oxygen tension was increased in a glucose-limited chemostat culture of *S. ruminantium*, production of propionate and succinate decreased whereas that of acetate progressively increased and that of lactate at first rose, only to decline at higher oxygen tensions (Fig. 1). When a culture of *Veillonella alcalescens* growing anaerobically on lactate and producing an equimolar ratio of acetate and propionate was aerated, the response was critically dependent on the degree of aeration imposed (de Vries *et al.*, 1978). At a low dissolved oxygen tension (<0.3 kPa), the efficiency of conversion of lactate into biomass increased twofold as the fermentation yielded more acetate and less propionate. At a higher oxygen tension (0.3 kPa) very little propionate was produced, and pyruvate was now the major product with acetate. At a still higher dissolved oxygen tension (15 kPa), growth and lactate utilisation both ceased as lactate dehydrogenase was rapidly inactivated.

Anaerobic cultures of *Propionibacterium pentosaceum* growing on lactate produced no propionate and less acetate when they were exposed to low concentrations of dissolved oxygen (20–55 μM); instead, pyruvate accumulated as oxygen was consumed (van Gent-Ruijters *et al.*, 1976). In this instance, the specific activities of enzymes such as fumarate reductase, lactate dehydrogenase, and NADH oxidase were lower in cells growing in the presence of oxygen than in its absence. Even so, the cessation of propionate production seemed to be attributable to the shortage of NADH, since lactate oxidation was coupled with the reduction of flavoprotein, which could contribute to the generation of NADH only via reversed electron transport. It was therefore suggested that in the presence of oxygen, the reduced flavoprotein was "directly oxidised" and was hence unavailable for NADH generation, and that pyruvate oxidation also did not occur (van Gent-Ruijters *et al.*, 1976). The inability of this organism to grow at still

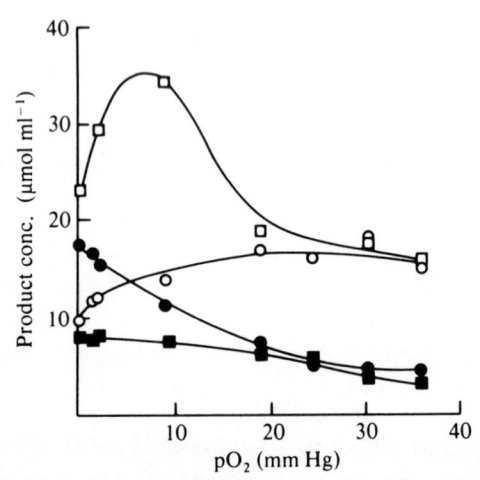

Fig. 1. Steady-state values for the production of acetate (○), propionate (●), lactate (□), and succinate (■) in chemostat cultures of *Selenomonas ruminantium* WPL 151/1. Each point represents the mean of three samples. From Samah and Wimpenny (1982).

higher dissolved oxygen tensions (for example, on agar plates in air) might, however, be due to inhibition of cytochrome biosynthesis (de Vries *et al.*, 1972). The obligately anaerobic character of the purple sulphur, phototrophic bacterium *Chromatium D* might be explicable in similar terms, for although exposure to air caused no immediate effect on the viability or motility of this organism, it inhibited its production of essential photosynthetic pigments (Hurlbert, 1967). In other cases, such variations in response to exposure to oxygen are displayed by distinct strains of a single species that different mechanisms of oxygen toxicity may be operative. Such apparently was the situation with the six strains of *Bifidobacterium* examined by de Vries and Stouthamer (1969), who found that they divided into three groups on the basis of their mode of interaction with oxygen.

It is noteworthy that abrupt exposure to oxygen can provoke metabolic changes markedly different from those which occur during the course of a gradual transition to the same final oxygen tension. Thus, in batch culture or washed suspension, *Propionibacterium shermanii* responded to sudden exposure to oxygen by the cessation of acetate and propionate production (Schwartz *et al.*, 1976). Yet, when in chemostat culture this organism was exposed to a serial succession of relatively small incremental increases in oxygen tension, its acetate production was enhanced (Pritchard *et al.*, 1977). Only when the dissolved oxygen concentration reached a just measurable level (at an atmospheric pO_2 of about 66 kPa) did it cause loss of viability with consequent "washout" of the culture.

Oxygen and the Carbon Metabolism of Obligate Aerobes

Even obligate aerobes are subject to oxygen toxicity at elevated oxygen tensions (so-called hyperbaric oxygen toxicity). Indeed, it seems that life in general is not possible in oxygen tensions greater than 5–10 times that present in "normal" air, so that for some obligate aerobes oxygen is toxic at 1 atm (101.3 kPa) pressure. Discounting the mutagenic action of oxygen, which may be manifested even at 20 kPa pressure (Bruyninckx *et al.*, 1978), the prime causes of hyperbaric oxygen tension are as disputed, and hence probably as diverse, as the cause of the hypersensitivity to oxygen of obligate anaerobes. In the case of bacteria, not surprisingly, the facultative aerobe *E. coli* K-12 has been most studied, following the discovery that its exposure to hyperbaric oxygen had a consequential bacteriostatic rather than bactericidal effect and that various nutritional supplements, especially of branched chain amino acids, provided some temporary protection. Oxygen inactivation of dihydroxyacid dehydratase appeared to explain the relief provided by L-valine, and the oxygen-provoked inhibition of the synthesis of specific amino acids, and thus of protein, could explain the observed triggering of the stringent response via rapid production of ppGpp. Additional effects including decreased synthesis of NAD (caused by inhibition of PRPP synthetase and/or of phosphoribosylquinolate transferase), inhibition of fructose 1,6-diphosphatase, and impairment of RNA synthesis have also been invoked as specific sites of hyperbaric oxygen toxicity (Brown, 1979). However, the range of oxygen tensions that may be encountered by an obligate aerobe (below that at which oxygen becomes toxic) remains wide, and the prevailing oxygen concentration can determine not only the rate of growth of the organism but also the manner of its carbon metabolism (Dawes, 1978).

The chief distinguishing feature and common denominator of obligately aerobic chemotrophs is their reliance on aerobic respiration for free energy conservation. The use of the O_2/H_2O redox couple ($E^{0'}$, +816 mV) as the terminal oxidant means that the oxidation of NADH ($E^{0'}$ of the NAD/NADH couple is -320 mV) is a highly exergonic reaction ($\Delta E^{0'}$ of 1.14 V, $\Delta G^{0'}$ of -220 kJ mol^{-1} of electron pairs) which is quite sufficient to generate 3 moles of ATP per mole of NADH oxidised under level flow conditions. The bulk of the NADH so utilised is likely to be derived from the activity of the tricarboxylic acid (TCA) cycle, whose operation is another feature of most respiratory organisms. Thus, complete oxidation of the organic source of carbon and energy (to CO_2 and H_2O) is generally possible, though some aerobes, either because they do not possess a complete TCA cycle or because the rate at which they consume an organic substrate exceeds the rate at which it can be totally "combusted," may excrete partially oxidised products such as acetate and pyruvate (Andersen and von Meyenburg, 1980). The rate of combustive oxidation of acetyl CoA via the TCA cycle may be regulated both by coarse controls on the synthesis of key enzymes of the cycle and by fine controls imposed upon the activities of these enzymes (Weitzmann, 1981). Similarly, rates of respiratory electron transport can be determined by controls on the synthesis of components of the chain of redox agents; also regulated may be the efficiency of electron transport coupling to ATP generation (Jones, 1977).

In species of *Azotobacter* fixing dinitrogen, the need to protect the oxygen-sensitive nitrogenase may mean that the organism has to indulge in "wasteful" and rapid consumption of the organic respiratory substrate so as to scavenge oxygen and moderate its intracellular concentration (Robson and Postgate, 1980). This respiratory protection of its nitrogenase is made possible by the organism's terminally branched electron transport chain, which is capable of effecting very high rates of oxygen utilisation; for example, *Azotobacter vinelandii* may consume as much as 4 ml O_2 hr^{-1} mg^{-1} dry weight of cells. The maximal rates of oxygen consumption by other aerobic microbes can also be surprisingly high, and although the threshold value to which the dissolved oxygen concentration must fall before its provision limits the respiratory rate (the so-called critical oxygen concentration) varies from species to species, it is generally of the order of 1% of its concentration in media saturated with air. Thus, unless their media are continuously and well aerated, dense cultures of aerobic bacteria would very quickly render their environment anaerobic, with possibly dire consequences (Smith, 1981).

A dramatic example of a shift in carbon metabolism provoked by curtailment of the oxygen supply to an obligate aerobe is the massive synthesis of poly-β-hydroxybutyrate polymer by *Azotobacter beijerinckii* growing on glucose as a carbon and energy source (Fig. 2) when, under conditions of limited aeration, up to 50% of the dry biomass may consist of the polymer. Dawes and his colleagues (Senior *et al.*, 1972) convincingly explained this as being due to the reductive stage in polymer synthesis serving as an alternative electron sink when oxygen was not available, thus permitting continued growth of the organisms under these suboptimal conditions. Although the specific activities of key enzymes of poly-β-hydroxybutyrate metabolism and of certain TCA cycle enzymes in *A. beijerinckii* responded predictably to changes in oxygen tension (Jackson and Dawes, 1976), the specific activities of enzymes of the "preliminary" Entner-Doudoroff pathway of glucose catabolism were not affected (Stephenson *et al.*, 1978; Carter and Dawes, 1979). Interestingly, imposition of an abrupt oxygen limitation on a nitrogen-limited chemostat culture of *A. beijerinckii* (dilution rate, 0.1 hr^{-1}) caused the NADH/NAD ratio first to rise from its aerobic value of 0.4 to 1.1 within 30 min, and then to fall back to 0.55 over the next 60 min, coincident with the rapid synthesis of poly-β-hydroxybutyrate (Dawes, 1981). The glucose-6-phosphate dehydrogenase, citrate synthase, and isocitrate dehydrogenase of this organism were all powerfully inhibited by NAD(P)H, so that under conditions of oxygen limitation, when the concentration of these reduced coenzymes increased, glucose metabolism and the operation of the TCA cycle would decrease. Under these circumstances, diversion of some acetyl CoA to poly-β-hydroxybutyrate synthesis partially alleviated the problem by engendering reoxidation of the reduced pyridine nucleotide.

Pseudomonas aeruginosa, though not a strict aerobe, since it can grow anaerobically by nitrate respiration, is nevertheless another example of an obligately respiratory bacterium which uses glucose via the Entner-Doudoroff and TCA cycle pathways. It can also metabolise glucose by a more direct oxidative route via gluconate and 2-oxogluconate, which are produced in the periplasm for transport into the cell. Oxygen limitation causes glucose metabolism to proceed by the intracellular phosphorylative route in preference to the periplasmic oxidative pathway. During the growth of *P. aeruginosa* with limited aeration, there was repression of synthesis of several TCA cycle dehydrogenases (for 2-oxoglutarate, succinate, and malate), and also of citrate synthase, pyruvate dehydrogenase, and glucose-6-phosphate dehydrogenase: Levels of isocitrate dehydrogenase and hexokinase increased, but no change was noted in the specific activities of the enzymes of the Entner-Doudoroff pathway (Mitchell and Dawes, 1982).

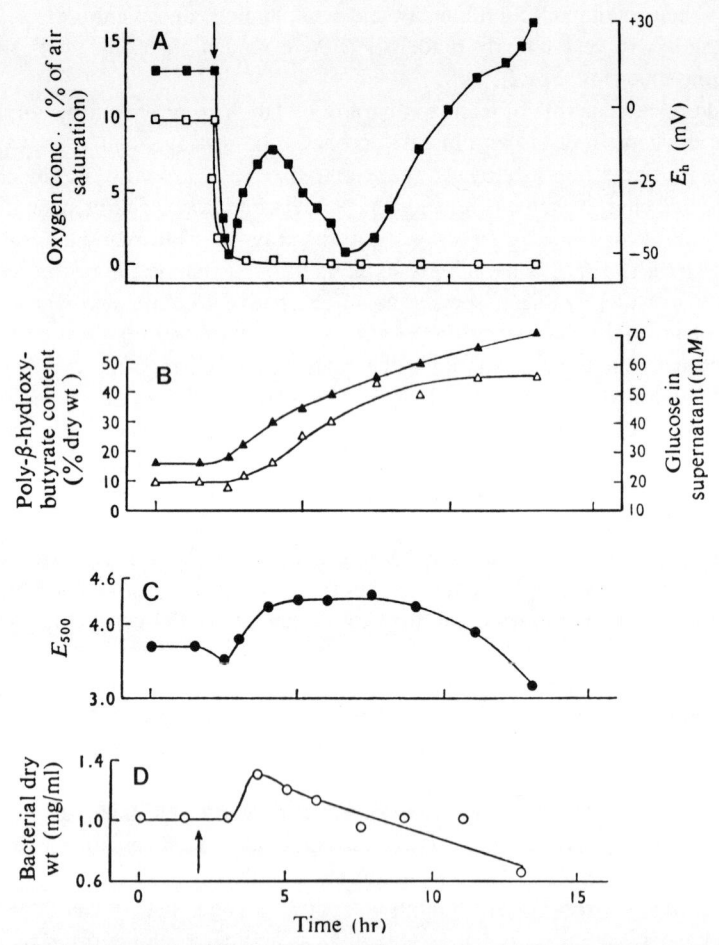

Fig. 2. Effect of the imposition of oxygen limitation on a nitrogen-limited chemostat culture of *Azotobacter beijerinckii*. The nitrogen-limited culture growth rate was 0.233 hr^{-1}, and at the point indicated by the arrow in A and D the oxygen supply rate was decreased from 100 to 25 ml min^{-1}, imposing an oxygen limitation; 40 min later the oxygen supply rate was decreased further to 15 ml min^{-1}. (A) □, Culture dissolved oxygen concentration; ■, redox potential *in situ*; (B) △, poly-β-hydroxybutyrate content; ▲, culture glucose concentration; (C) ●, culture turbidity (E_{500}); (D) ○, bacterial dry weight. From Senior *et al.* (1972).

The carbon metabolism of aerobic hydrogen bacteria responds in a particularly informative manner to shifts in environmental oxygen tension (Bowien and Schlegel, 1981). In general, the autotrophic growth of these bacteria is much more sensitive to inhibition by high oxygen tensions than is their alternative

3. OXYGEN TENSION AND MICROBIAL METABOLISM OF ORGANIC CARBON

heterotrophic growth. Thus, the growth rates of *Alcaligenes eutrophus* and *Xanthobacter autotrophicus* were unaffected over a 10-hr period by exposure to pure oxygen, but their autotrophic growth rates either ceased or decreased markedly within 2–4 hr in an atmosphere containing 40% O_2 v/v (Schlegel, 1976). Two enzymes required by *A. eutrophus* for autotrophic growth are likely targets of this oxygen toxicity. The ribulose bisphosphate carboxylase of this as of other organisms may, at high oxygen tensions, function as an oxygenase, resulting in loss of the acceptor molecule for carbon dioxide and the operation of an energy-wasteful, futile cycle of organic carbon metabolism. Then again, the NAD-reducing hydrogenase of *A. eutrophus* H16, though stable in cells maintained in air or hydrogen, catalysed its own inactivation when the cells were simultaneously exposed to oxygen and hydrogen—an example of autogenous inactivation (Schneider and Schlegel, 1981). Equally interesting was the finding that this obligate aerobe excreted typical products of glucose fermentation (for example, ethanol, lactate, and butanediol) when it was grown in conditions of limited aeration and when the respiratory rate was only 3–15% of the normal maximum (Schlegel and Vollbrecht, 1980). The dehydrogenases responsible for the formation of these reduced end products were not synthesised when the oxygen supply was sufficient to allow maximum respiration and the nature of the excreted metabolite was related to the relative respiration rate, defined as the ratio ($\times$ 100) of the restricted rate of respiration divided by the maximum cellular respiratory rate. At high relative respiration rates, the products excreted by *A. eutrophus* were normal intermediates in the metabolism of strict aerobes, for example 2-oxoglutarate, *cis*-aconitate, and succinate. Though a low relative respiration rate was not required for the induction of synthesis of the NAD-dependent hydrogenase of this organism, Vollbrecht *et al.* (1979) suggested that the relative respiration rate determined the prevailing NADH/NAD ratio, which in turn was the "primary metabolic signal" which elicited derepression of synthesis of each of the fermentative dehydrogenases.

Through the metabolic changes that they provoke, abrupt shifts to high (or low) oxygen tensions, especially in chemostat culture, can often prove highly informative. Yet in some instances, as in the cultures of *Chlorella* studied by Pirt and Pirt (1980), an organism which cannot accommodate a sharp increase in oxygen concentration can sometimes be adapted to grow at the same higher oxygen level if this is attained gradually (see above, p. 65).

Oxygen and the Carbon Metabolism of Facultative Anaerobes

Obligate anaerobes and aerobes are specialised organisms limited by their abhorrence of, or dependence on, oxygen. In contrast, the facultative anaerobe is a versatile entrepreneur, capable of growing without oxygen but eager to exploit it, when available, as the preferred respiratory oxidant. It is this facility, founded on the preservation of both anaerobic and aerobic means of free energy conservation adequate to their needs for growth and multiplication, that led Gest (1981) to suggest that such organisms by renamed "amphiaerobes." He defined an amphiaerobe as "an organism or cell that can use oxygen as the terminal electron acceptor for energy conservation, or some alternative energy transduction process that is independent of oxygen." Their ability to switch successfully between oxygen-dependent and oxygen-independent modes of free energy conservation makes such organisms especially informative experimental subjects in studies designed to reveal the various ways in which the synthesis of the necessary components of the anaerobic and aerobic ATP-generating pathways is controlled and the means whereby their operation is regulated.

The facultative anaerobe which changes from fermentation to aerobic respiration of its limited supply of a carbon and energy source will most likely display a higher biomass yield accompanied by a decreased yield of the partially oxidised products of the fermentation. Thus, in the case of *Saccharomyces cerevisiae* growing on glucose, the biomass yield is fivefold greater under aerobic than anaerobic conditions when only about 10% (w/w) of the glucose utilised is converted into biomass, the rest being largely converted into ethanol, glycerol, and pyruvate (Fiechter et al., 1981). When even a small amount of oxygen is supplied to the yeast, glycerol production is much decreased, since in the presence of oxygen, as of some alternative oxidants, excess NADH can be disposed of by a means other than the reduction of dihydroxyacetone phosphate (Oura, 1977; see also Markham, 1969). Such adaptation to aerobic growth conditions will necessitate the synthesis of those components of the aerobic respiratory apparatus—specific cytochromes, enzymes of the TCA cycle—whose production is repressed in the absence of oxygen. Sometimes a switch from anaerobic to aerobic conditions, conversely, will most evidently be accompanied by cessation of synthesis of some key component(s) of the anaerobic mechanism of free energy conservation, as in the repression of synthesis of photosynthetic pigments in members of the *Athiorhodaceae* when these are transferred from anaerobic/light to aerobic/dark growth conditions (Cohen-Bazire et al., 1957). Such changes were initially studied with batch cultures of facultative anaerobes subjected to different levels of aeration, but have subsequently been better examined in chemostat cultures growing at fixed rates under conditions wherein the dissolved oxygen tension is varied and the carbon and energy source is growth limiting.

An informative series of such experiments was undertaken with *Klebsiella aerogenes* by Pirt and his colleagues (Harrison, 1972, 1973, 1976). When the dissolved oxygen tension was progressively changed in a glucose-limited chemostat culture ($D = 0.16$ hr^{-1}, pH 6.0) three distinct phases of organismic response were discernible. In the aerobic phase, when oxygen was manifestly present in excess (at 1.3–21.3 kPa), the specific respiratory rate of the culture

(q_{O_2}, expressed as mmol O_2 consumed g^{-1} organisms hr^{-1}) was independent of the dissolved oxygen concentration. Within this range of oxygen tensions, no significant changes were noted in the metabolic products excreted by the cells, and all of the glucose carbon was accounted for as biomass and carbon dioxide. At the other extreme, in the phase of limited aeration (when the dissolved oxygen tension, though stable, was less than 0.13 kPa, and thus too small to be measured with a conventional oxygen electrode), the q_{O_2} of the culture was dependent on the rate of oxygen supply. As this rate diminished, the biomass yield progressively decreased, whereas production of reduced end products, such as ethanol and butanediol, increased. The third "transitional" phase was inherently unstable in that the dissolved oxygen tension fluctuated between 0.1 and 1.7 kPa, even though the oxygen supply rate was maintained constant. In this phase the q_{O_2} of the culture was markedly increased (Fig. 3). To explain this, Harrison and Pirt (1967) suggested that when the oxygen concentration fell below a certain critical value, there was a switch to a higher respiratory rate associated with lower growth efficiency. Under wholly anaerobic conditions, *K. aerogenes* formed ethanol, formic acid, butanediol, acetoin, acetic acid, and carbon dioxide. As the dissolved oxygen tension was increased, ethanol and formic acid, then butanediol and acetoin, and finally acetic acid ceased to be produced, whereas there was an increase in the amount of carbon dioxide formed.

In glucose-limited cultures of *E. coli,* both the actual and potential respiration rates were higher under conditions of oxygen limitation than excess oxygen provision (Harrison, 1972). Though it was suggested that feedback regulation via adenine nucleotide levels might, under these conditions, enhance the supply of substrate for respiration, considerable interest was also expressed in the derepression of synthesis of novel cytochrome(s) at low oxygen tensions. For example, the synthesis of cytochrome *d* has been shown to be stimulated in a number of facultative anaerobes when their supply of oxygen was restricted (Rice and Hempfling, 1978) or their ability to utilise oxygen was in some way curtailed (Poole and Chance, 1981).

McPhedran *et al.* (1961) found that the synthesis of both ethanol and glycerol dehydrogenases was repressed by oxygen in *K. aerogenes,* and Pichinoty (1962) reported that hydrogenase and formate hydrogenlyase activities were absent from aerobic cultures of *E. coli, K. aerogenes, Proteus vulgaris,* and *Salmonella oranienburg.* In an extended series of such studies of the effects of aeration on *E. coli's* content of fermentative and TCA cycle enzymes, Wimpenny and his colleagues (Gray *et al.*, 1966a,b; Wimpenny, 1969a,b; Cole, 1976) substantiated their view that two classes of dehydrogenase could be distinguished since their syntheses were regulated by oxygen in opposite ways. Aeration caused repression of synthesis of those dehydrogenases which generally reduced their substrates and which normally operated during anaerobic fermentation. In contrast, synthesis of the TCA cycle dehydrogenases and flavoprotein dehydrogenases

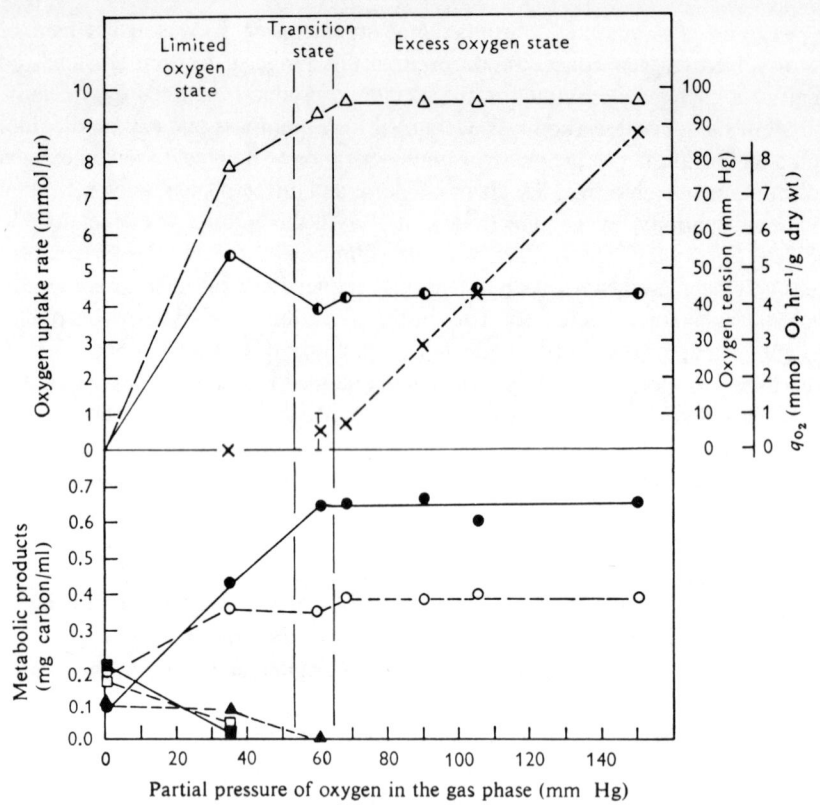

Fig. 3. *Klebsiella aerogenes* NCIB 8017. The fate of glucose at different oxygen tensions with glucose-limited growth at pH 6.0, dilution rate 0.16 hr^{-1}. Glucose concentration in entering medium, 2.6 mg ml^{-1}. X- - -X, Dissolved oxygen tension; △- - -△, oxygen uptake rate; ◐——◐, respiration rate (q_{O_2}). Products: ●———●, organism dry weight; ○- - -○, CO_2; ■———■, 2,3-butanediol; □- - -□, ethanol; ▲- - -▲, volatile acid. $\overline{\underline{\text{I}}}$ denotes the amplitude of oscillations in oxygen tension. Glucose utilization was over 97% of that supplied under all conditions. From Harrison and Pirt (1967).

feeding electrons into the respiratory electron transport chain was enhanced by the presence of oxygen. In a glucose-limited chemostat culture of *E. coli* K-12, the gradual increase in the dissolved oxygen tension was reflected in an increase in the culture's E_h value, making it possible to distinguish between differing levels of aeration even when the oxygen tension was too small to be measured by normal means. As the culture E_h rose with aeration (to +100 to +200 mV), the hydrogenase activity of the cells fell to zero. Cytochrome *b* was maximally synthesised between +50 and +150 mV, and TCA cycle enzymes were maximally produced at +250 to +300 mV (Wimpenny and Necklen, 1968).

Several of the observations made in these and earlier studies have been mirrored in a more recent account of the regulation of respiratory and fermentative modes of growth of the facultative anaerobe *Citrobacter freundii* (Keevil *et al.*, 1979a). During anaerobic growth on glucose, cultures of this organism accumulated substantial amounts of acetate, ethanol, formate, pyruvate, lactate, and succinate. During aerobic batch growth with excess glucose, or in the course of similar anaerobic growth in the presence of nitrate, partial oxidation products of glucose, especially acetate and pyruvate, were accumulated in large amounts. Succinate also accumulated during anaerobic growth with nitrate or under limited aeration, but was replaced by malate, fumarate, and isocitrate when the culture was well aerated. The rates of growth under these conditions are shown in Fig. 4, and the distribution of glucose carbon between the various products is summarised in Table 1. In glucose-limited chemostat culture, a shift from anaerobic to aerobic conditions caused a fivefold increase in biomass yield (Y_{glucose} in-

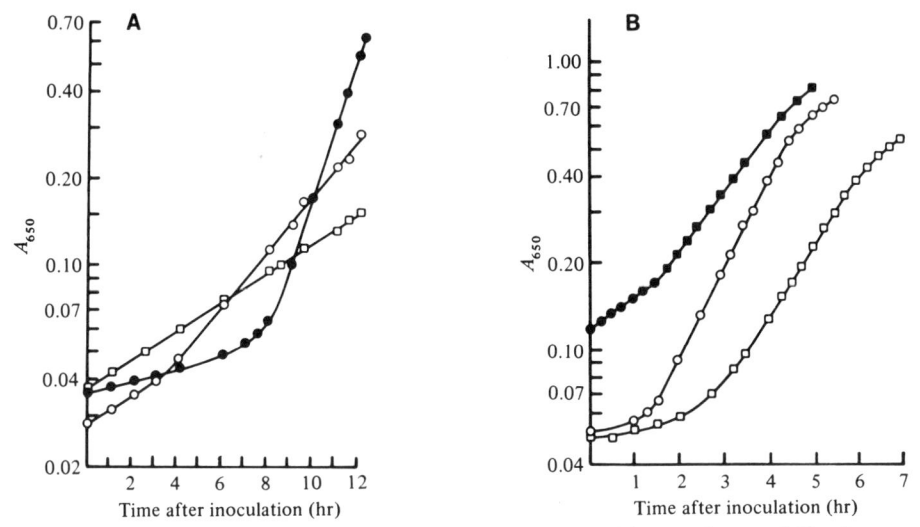

Fig. 4. Effect of inorganic electron acceptors on the anaerobic growth rate of *Citrobacter freundii*. In A bacteria were first grown anaerobically in 100 ml of supplemented minimal medium, pH 6.5, at 37°C for 16 hr. The exponentially growing culture provided a 1% (v/v) inoculum for 100 ml of similar medium, either in 100-ml conical flasks for anaerobic growth (□) or anaerobic growth with 50 mM KNO$_3$ (○), or in sterile 250-ml conical flasks for aerobic growth in a shaking water bath (●). In B 20-ml inocula in aerated nutrient broth were transferred to 500 ml of supplemented minimal medium, incubated for 16 hr at 37°C, and then transferred into 10 litres of similar medium in the stirred fermenter. □, Anaerobic culture without nitrate; ○, anaerobic culture with 80 mM nitrate; ●, aerated culture sparged with 1 litre O$_2$ min^{-1}; ■, aerated culture sparged with 5 litre O$_2$ min^{-1}. In both A and B, the initial glucose concentration was 20 mM. From Keevil *et al.* (1979a).

Table 1. *Analysis of soluble fermentation products in culture supernatants of* Citrobacter freundii *grown in 10-litre batch cultures with or without inorganic electron acceptors*[a,b]

Fermentation product	Terminal electron acceptor during growth			
	None	NO_3^- (80 mM)	O_2 (11 min^{-1})	O_2 (5 liter min^{-1})
Formate	6.0	3.0	0.1	0.1
Acetate	17.8	32.3	28.0	17.7
Ethanol	16.8	1.7	0.7	0.1
Lactate	10.0	3.0	1.5	1.0
Pyruvate	12.5	10.0	18.5	16.7
Fumarate	0.1	0.1	0.1	2.4
Succinate	13.0	10.0	14.5	1.0
Malate	0.5	0.5	0.5	8.0
Isocitrate	0.8	0.1	2.0	5.0
Citrate	0.4	0.9	1.1	1.6
Unidentified	1.8	1.9	3.5	6.6

[a] Reproduced with permission from Keevil et al. (1979a).
[b] Concentrations are expressed as the percentage of glucose-carbon metabolized.

creasing from 11.1 to 60.0 g dry weight cells per mole of glucose utilised), with a corresponding decrease in the yield of excreted metabolites other than carbon dioxide (Tables 2 and 3). It is particularly noteworthy that acetate, pyruvate, and malate were not excreted during aerobic growth under conditions of glucose limitation, though they were major products of sulphate-limited aerobic growth when glucose was present in excess. Although aeration caused a marked change in the cytochrome complement and NADH oxidase activity of the cells, the availability of oxygen or nitrate appeared not to determine the rates at which most of the TCA cycle enzymes were synthesised. An exception was 2-oxoglutarate dehydrogenase, whose synthesis was induced by a very low concentration of oxygen (Keevil et al., 1979b).

Whether it is a fermentative or an anaerobic respiratory mode of ATP generation that is displaced in favour of aerobic respiration when oxygen is made available to an amphiaerobe, particularly important is the regulation of the synthesis and activity of the terminal enzyme of the anaerobic pathway—for example, in instances of anaerobic respiration, of fumarate reductase (Kröger, 1977) or nitrate reductase (Stouthamer, 1976). The general principle which appears to govern the organism's reaction to changes in availability of potential electron acceptors has been expressed by Stouthamer (1976) as follows: "the amount of energy released by catabolism is as large as possible since in all cases the formation of a reductase for a certain substrate is repressed under conditions at

Table 2. Carbon balances for Citrobacter freundii during growth in continuous culture[a,b]

			Terminal electron acceptor during growth			
	None	NO_2^- (20 mM)	NO_3^- (20 mM)	O_2 (11 min^{-1})	None	O_2 (4 liter air min^{-1})
Growth-limiting nutrient	Sulphate	Sulphate	Sulphate	Sulphate	Glucose	Glucose
Glucose consumed (mM)	57.8	58.7	57.4	54.9	30	30
Culture turbidity (A_{650})	0.98	1.20	1.15	1.10	0.83	4.5
$Y_{glucose}$ [g dry wt (mol glucose)$^{-1}$]	6.8	8.1	8.0	8.0	11.1	60.0
Carbon balance (% glucose-carbon consumed)						
Cell carbon	8.0	12.0	12.2	14.0	11.0	58.2
Soluble fermentation products	71.5	70.0	68.3	56.1	68.0	5.4
Volatile products (calculated)	20.5	18.0	19.5	29.9	21.0	36.4

[a] Reproduced with permission from Keevil et al. (1979a).
[b] When sulphate was the limiting nutrient, the glucose concentration was 60 mM; the glucose concentration for carbon-limited cultures was 30 mM.

Table 3. Analysis of soluble fermentation products of Citrobacter freundii after growth in continuous culture[a,b]

	Terminal electron acceptor during growth					
Growth-limiting nutrient Fermentation product	None Sulphate	NO_2^- (20 mM) Sulphate	NO_3^- (20 mM) Sulphate	O_2 (11 min^{-1}) Sulphate	None Glucose	O_2 (4 liter air min^{-1}) Glucose
Formate	0.4	0.5	0.8	0.4	1.0	<0.1
Acetate	21.0	29.1	28.8	17.4	25.0	0.1
Ethanol	30.2	14.9	10.7	0.5	18.7	0.1
Lactate	2.1	9.4	10.9	3.2	1.0	<0.1
Pyruvate	0.3	0.1	0.1	22.0	0.1	<0.1
Succinate	7.7	12.4	10.0	0.3	13.4	<0.1
Malate	0.1	2.6	2.0	11.0	8.5	<0.1
Isocitrate	0.1	0.3	0.1	4.0	0.1	<0.1
Citrate	0.1	0.5	0.5	3.7	0.1	<0.1
Unidentified	9.8	2.3	6.5	3.6	8.9	5.2

[a] Reproduced with permission from Keevil et al. (1979a).
[b] Concentrations are expressed as the percentage of glucose-carbon metabolized.

which a hydrogen acceptor with a higher energy-yielding potential is also present." Or as Wimpenny (1969b) put it: "For reasons of energy transfer there exists a sort of 'pecking order' in the choice of electron acceptors that versatile organisms can employ. For example, oxygen is 'preferred' to nitrate, which in turn is 'preferred' to organic electron acceptors." In this regard, it is interesting that below a critical dissolved oxygen tension of 2 kPa, *Klebsiella* K312 utilises both oxygen and nitrate as electron acceptors (Dunn *et al.*, 1979). This oxygen tension correlates well with that critical oxygen pressure which marks the switchover from oxidative to fermentative metabolism in cultures of *K. aerogenes* (Harrison, 1973). In the denitrifying organism *Paracoccus denitrificans,* it would seem that the reduction of nitrate is controlled by the amount of reduction of one or more components of the respiratory chain, which in turn is determined by the rate of electron transfer to oxygen (Alefounder *et al.*, 1981).

One must also be prepared for the singular, sometimes unpredictable response of a microorganism to oxygen. Thus, although the major consequence of exposure to air of a growing culture of *Rhodopseudomonas sphaeroides* is the repression of synthesis of its photosynthetic apparatus and its shift to a heterotrophic mode of carbon metabolism, another less evident consequence of aeration is repression of synthesis of its fructose 1,6-bisphosphate aldolase. This is of little consequence to the metabolism of glucose, which is utilised via the Entner-Doudoroff pathway both aerobically in the dark and anaerobically in the light. However, when *R. sphaeroides* is growing anaerobically in the light, fructose is utilised chiefly via the Embden-Meyerhof pathway, so that the imposition of aerobic conditions renders necessary a switch to the Entner-Doudoroff pathway as the major route of fructose utlisation (Conrad and Schlegel, 1977). In contrast, *R. capsulata* catabolises glucose via the Entner-Doudoroff pathway and fructose via the Embden-Meyerhof pathway irrespective of whether growth is proceeding aerobically or anaerobically (phototrophically).

It is evident therefore that when an aerobically growing culture of an amphiaerobe is deprived of oxygen (or conversely, when a culture which has been growing anaerobically is aerated), profound changes occur in its mode of carbon metabolism as a result of the repression and/or inhibition of certain enzymes and the induced synthesis and/or activation of others. For example, following removal of oxygen from a culture of *Escherichia coli* growing in a glucose minimal medium, although the rate of glucose consumption increases about 3-fold (Krebs, 1972), the fraction of this glucose that is utilised via the hexose monophosphate pathway decreases from 25% to about 7% with the Embden-Meyerhof pathway playing a correspondingly enhanced role. Since the pyruvate dehydrogenase complex is repressed and inhibited under anaerobic conditions (Schwartz *et al.*, 1968; Yamamoto and Ishimoto, 1975), the pyruvate produced by glycolysis is metabolised via the action of pyruvate formate lyase, which under anaerobic conditions is derepressed about 6-fold (Pecher *et al.*, 1982) and activated by a complex mechanism (Knappe and Schmitt, 1976). With the activity of the TCA cycle minimized as a result of repression of synthesis plus inhibition of a number of its component enzymes, acetyl CoA is converted to acetate by the action of phosphotransacetylase and acetate kinase. Therefore, one manifest consequence of the regulation of the carbon metabolism of the organism that is wrought by oxygen is that the protein "profile" of anaerobically grown cells of an amphiaerobe will differ from that displayed by aerobically grown cells. Smith and Neidhardt (1983a,b) measured the cellular concentrations in *E. coli* K-12 of 170 polypeptides following growth of the organism on glucose as sole source of carbon and energy both aerobically and anaerobically with or without nitrate. They found that 18 of these proteins achieved their highest intracellular levels (1.8- to 11-fold enhanced) in anaerobically grown organisms. Nitrate antagonized the anaerobic induction of synthesis of all but one of these polypeptides (Smith and Neidhardt, 1983a). On the other hand, 19 polypeptides including proteins of the pyruvate dehydrogenase and 2-oxo-glutarate complexes, several TCA cycle enzymes, manganese superoxide dismutase, and tetrahydropteroyltriglutamate transmethylase, attained their highest levels during aerobic growth. The synthesis of most of these was also induced during anaerobic growth in the presence of nitrate (Smith and Neidhardt, 1983b). In their consideration of the possible nature of these pleiotropic responses to the presence or absence of oxygen, Smith and Neidhardt (1983b) coined the somewhat inelegant term *stimulon* to signify "a set of genes (or a set of regulons) whose products are increased in response to a common environmental stimulus, irrespective of molecular mechanisms." Thus, the 18 anaerobiosis-enhanced polypeptides and the 19 aerobically induced polypeptides would be the structural protein products of the anaerobic and aerobic stimulons of *E. coli* K-12. The situation is, however, complicated by the fact that many of the 18 anaerobiosis-induced proteins appeared to be subject to regulation by stimuli other than the absence of oxygen, whilst at least one subset of the 19-member set of aerobically induced proteins shared sensitivity to a number of other regulatory signals.

Oxygen and the Efficiency of Microbial Growth

The oxygen supply to a microorganism can evidently determine in several ways the routes whereby it catabolises organic substrates and generates ATP in the course of metabolising those substrates which serve as sources of free energy. At the same time, biosynthesis remains essentially anaerobic, even in most aerobic organisms. It is therefore pertinent to consider how close is the linkage between the efficiency of carbon source conversion into biomass and the oxygen demand displayed by an aerobe. This is a subject of considerable concern to the microbial technologist who may wish to maximise cell yield whilst minimising the considerable cost of oxygenation (aeration) of large-volume cultures. Tempest and Neijssel (1981) have pointed out that microorganisms possess the capacity to dissociate catabolism from anabolism to varying degrees under different growth conditions. In the case of aerobic respiratory growth, variable stoichiometry between the rates of oxygen reduction and ATP formation may be one consequence of the possession of branched electron transport pathways, and "respiratory control" may additionally be circumvented by the operation of energy-spilling reactions or futile cycles. As an example, Tempest and Neijssel (1981) instanced methanol-utilising bacteria when a 25% drop in carbon conversion efficiency was accompanied by a 60% increase in the oxygen demand with a proportional increase in heat output.

This theme, namely, that oxygen uptake cannot be stoichiometrically related to ATP generation and biomass production, has been developed, amongst others, by Harder and his colleagues (1981). They have postulated that in terms of free energy conversion, rapidly multiplying aerobic microorganisms maintained under conditions of balanced growth do not work at maximum efficiency but in a manner which generates the maximum rate of biomass production. The efficiency of free energy transduction in such organisms may be close to 50%. Since the theoretical Y_{ATP}^{max} value for growth of a microbe on a known carbon and energy source can be calculated (Stouthamer, 1979), a fairly accurate prediction of the growth yield of an aerobic organism growing on a particular substrate can be made if one knows the Y_{ATP}^{max} value and the maintenance energy requirement and assumes an overall P/O ratio for $NADH_2$ oxidation based on approximately 50% efficiency of free energy transduction. A variable degree of coupling (slippage) within membranous ionmotive sources and sinks would constitute an important general mechanism whereby such free energy may be squandered.

It is to such considerations that we must turn for an explanation of the following fact: The increase in biomass yield (per mole of substrate utilised) that is achieved following transfer of a facultative anaerobe from anaerobic to aerobic conditions of growth is generally substantially less than would be anticipated from the enhanced specific yield of ATP derivable from the substitution of aerobic respiration (and a functional TCA cycle) for fermentation.

Oxygen as a Metabolic Regulator

The literature abounds with examples of effects of changes in oxygen tension on the metabolism of various microbes. In many of these reports, the account of the oxygen effect verges on the anecdotal, in some the effects have been analysed and quantified, but only in very few is there any suggestion regarding the regulatory mechanism triggered by enhancement or curtailment of the oxygen supply.

In each case, it is of prime importance to determine whether the observed effect is provoked uniquely by oxygen or by other electron acceptors as well. In the latter situation, some sort of "redox governor" hypothesis is generally invoked to explain the phenomenon. According to such a hypothesis, oxygen and alternative electron acceptors act by determining the state of oxidation of some key redox couple in the cell, which might be a membrane-integrated component of the respiratory electron transport chain or a soluble redox couple whose condition reflects the "oxidation–reduction state" of the cytosol. For example, in *Staphylococcus aureus,* anaerobiosis led to a 10-fold increase in NAD-linked lactate dehydrogenase activity, but in haemin-less mutants of this organism grown in the absence of haemin, lactate dehydrogenase was present at equally high specific activity levels following aerobic or anaerobic growth. Restoration of haemin to these mutants caused repression of synthesis of lactate dehydrogenase to be elicited under aerobic conditions, which provoked Garrard and Lascelles (1968) to suggest that the control was exercised by "reducing conditions inside the cell." Similarly, in a mutant strain of *Escherichia coli* unable to synthesise ubiquinone, high levels of ethanol and lactate dehydrogenases were present even in cells which had been grown aerobically (Jones and Lascelles, 1967). Broadly similar repression of synthesis of dehydrogenase enzymes, hydrogenase, and cytochrome *b* was noted when in cultures of *E. coli* appropriate E_h values were established using nitrate or nitrite in place of oxygen (Wimpenny, 1969a). This suggested that the prevailing intracellular NADH/NAD ratio might be important in determining whether such proteins were synthesised. Yet when Wimpenny and Firth (1972) actually measured the NADH/NAD ratio in cells of *E. coli* grown anaerobically and at various levels of aeration, they found that this ratio (and hence the intracellular E_h) remained fairly constant, increasing by only 30 mV when the culture E_h was raised by aeration from -350 to $+15$ mV.

The redox governor hypothesis was actually first proposed by Cohen-Bazire *et al.* (1957), who suggested that the rates of bacteriochlorophyll and carotenoid synthesis in nonsulphur purple bacteria (both of which decreased as light intensity or oxygen tension was increased) were inversely related to the state of oxidation of some component of the respiratory electron transport system. Marrs and Gest (1973), however, discovered that in a mutant strain of *Rhodopseudomonas capsulata* which lacked cytochrome oxidase activity, the sensitivity to oxygen of bacteriochlorophyll formation was essentially unaltered. They proposed that oxygen directly inactivated some factor (F^{bchl}) involved in bacteriochlorophyll synthesis in this organism; reactivation of this factor necessitated an electron flow from the electron transport system. This effect of oxygen was immediate, and was quite distinct from the known ability of oxygen to repress the synthesis of certain enzymes concerned with bacteriochlorophyll formation. Transfer to a R' plasmid (pRPS 404) of all of the genetic information required for development of the photosynthetic apparatus of *Rps. capsulata* has recently rendered it possible to examine how the concentrations of mRNA coding for antenna and reaction centre proteins vary in response to changes in oxygen tension. Messenger RNA transcripts corresponding to the genes for certain pigment-binding polypeptides increased some 40-fold following a shift from high- to low-oxygen tension. Transcripts hydridizing to genes involved in the synthesis of bacteriochlorophyll increased to a much lesser extent, whilst the transcription of several genes implicated in carotenoid biosynthesis was seemingly not affected by oxygen (Clark *et al.*, 1984). Although in this study it was not possible to distinguish between increased transcription of the genes or decreased degradation of their mRNA transcripts, evidence was obtained suggesting the involvement in the organism's response to oxygen of a regulatory genetic element located on the chromosome of *Rps. capsulata* adjacent to the *rxc*A locus wherein mutations affect the synthesis of reaction centre- and light harvesting-polypeptides.

Fine control of the activity of a preexisting enzyme system is apparently exerted indirectly by oxygen in the case of the energy-dependent inactivation of citrate lyase in *Enterobacter aerogenes*. Inactivation of this enzyme (by deacetylation) occurred not only in cells exposed to aerobic conditions but also when the organism was supplied anaerobically with nitrate. Kulla and Gottschalk (1977) proposed that the inactivation was dependent on the operation of electron transport processes in the cell membrane which generated a citrate lyase deacetylating activity. The L-1,2-propanediol oxidoreductase of *E. coli* K-12 which is functional during anaerobic growth of this organism on L-fucose was also inducibly synthesised during aerobic growth on fucose but under these conditions the protein was either catalytically inert or was rapidly inactivated after synthesis (Chen and Lin, 1984). In the cyanobacterium *Anabaena* sp. strain PCC7120, the glucose 6-phosphate dehydrogenase, which is the regulatory enzyme of the oxidative pentose pathway, was deactivated by reduced thioredoxin and reactivated by exposure to oxidised glutathione or H_2O_2 (Udvardy *et al.*, 1984). Indirect fine control of the activity of a key enzyme or enzymes, but of a rather different type, appears to lie at the heart of the classic Pasteur effect in yeast. This is manifested in cultures of *Saccharomyces cerevisiae* as a decreased rate of glucose utilisation in aerobic as opposed to anaerobic conditions. The most reasonable explanation of the means whereby the availability of oxygen controls the rate of glycolysis in the yeast, is that this involves feedback regulation by adenine nucleotides of the activity of Embden-Meyerhof pathway enzymes, especially phosphofructokinase (Ramaiah, 1974; Sols, 1976).

Direct interaction between oxygen and key enzymes, which results in their inactivation, is not an uncommon phenomenon, but the mechanisms of such inactivation are not always obvious. The list of such oxygen-sensitive enzymes is quite extensive, including, for example, nitrogenases (Robson and Postgate,

1980), the gramicidin S synthetase of *Bacillus brevis* (Friebel and Demain, 1977), the L(+)-lactate dehydrogenase of *Desulfovibrio desulfuricans* (Stams and Hamsen, 1982), or the glutamine phosphoribosylpyrophosphate amidotransferase of *Bacillus subtilis* (Turnbough and Switzer, 1975). Subtly different is the enzyme which catalyses its own inactivation when supplied with oxygen and an electron donor. An example of such autogenous inactivation is provided by the NAD-reducing hydrogenase of *Alcaligenes eutrophus* (see p. 69). It seems that when the enzyme encounters hydrogen, oxygen and NADH simultaneously, it produces superoxide anions and so brings about its own inactivation (Schneider and Schlegel, 1981).

When the actual production of an enzyme or another protein is determined by whether or not the organisms are supplied with oxygen, the major question (so far unanswered in most cases) is, how directly does oxygen regulate the transcription of the genes which specify those proteins? The facultative anaerobe *Bacillus licheniformis* operates two pathways of arginine catabolism. In well-aerated cultures, the amino acid is utilised via the "arginase route" and the catabolic ornithine carbamoyltransferase is present only in low specific activity in the cells. In anaerobic culture the arginine deiminase pathway is followed, with the cells containing only low levels of arginase and ornithine transaminase. Nitrate (added to the anaerobic culture) has the same effect as oxygen, for in its presence the synthesis of the enzymes of the arginase pathway is derepressed, whereas formation of the enzymes of the deiminase route is repressed. It would appear that nitrate and oxygen elicit this controlled response via some common respiration-mediated mechanism (Broman *et al.*, 1978).

Not unnaturally, the means whereby enzymes of anaerobic electron transport are repressed during aerobiosis (the "oxygen effect" of Pichinoty, 1965) has been the subject of considerable interest, with numerous studies being undertaken of the control exercised by oxygen over the induced synthesis of respiratory nitrate reductase in *Escherichia coli*. At least three hypotheses have been favoured at various times. Pichinoty (1965) tended to think in terms of oxygen itself acting as the corepressor. Another suggestion was that the repressor is sensitive in some way to the intracellular redox potential (Wimpenny, 1969a). In some later studies, the emphasis tended to shift from control at the level of transcription of the structural gene to regulation of the production of active enzyme by interference with its proper assembly into the cytoplasmic membrane, a process which might be subject to regulation by electron transport to oxygen also accomplished at that membrane (Stouthamer, 1976). Evidence suggests that there may indeed be two ways whereby oxygen might control production of nitrate reductase in an active form: by inhibition (1) of transcription of the *chl* C gene and (2) of enzyme assembly (Kaprálek *et al.*, 1982). Fimmel and Haddock (1979) used *chl* C–lac fusions to determine the means of regulation of the *chl* C gene in *E. coli* K-12. Their findings implicated the promoter/operator region as the likely site of oxygen action. However, evidence also obtained with *E. coli* suggested that exposure of intact cells to oxygen prevented insertion of soluble nitrate reductase into the cell membrane (Hackett and MacGregor, 1981).

In the course of their study of mutants of *Escherichia coli* unable to use fumarate as an anaerobic electron acceptor, Lambden and Guest (1976) described a regulatory *fnr* gene, located at 29.3 min on the *E. coli* linkage map, mutations within which proved to be pleiotropic causing deficiencies in fumarate and nitrate reduction and in hydrogenase activity. Similarities in map position suggested that *fnr* was identical to the *nir*A and *nir*R genes (Newman and Cole, 1978; Chippaux *et al.*, 1978), a supposition that is strengthened by the multiple deficiencies in *fnr, nir*A and *nir*R mutants of nitrate, nitrite and fumarate reductases, hydrogenase, a formate dehydrogenase, formate hydrogen-lyase, and the anaerobic cytochrome c_{552}. Study of the kinetics of transcription of the nitrate reductase genes, performed in strains in which *lac* structural genes were fused to the promoter of the nitrate reductase operon (Chippaux *et al.*, 1981) confirmed the view that the *fnr* (*nir*R) gene product acted at the level of transcription of the structural genes under its control. The most plausible model for its mode of action proposed that the *fnr* gene product exerted positive control responsive in some manner to the prevailing "redox state" of the cell. When the nucleotide sequence of the cloned *fnr* gene was determined, not only was it possible to confirm that its product was a Fnr protein of MW 27,947, but also it was evident that the primary structure of this protein contained regions of homology with several other transcriptional regulator proteins (Shaw and Guest, 1982a,b). In particular, Fnr protein possessed three regions of sequence homology with the cyclic AMP receptor protein (CRP) which mediates catabolite repression in *E. coli* by binding to specific regulatory sites of catabolite-sensitive genes (Shaw *et al.*, 1983). The regions of homology were found in both the DNA-binding and the nucleotide-binding domains of CRP, though the amino acid residues that make specific contacts with cAMP were not conserved in Fnr. This suggested that Fnr might function in a manner akin to CRP in its activation of the transcription of genes specifically concerned with anaerobic metabolism, in which case, interaction with DNA could involve Fnr binding some effector molecule(s) whose availability is determined by the redox state of the cell. Yet recently, both cAMP and adenylate cyclase (but not CRP) have been reported to be essential for the expression of the fumarate reductase system of *E. coli* (Unden and Guest, 1984). It would thus appear that in *E. coli* the high intracellular concentration of cAMP that signals energy deprivation activates the transcription of genes controlled by CRP and Fnr. The actual control system that responds to the redox state remains unidentified, unless it is that Fnr itself is a redox-sensitive protein. Such a notion is encouraged by the finding that Fnr differs from CRP in possessing a small terminal domain containing a cluster of three cysteine residues (Unden and Guest, 1984). In summary then, the picture emerges of anaerobic respiratory systems in *E. coli* being under multiple control involving specific repressor proteins responding to the terminal electron acceptor (fumarate or nitrate) and a general regulatory protein Fnr responsive to cAMP level and possibly itself redox-sensitive (for further information see Stewart, 1982; Cole, 1984; Pascal *et al.*, 1982).

Since an increase in *fnr* gene dosage had a differential effect on the expression of structural genes encoding components of the anaerobic respiratory enzymes (fumarate and nitrate reductases) and those purely fermentative enzymes whose activities increase under anoxic conditions (Shaw and Guest, 1982a), it is conceivable that additional regulatory circuits may control the expression of the latter class of enzymes (Pecher *et al.*, 1983). Using *Mu d (Ap lac)*-induced mutations in the formate hydrogen-lyase branch of the glucose fermentation pathway in *E. coli*, Pecher *et al.* (1983) showed that redox control of formate dehydrogenase and hydrogenase synthesis by oxygen, nitrate, or fumarate occurred at the level of transcription and that the degree of repression is somehow correlated with the magnitude of the redox potential of the terminal electron-accepting couple. These workers concluded that the effect of oxygen, nitrate, and fumarate on the synthesis of hydrogenase, formate dehydrogenase (and the nitrogenase of *Klebsiella pneumoniae*) could have a common basis. The effect of Fnr was not however examined by these workers. Interestingly, however, Fnr proved to be a positive regulator of synthesis of the anaerobic *sn*-glycerol-3-phosphate dehydrogenase of *E. coli* (Kuritzkes *et al.*, 1984) so that when this organism grows anaerobically on glycerol plus fumarate, Fnr is a regulator of the levels of both the primary electron donating- and terminal electron accepting-elements of the anaerobic respiratory chain.

The precise mechanism of control of expression of anaerobic stimulons in other bacteria may, of course, not be identical with that operative in the enterobacteria. For example, as was pointed out by Shaw *et al.* (1983), in *Rhizobium japonicum* the intracellular concentration of cGMP seems to be under redox control (Lim *et al.*, 1979) and in this organism cGMP inhibits expression of nitrogenase, nitrate reductase, and hydrogenase.

The genes specifying enzymes of nitrogen assimilation in *Klebsiella pneumoniae* can be divided into three classes depending on the manner in which their expression is regulated by exposure of the organism to oxygen. The transcription of several genes is affected slightly or not at all (for example, *gln* A specifying glutaminase, or *ure* A specifying urease). Others (such as *nif* specifying the nitrogenase system necessary for dinitrogen fixation) are repressed, whereas some are expressed only under aerobic conditions, including *hut* specifying a catabolic histidase (Goldberg and Hanau, 1980). One proposal as to how oxygen may control gene transcription concerns the aerobic repression of synthesis of nitrogenase in *K. pneumoniae* (Buchanan-Wollaston *et al.*, 1981). In this model, it was proposed that the protein specified by a regulatory *nif* L gene is a repressor of transcription acting at the promoters of all *nif* operons (save the regulatory *nif* LA operon). Under conditions in which *nif* is expressed, the *nif* L protein is inactivated by interaction with a regulatory metabolite. When such derepressed cells are exposed to oxygen (as also occurs when a source of fixed nitrogen is provided), the level of the regulatory metabolite is decreased, *nif* L repressor protein is "liberated", RNA polymerase possibly assumes a different structural form, and expression of *nif* is halted. According to this suggestion, the repressive effect of oxygen on the synthesis of nitrogenase is effected at second hand by its causing, via some undisclosed physiological means, the critical decrease in the intracellular concentration of the regulatory metabolite. It is significant that a functional *fnr* gene is believed to be essential for the expression in *E. coli* of plasmid-borne nitrogen fixation genes derived from *K. pneumoniae* (Skotnicki and Rolfe, 1979).

It is evident that much work remains to be done on this fundamental problem of the mechanism(s) of oxygen control of microbial synthesis of proteins and other products. The findings to date suggest that once again we must be prepared to question simplistic solutions and expect that different mechanisms might operate in different organisms. The problem is not just of academic interest, as is clear from the manner in which bacterial synthesis of several important exoenzymes can be influenced by the degree of aeration imposed upon the culture. Production of L-asparaginase by cultures of *Erwinia aroidea* was little affected by the prevailing dissolved oxygen tension (Liu and Zajic, 1973), but *Serratia marcescens* produced this commercially valuable enzyme in maximum amount only when aeration was so limited that the dissolved oxygen concentration was undetectably small (Heinemann *et al.*, 1970). *Escherichia coli* A-1 also formed L-asparaginase in high yield only when the dissolved oxygen tension was virtually zero (Barnes *et al.*, 1977). Biosynthesis of hydrogen cyanide by *Pseudomonas aeruginosa* was triggered by lack of oxygen and was inhibited when normal aeration was restored, the latter effect being attributed to a combination of inactivation of the hydrogen cyanide synthase and repression of synthesis of this enzyme (Castric *et al.*, 1981). In the case of neuraminidase secretion by type III group B streptococci, the enzyme, which serves as a virulence factor, was maximally produced under conditions of "limited" oxygen supply (Caldwell and Lim, 1981), whereas production of an extracellular collagenase by *Vibrio alginolyticus* was enhanced by aeration (Hare *et al.*, 1981).

One should, of course, always be watchful in studies undertaken with batch cultures that any "oxygen effect" is not merely due to a change in growth rate provoked by enhanced aeration. Thus, although the cytochrome *d* content of *Arthrobacter crystallopoietes* was genuinely determined by the prevailing oxygen tension in the culture, the induction of the change in the shape of the organism from rod to sphere which could be provoked by oxygen deprivation was in fact caused by the substantially lower growth rate in the oxygen-limited culture (Faller and Schleifer, 1981). Such an observation emphasises the desirability of investigating the effects of changes in oxygen concentration in chemostat culture.

Some Ecological Implications

In nature, various microbes can inhabit locations in which the oxygen tension may be as low as zero or as high as those "supersaturating" levels found in dense algal and cyanobacterial mats (Eberly, 1964). Nor are all microorganisms passive in expressing their particular needs. Thus, of 24 species of highly motile bacteria that were examined by Baracchini and Sherris (1959), all of the aerobic and facultative aerobic species showed positive aerotaxis, whereas three anaerobic species of *Clostridium* demonstrated negative aerotaxis. An apparently even more ingenious mechanism is employed by the magnetic spirillum MS-1, which is an extreme microaerophile (optimal dissolved oxygen tension of 17 Pa). It has been proposed that the role of its magnetotaxis could be to direct the motile cell downwards in aquatic environments to regions of appropriately low oxygen tension (Blakemore, 1975; Escalente-Semerena *et al.*, 1980).

Studies on the microflora of aquatic sediments and on the oxygen relations of the constituent organisms have illustrated the complexity of the interrelations that can be established. When E_h is used as an indicator of the degree of oxidation of a sediment at various depths, values ranging from +700 mV (highly oxidised) to −300 mV (highly reduced) can be measured (Hambrick *et al.*, 1980), and the picture emerges of a minimally two-layered system—a relatively thin aerobic surface layer over an anaerobic base. Mackenzie and Wollast (1977) reported that the E_h in coastal marine sediments fell stepwise with depth, mirroring the microbial utilisation of major electron acceptors of decreasing midpoint potentials. Oxygen actually penetrated only a relatively small depth (about 5 mm) into the surface of most marine sediments (Revsbech *et al.*, 1980). It is likely that this surface layer may be one of the "natural" locations for microaerophiles, as well as for facultative aerobes, and it is supposed that the obligate anaerobes would be found beneath this zone. This, however, is too simple a view, for the surface zone itself is likely to be highly heterogeneous. In studies of estuarine sediments (Hallberg, 1968; Jørgensen, 1977; Laanbroek and Veldkamp, 1982), it has been found that the surface layer may contain a myriad of minute anaerobic "pockets" distinguishable by their blackening by the activity of sulphate-reducing bacteria. Around and between these reduced microniches, relatively aerobic tracts can harbour quite different microbes, including species of *Thiobacillus* that can reoxidise the sulphide ions generated nearby. Such aerobic capillaries might form, be obliterated, and be re-formed, for example, by the activities of the microfauna, and even the *Thiobacillus* can survive a temporary period of anaerobiosis by laying down an intracellular polyglucan during aerobic metabolism, which can thereafter be fermentatively catabolised when oxygen is in short supply (Beudeker *et al.*, 1981). Thus is the sulphur cycle sustained in the surface layer whilst a host of other organisms degrade the organic detritus to yield the substrates used fermentatively and acetogenically.

Even this two-dimensional view of the gradient of oxygen through the surface layer of the sediment is manifestly too simple. Wimpenny (1981) does well to remind us that natural microbial ecosystems are structured both in space and in time. Even the colony of an aerobic bacterium growing in air on the surface of an agar-solidified medium (Fig. 5) well demonstrates the restricted penetration of oxygen which can cause even such an apparently homogeneous structure to demonstrate differentiation in both biochemical and morphological characteristics (Wimpenny and Parr, 1979). Serial changes with time which may occur as a heterogeneous microbial community becomes progressively more limited in oxygen have been observed in many diverse situations. For example, Takai and Kamura (1966), studying the microbial succession which followed waterlogging of paddy soil samples, found that in an initial rapid phase oxygen was first wholly consumed, followed by utilisation of the nitrate and then by the formation

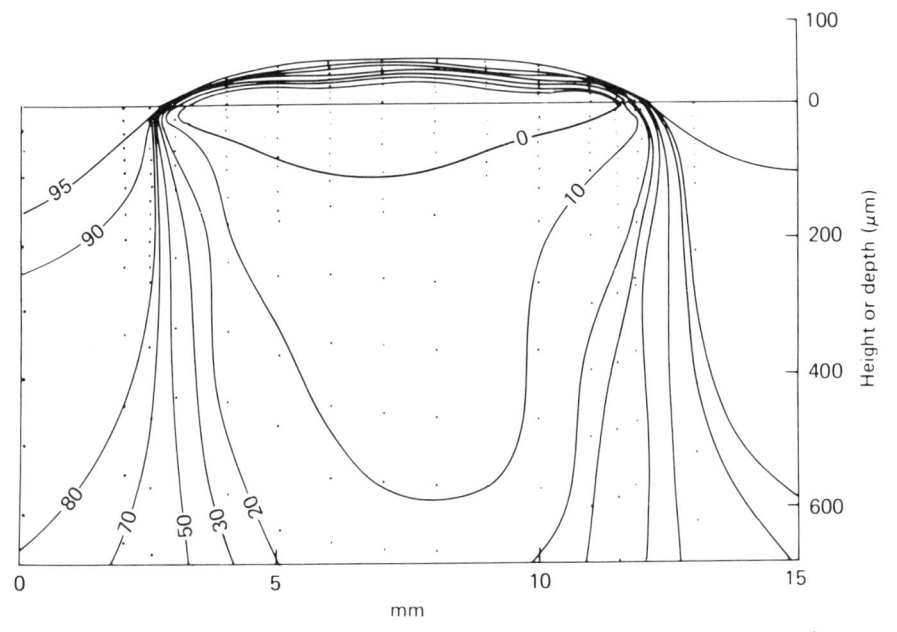

Fig. 5. The distribution of oxygen in and around a colony of *Bacillus cereus* growing at 37°C on a tryptone-soya agar. Isopleths as a percentage of the air-saturated value. Data were obtained with a Transidyne oxygen sensor and microelectrode. From Wimpenny (1981).

of Mn (II) and Fe (II) ions from MnO_2 and $Fe(OH)_3$. During this sequence the E_h fell from +600 to +300 mV. In the following slower phase during which the E_h dropped still further to −200 mV, organic acids accumulated, H_2S was produced, and ultimately methane was generated (see also Wakao and Furusaka, 1976).

It is only by seeking to unravel the tangled relations that exist between the components of such natural systems, in all their complexities, that we can hope eventually so to control the carbon metabolism of mixed cultures of anaerobes with facultative and obligate aerobes as to optimise the yield of a desired product with the facility that is currently possible only with single-species cultures.

References

Alefounder, P. R., McCarthy, J. E. G. and Ferguson, S. J. (1981). The basis of the control of nitrate reduction by oxygen in *Paracoccus denitrificans*. *FEMS Microbiol. Lett.* **12**, 321–326.

Andersen, K. B. and von Meyenburg, K. (1980). Are growth rates of *Escherichia coli* in batch cultures limited by respiration? *J. Bacteriol.* **144**, 114–123.

Atlas, R.M. (1981). Microbial degradation of petroleum hydrocarbons: an environmental perspective. *Microbiol. Rev.* **45**, 180–209.

Bannister, J. V. and Hill, H. A. O. (1982). Chemical reactivity of oxygen-derived radicals with reference to biological systems. *Biochem. Soc. Trans.* **10**, 68–69.

Baracchini, O. and Sherris, J. C. (1959). The chemotactic effect of oxygen on bacteria. *J. Pathol. Bacteriol.* **77**, 565–574.

Barnes, W. R., Dorn, G. L. and Vela, G. R. (1977). Effect of culture conditions on synthesis of L-asparaginase by *Escherichia coli* A-1. *Appl. Environ. Microbiol.* **33**, 257–261.

Beudeker, R. F., de Boer, W. and Kuenen, J. G. (1981). Heterolactic fermentation of intracellular polyglucose by the obligate chemolithotroph *Thiobacillus neapolitanus* under anaerobic conditions. *FEMS Microbiol. Lett.* **12**, 337–342.

Blakemore, R. P. (1975). Magnetotactic bacteria. *Science* **190**, 377–379.

Bloch, K. (1976). On the evolution of a biosynthetic pathway. *In* "Reflections on Biochemistry" (Eds. A. Kornberg, B. L. Horecker, L. Cornudella and J. Oró), pp. 143–150. Pergamon, Oxford.

Bowien, B. and Schlegel, H. G. (1981). Physiology and biochemistry of aerobic hydrogen-oxidizing bacteria. *Annu. Rev. Microbiol.* **35**, 405–452.

Broman, K., Lauwers, N., Stalon, V. and Wiame, J-M. (1978). Oxygen and nitrate in utilization by *Bacillus licheniformis* of the arginase and arginine deiminase routes of arginine catabolism and other factors affecting their synthesis. *J. Bacteriol.* **135**, 920–927.

Brown, D. E. (1970). Aeration in the submerged culture of micro-organisms. *In* "Methods in Microbiology" (Eds. J. R. Norris and D. W. Ribbons), Vol. 2, pp. 127–174. Academic Press, London.

Brown, O. R. (1979). Specific enzymatic sites and cellular mechanisms of oxygen toxicity. *In* "Biochemical and Clinical Aspects of Oxygen" (Ed. W. S. Caughey), pp. 755–766. Academic Press, New York.

Bruyninckx, W., Mason, H. S. and Morse, S. (1978). Oxygen is a mutagen at physiological oxygen tensions. *Nature (London)* **274**, 606–607.
Buchanan-Wollaston, V., Cannon, M. C., Beynon, J. L. and Cannon, F. C. (1981). Role of the *nif* A gene product in the regulation of *nif* expression in *Klebsiella pneumoniae*. *Nature (London)* **194**, 776–778.
Caldwell, J. B. and Lim, D. Y. (1981). The Effects of oxygen concentration on neuraminidase production by Type III Group B streptococci. *Curr. Microbiol.* **5**, 175–178.
Carlsson, J. and Carpenter, V. S. (1980). The *rec* A+ gene product is more important than catalase and superoxide dismutase in protecting *Escherichia coli* against hydrogen peroxide toxicity. *J. Bacteriol.* **142**, 319–321.
Carter, I. S. and Dawes, E. A. (1979). Effects of oxygen concentration and growth rate on glucose metabolism, polyβ-hydroxybutyrate biosynthesis and respiration of *Azotobacter beijerinckii*. *J. Gen. Microbiol.* **110**, 393–400.
Castric, K. F., McDevitt, D. A. and Castric, P. A. (1981). Influence of aeration on hydrogen cyanide biosynthesis by *Pseudomonas aeruginosa*. *Curr. Microbiol.* **5**, 223–226.
Chen, Y.-M. and Lin, E. C. C. (1984). Post-transcriptional control of L-1,2-propanediol oxido-reductase in the L-fucose pathway of *Escherichia coli* K-12. *J. Bacteriol.* **157**, 341–344.
Chippaux, M., Giudici, D., Abou-Jaoudé, A., Casse, F. and Pascal, M. C. (1978). A mutation leading to the total lack of nitrite reductase activity in *Escherichia coli* K-12. *Mol. Gen. Genet.* **160**, 225–229.
Chippaux, M., Bonnefoy-Orth, V., Ratouchniak, J. and Pascal, M. C. (1981). Operon fusions in the nitrate reductase operon and study of the control gene *nir R* in *Escherichia coli*. *Mol. Gen. Gene.* **182**, 477–479.
Clark, W. G., Davidson, E. and Marrs, B. L. (1984). Variation in levels of mRNA coding for antenna and reaction center polypeptides in *Rhodopseudomonas capsulata* in response to changes in oxygen concentration. *J. Bacteriol.* **157**, 945–948.
Cocucci, M. C., Belloni, G. and Gianini, L. (1975). Oxygen pressure, fatty acid composition and ergosterol level in *Rhodotorula gracilis*. *Arch. Microbiol.* **105**, 17–20.
Cohen-Bazire, G., Sistrom, W. R. and Stanier, R. Y. (1957). Kinetic studies of pigment synthesis by non-sulfur purple bacteria. *J. Cell. Comp. Physiol.* **49**, 25–68.
Cole, J. A. (1976). Microbial gas metabolism. *Adv. Microb. Physiol.* **14**, 1–92.
Cole, S. T. (1984). Molecular and genetic aspects of the fumarate reductase of *Escherichia coli*. *Biochem. Soc. Trans.* **12**, 237–238.
Conrad, R. and Schlegel, H. G. (1977). Influence of aerobic and phototrophic growth conditions on the distribution of glucose and fructose carbon into the Entner-Doudoroff and Embden-Meyerhof pathways in *Rhodopseudomonas sphaeroides*. *J. Gen. Microbiol.* **101**, 277–290.
Dagley, S. (1978). Pathways for the utilisation of organic growth substances. *In* "The Bacteria" (Eds. L. N. Ornston and J. R. Sokatch), Vol. 6, pp. 305–388. Academic Press, New York.
Dalton, H. (1981). Methane monooxygenases from a variety of microbes. *In* "Microbial Growth on C_1 Compounds" (Ed. H. Dalton), pp. 1–10. Heyden, London.
Dawes, E. A. (1978). Oxygen and the enzymes of obligate aerobes. *Biochem. Soc. Trans.* **6**, 363–367.
Dawes, E. A. (1981). Carbon metabolism. *In* "Continuous Cultures of Cells" (Ed. P. H. Calcott), Vol. 2, pp. 1–38. CRC Press, Boca Raton, Florida.
de Vries, W. and Stouthamer, A. H. (1969). Factors determining the degree of anaerobiosis of *Bifidobacterium* strains. *Arch. Mikrobiol.* **65**, 275–287.

de Vries, W., van Wijck-Kapteyn, W. M. C. and Stouthamer, A. H. (1972). Influence of oxygen on growth, cytochrome synthesis and fermentation pattern in propionic acid bacteria. *J. Gen. Microbiol.* **71**, 515–524.
de Vries, W., Donkers, C., Boellaard, M. and Stouthamer, A. H. (1978). Oxygen metabolism by the anaerobic bacterium *Veillonella alcalescens*. *Arch. Microbiol.* **119**, 167–174.
Dunn, G. M., Herbert, R. A. and Brown, C. M. (1979). Influence of oxygen tension on nitrate reduction by a *Klebsiella* sp. growing in chemostat culture. *J. Gen. Microbiol.* **112**, 379–383.
Eberly, W. R. (1964). Further studies on the metalimnetic oxygen maximum with special reference to its occurrence throughout the world. *Invest. Indiana Lakes Streams* **6**, 103–139.
Ellefson, W. L. and Wolfe, R. S. (1981). Biochemistry of methylreductase and evolution of methanogens. *In* "Microbial Growth on C_1 Compounds" (Ed. H. Dalton), pp. 171–180. Heyden, London.
Escalante-Semerena, J. C., Blakemore, R. P. and Wolfe, R. S. (1980). Nitrate dissimilation under microaerophilic conditions by a magnetic spirillum. *Appl. Environ. Microbiol.* **40**, 429–430.
Evans, W. C. (1977). Biochemistry of the bacterial metabolism of aromatic compounds in anaerobic environments. *Nature (London)* **270**, 17–22.
Faison, B. D. and Kirk, T. K. (1982). Effects of molecular oxygen on lignolytic activity in *Phanerochaete chrysosporium*. *Abstr. Annu. Meet. Am. Soc. Microbiol.* Abstract K125, p. 157.
Faller, A. H. and Schleifer, K.-H. (1981). Effects of growth phase and oxygen supply on the cytochrome composition and morphology of *Arthrobacter crystallopoietes*. *Curr. Microbiol.* **6**, 253–258.
Fee, J. A. (1981). A comment on the hypothesis that oxygen toxicity is mediated by superoxide. *In* "Oxygen and Life. Second BOC Priestley Conference," pp. 77–97. Royal Society of Chemistry, London.
Fiechter, A., Fuhrmann, G. F. and Käppeli, O. (1981). Regulation of glucose metabolism in growing yeast cells. *Adv. Microb. Physiol.* **22**, 123–183.
Fimmel, A. L. and Haddock, B. A. (1979). Use of *chl C—lac* fusions to determine regulation of gene *chl* C in *Escherichia coli* K-12. *J. Bacteriol.* **138**, 726–730.
Forney, L. J. and Reddy, C. A. (1982). The apparent involvement of oxygen radicals in lignin degradation by *Phanerochaete chrysosporium*. *Abstr. Annu. Meet. Am. Soc. Microbiol.* Abstract K126, p. 157.
Fridovich, I. (1981). Superoxide radical and superoxide dismutases. *In* "Oxygen and Living Processes" (Ed. D. L. Gilbert), pp. 250–272. Springer-Verlag, Berlin and New York.
Friebel, T. E. and Demain, A. L. (1977). Oxygen-dependent inactivation of gramicidin S synthetase in *Bacillus brevis*. *J. Bacteriol.* **130**, 1010–1016.
Garrard, W. and Lascelles, J. (1968). Regulation of *Staphylococcus aureus* lactate dehydrogenase. *J. Bacteriol.* **95**, 152–156.
Gest, H. (1981). Evolution of the citric acid cycle and respiratory energy conversion in prokaryotes. *FEMS Microbiol. Lett.* **12**, 209–215.
Goldberg, R. B. and Hanau, R. (1980). Regulation of *Klebsiella pneumoniae hut* operons by oxygen. *J. Bacteriol.* **141**, 745–750.
Gray, C. T., Wimpenny, J. W. T., Hughes, D. E. and Mossman, M. R. (1966a). Regulation of metabolism in facultative bacteria. 1. Structural and functional changes in *Escherichia coli* associated with shifts between the aerobic and anaerobic states. *Biochim. Biophys. Acta* **117**, 22–32.

Gray, C. T., Wimpenny, J. W. T. and Mossman, M. R. (1966b). Regulation of metabolism in facultative bacteria. II. Effects of aerobiosis, anaerobiosis and nutrition on the formation of Krebs cycle enzymes in *Escherichia coli*. *Biochim. Biophys. Acta* **117**, 33–41.
Hackett, C. S. and MacGregor, C. H. (1981). Synthesis and degradation of nitrate reductase in *Escherichia coli*. *J. Bacteriol.* **146**, 352–359.
Hallberg, R. O. (1968). Some factors of significance in the formation of sedimentary metal sulphides. *Stockholm Contrib. Geol.* **15**, 39–66.
Hambrick, G. A., De Laune, R. D. and Patrick, W. H. (1980). Effect of estuarine sediment pH and oxidation–reduction potential on microbial hydrocarbon degradation. *Appl. Environ. Microbiol.* **40**, 365–369.
Hänel, F., Hillinger, M. and Gräfe, U. (1981). Effect of oxygen limitation on cellular L-lysine pool and lipid spectrum of *Corynebacterium glutamicum*. *Biotechnol. Lett.* **3**, 461–464.
Harder, W., van Dijken, J. P. and Roels, J. A. (1981). Utilization of energy in methylotrophs. *In* "Microbial Growth on C_1 Compounds" (Ed. H. Dalton), pp. 258–269. Heyden, London.
Hare, P., Long, S., Robb, F. T. and Woods, D. R. (1981). Regulation of exoprotease production by temperature and oxygen in *Vibrio alginolyticus*. *Arch. Microbiol.* **130**, 276–280.
Harrison, D. E. F. (1972). Physiological effects of dissolved oxygen tension and redox potential on growing populations of micro-organisms. *J. Appl. Chem. Biotechnol.* **22**, 417–440.
Harrison, D. E. F. (1973). Growth, oxygen and respiration. *CRC Crit. Rev. Microbiol.* **2**, 185–228.
Harrison, D. E. F. (1976). The regulation of respiration rate in growing bacteria. *Adv. Microb. Physiol.* **14**, 243–313.
Harrison, D. E. F. and Pirt, S. J. (1967). The influence of dissolved oxygen concentration on the respiration and glucose metabolism of *Klebsiella aerogenes* during growth. *J. Gen. Microbiol.* **46**, 193–211.
Heinemann, B., Howard, A. J. and Palocz, H. J. (1970). Influence of dissolved oxygen levels on production of L-asparaginase and prodigiosin by *Serratia marcescens*. *Appl. Microbiol.* **19**, 800–804.
Hoshino, E., Frölander, F. and Carlsson, J. (1978). Oxygen and the metabolism of *Peptostreptococcus anaerobius* VPI 4330-1. *J. Gen. Microbiol.* **107**, 235–248.
Hurlbert, R. E. (1967). Effect of oxygen on viability and substrate utilization in *Chromatium*. *J. Bacteriol.* **93**, 1346–1352.
Jackson, F. A. and Dawes, E. A. (1976). Regulation of the TCA cycle and polyβ-hydroxybutyrate metabolism in *Azotobacter beijerinckii* grown under nitrogen or oxygen limitation. *J. Gen. Microbiol.* **97**, 303–312.
Jones, C. W. (1977). Aerobic respiratory systems in bacteria. *Sym. Soc. Gen. Microbiol.* **27** 23–59.
Jones, R. G. W. and Lascelles, J. (1967). The relationship of 4-hydroxybenzoic acid to lysine and methionine formation in *Escherichia coli*. *Biochem. J.* **103**, 709–713.
Jørgensen, B. B. (1977). Bacterial sulfate reduction within reduced microniches of oxidized marine sediments. *Mar. Biol. (Berlin)* **41**, 7–17.
Káprálek, F., Jechová, E. and Otavová, M. (1982). Two sites of oxygen control in induced synthesis of respiratory nitrate reductase in *Escherichia coli*. *J. Bacteriol.* **149**, 1142–1145.
Keevil, C. W., Hough, J. S. and Cole, J. A. (1979a). Regulation of respiratory and

fermentative modes of growth of *Citrobacter freundii* by oxygen, nitrate and glucose. *J. Gen. Microbiol.* **113**, 83–95.
Keevil, C. W., Hough, J. S. and Cole, J. A. (1979b). Regulation of 2-oxoglutarate dehydrogenase synthesis in *Citrobacter freundii* by traces of oxygen in commercial nitrogen gas and by glutamate. *J. Gen. Microbiol.* **114**, 355–359.
Kirby, T. W. Lancaster, J. R. and Fridovich, I. (1981). Isolation and characterisation of the iron-containing superoxide dismutase of *Methanobacterium bryantii*. *Arch. Biochem. Biophys.* **210**, 140–148.
Knappe, J. and Schmitt, T. (1976). A novel reaction of S-adenosyl-L-methionine correlated with the activation of pyruvate formate-lyase. *Biochem. Biophys. Res. Commun.* **71**, 1110–1117.
Knoell, H.-E. (1981). Stand-by position of the dioxygen-dependent ubiquinone 8 synthesis apparatus in anaerobically grown *Escherichia coli* K-12. *FEMS Microbiol. Lett.* **10**, 59–62.
Krebs, H. A. (1972). The Pasteur effect and the relations between respiration and fermentation. In "Essays in Biochemistry" (Eds. P. N. Campbell and F. Dickens), Vol. 8, pp. 1–34. Academic Press, New York.
Krinsky, N. I. (1979). Carotenoid protection against oxidation. *Pure Appl. Chem.* **51**, 649–660.
Kröger, A. (1977). Phosphorylative electron transport with fumarate and nitrate as terminal hydrogen acceptors. *Symp. Soc. Gen. Microbiol.* **27**, 61–93.
Kulla, H. and Gottschalk, G. (1977). Energy-dependent inactivation of citrate lyase in *Enterobacter aerogenes*. *J. Bacteriol.* **132**, 764–770.
Kuritzkes, D. R., Zhang, X.-Y. and Lin, E. C. C. (1984). Use of Φ (*glp-lac*) in studies of respiratory regulation of the *Escherichia coli* anaerobic *sn*-glycerol-3-phosphate dehydrogenase (*glp A B*). *J. Bacteriol.* **157**, 591–598.
Laanbroek, H. J. and Veldkamp, H. (1982). Microbial interactions in sediment communities. *Philos. Trans. R. Soc. London, Ser. B* **297**, 533–550.
Lambden, P. R. and Guest, J. R. (1976). Mutants of *Escherichia coli* K-12 unable to use fumarate as an anaerobic electron acceptor. *J. Gen. Microbiol.* **97**, 145–160.
Lim, S. T., Hennecke, H. and Scott, D. B. (1979). Effect of cyclic guanosine 3′,5′-monophosphate on nitrogen fixation in *Rhizobium japonicum*. *J. Bacteriol.* **139**, 256–263.
Liu, F. S. and Zajic, J. E. (1973). Effect of oxygen-transfer rate on production of L-asparaginase by *Erwinia aroideae*. *Can. J. Microbiol.* **19**, 1153–1158.
Lloyd, D., James, K., Williams, J. and Williams, N. (1981). A membrane-covered probe for oxygen measurements in the nanomolar range. *Anal. Biochem.* **116**, 17–21.
McFadden, B. A. (1973). Autotrophic carbon dioxide assimilation and the evolution of ribulose diphosphate carboxylase. *Bacteriol. Rev.* **37**, 289–319.
Mackenzie, F. T. and Wollast, R. (1977). Thermodynamic and kinetic controls of global chemical cycles of the elements. In "Chemical Cycles and their Alterations by Man" (Ed. W. Stumm), pp. 45–59. Dahlem Konferenzen, Berlin.
Maclennan, D. G. and Pirt, S. J. (1966). Automatic control of dissolved oxygen concentration in stirred microbial cultures. *J. Gen. Microbiol.* **45**, 289–302.
McPhedran, P., Sommer, B. and Lin, E. C. C. (1961). Control of ethanol dehydrogenase levels in *Aerobacter aerogenes*. *J. Bacteriol.* **81**, 852–857.
Markham, E. (1969). The role of oxygen in brewery fermentations. *Wallerstein Lab. Commun.* **32**, 5–12.
Marrs, B. and Gest, H. (1973). Regulation of bacteriochlorophyll synthesis by oxygen in respiratory mutants of *Rhodopseudomonas capsulata*. *J. Bacteriol.* **114**, 1052–1057.

Mattoon, J. R., Lancashire, W. E., Sanders, H. K., Carvajal, E., Malamud, D. R., Braz, G. R. C. and Panek, A. D. (1979). Oxygen and catabolite regulation of hemoprotein biosynthesis in yeast. *In* "Biochemical and Clinical Aspects of Oxygen" (Ed. W. S. Caughey), pp. 421–435. Academic Press, New York.

Mitchell, C. G. and Dawes, E. A. (1982). The role of oxygen in the regulation of glucose metabolism, transport and the TCA cycle in *Pseudomonas aeruginosa*. *J. Gen. Microbiol.* **128**, 49–59.

Morris, J. G. (1975). The physiology of obligate anaerobiosis. *Adv. Microb. Physiol.* **12**, 169–246.

Morris, J. G. (1979). Nature of oxygen toxicity in anaerobic micro-organisms. *In* "Strategies of Microbial Life in Extreme Environments" (Ed. M. Shilo), Dahlem Conf. Life Sci. Res. Rep. 13, pp. 149–162. Verlag Chemie, Weinheim.

Newman, B. M. and Cole, J. A. (1978). The chromosomal location and pleiotropic effects of mutations of the *nir A*⁺ gene of *Escherichia coli* K-12; the essential role of *nir A*⁺ in nitrite reduction and in other anaerobic redox reactions. *J. Gen. Microbiol.* **106**, 1–12.

O'Brien, R. W. and Morris, J. G. (1971). Oxygen and the growth and metabolism of *Clostridium acetobutylicum*. *J. Gen. Microbiol.* **68**, 307–318.

Oura, E. (1977). Reaction products of yeast fermentations. *Process Biochem.* **12**, 19–21 and 35.

Pascal, M.-C., Burini, J.-F., Ratouchniak, J. and Chippaux, M. (1982). Regulation of the nitrate reductase operon: Effect of mutations in *Chl A, B, D,* and E genes. *Mol. Gen. Genet.* **188**, 103–106.

Pecher, A., Blaschkowski, H. P., Knappe, K. and Böck, A. (1982). Expression of pyruvate formate-lyase from the cloned structural gene. *Arch. Microbiol.* **132**, 365–371.

Pecher, A., Zinoni, F. Jatisatienr, C., Wirth, R., Hennecke, H. and Böck, A. (1983). On the redox control of synthesis of anaerobically induced enzymes in Enterobacteriaceae. *Arch. Microbiol.* **136**, 131–136.

Perry, J. J. (1979). Microbial co-oxidations involving hydrocarbons. *Microbiol. Rev.* **43**, 59–72.

Pichinoty, F. (1962). Inhibition par l'oxygène de la biosynthèse et de l'activité de l'hydrogenase et de l'hydrogènelyase chez les bactéries anaérobies facultatives. *Biochim. Biophys. Acta* **64**, 111–119.

Pichinoty, F. (1965). L'effet oxygène et la biosynthèse des enzymes d'oxydoreduction bacteriens. *Colloq. Int. C.N.R.S.* **124**, 507–522.

Pirt, M. W. and Pirt, S. J. (1980). The influence of carbon dioxide and oxygen partial pressures on *Chlorella* growth in photosynthetic steady-state cultures. *J. Gen. Microbiol.* **119**, 321–326.

Poole, R. K. and Chance, B. (1981). The reaction of cytochrome o in *Escherichia coli* K-12 with oxygen. Evidence for a spectrally and kinetically distinct cytochrome o in cells from oxygen-limited cultures. *J. Gen. Microbiol.* **126**, 277–287.

Pritchard, G. G., Wimpenny, J. W. T., Morris, H. A., Lewis, M. W. A. and Hughes, D. E. (1977). Effects of oxygen on *Propionibacterium shermanii* grown in continuous culture. *J. Gen. Microbiol.* **102**, 223–233.

Ramaiah, A. (1974). Pasteur effect and phosphofructokinase. *Curr. Top. Cell. Regul.* **8**, 298–345.

Revsbech, N. P., Jørgensen, B. B. and Blackburn, T. H. (1980). Oxygen in the sea bottom measured with a microelectrode. *Science* **207**, 1355–1356.

Rice, C. W. and Hempfling, W. P. (1978). Oxygen-limited continuous culture and respiratory energy conservation in *Escherichia coli*. *J. Bacteriol.* **134**, 115–124.

Robson, R. L. and Postgate, J. R. (1980). Oxygen and hydrogen in biological nitrogen fixation. *Annu. Rev. Microbiol.* **34**, 183–207.

Roodyn, D. B. and Wilkie, D. (1968). "The Biogenesis of Mitochondria." Methuen, London.

Samah, O. A. and Wimpenny, J. W. T. (1982). Some effects of oxygen on the physiology of *Selenomonas ruminantium* WPL 151/1 grown in continuous culture. *J. Gen. Microbiol.* **128**, 355–360.

Schatzmann, H. (1975). Anaerobes Wachstum von *Saccharomyces cerevisiae*. Thesis No. 5504, ETH Zürich.

Schlegel, H. G. (1976). The physiology of hydrogen bacteria. *Antonie van Leeuwenhoek* **42**, 181–201.

Schlegel, H. G. and Vollbrecht, D. (1980). Formation of the dehydrogenases for lactate, ethanol and butanediol in the strictly aerobic bacterium *Alcaligenes eutrophus*. *J. Gen. Microbiol.* **117**, 475–481.

Schneider, K. and Schlegel, H. G. (1981). Production of superoxide radicals by soluble hydrogenase from *Alcaligenes eutrophus* H16. *Biochem. J.* **193**, 99–107.

Schönheit, P., Keweloh, H. and Thauer, R. K. (1981). Factor F_{420} degradation in *Methanobacterium thermoautotrophicum* during exposure to oxygen. *FEMS Microbiol. Lett.* **12**, 347–349.

Schwartz, A. C., Mertens, B., Voss, K. W. and Hahn, H. (1976). Inhibition of acetate and propionate formation upon aeration of resting cells of the anaerobic *Propionibacterium shermanii*: evidence of the Pasteur reaction. *Z. Allg. Mikrobiol.* **16**, 123–131.

Schwartz, E. R., Old, L. O. and Reed, L. J. (1968). Regulatory properties of pyruvate dehydrogenase from *Escherichia coli*. *Biochem. Biophys. Res. Commun.* **31**, 495–500.

Scott, D., Brannan, J. and Higgins, I. J. (1981). The effect of growth conditions on intracytoplasmic membranes and methane mono-oxygenase activities in *Methylosinus trichosporium* OB 3b. *J. Gen. Microbiol.* **125**, 63–72.

Senior, P. J., Beech, G. A., Ritchie, G. A. F. and Dawes, E. A. (1972). The role of oxygen limitation in the formation of polyβ-hydroxybutyrate during batch and continuous culture of *Azotobacter beijerinckii*. *Biochem. J.* **128**, 1193–1201.

Shaw, D. J. and Guest, J. R. (1982a). Amplification and product identification of the *fnr* gene of *Escherichia coli*. *J. Gen. Microbiol.* **128**, 2221–2228.

Shaw, D. J. and Guest, J. R. (1982b). Nucleotide sequence of the *fnr* gene and primary structure of the Fnr protein of *Escherichia coli*. *Nucleic Acids Res.* **10**, 6119–6130.

Shaw, D. J., Rice, D. W. and Guest, J. R. (1983). Homology between CAP and Fnr, a regulator of anaerobic respiration in *Escherichia coli*. *J. Mol. Biol.* **166**, 241–247.

Skotnicki, M. L. and Rolfe, B. G. (1979). Pathways of energy metabolism required for phenotypic expression of nif^+_{Kp} genes in *Escherichia coli*. *Aust. J. Biol. Sci.* **32**, 637–649.

Smith, M. W. and Neidhardt, F. C. (1983a). Proteins induced by anaerobiosis in *Escherichia coli*. *J. Bacteriol.* **154**, 336–343.

Smith, M. W. and Neidhardt, F. C. (1983b). Proteins induced by aerobiosis in *Escherichia coli*. *J. Bacteriol.* **154**, 344–350.

Smith, S. R. L. (1981). Some aspects of ICI's single cell protein process. *In* "Microbial Growth on C_1 Compounds" (Ed. H. Dalton), pp. 342–348. Heyden, London.

Sols, A. (1976). The Pasteur effect in the allosteric era. *In* "Reflections on Biochemistry" (Eds. A. Kornberg, B. L. Horecker, L. Cornudella and J. Oró), pp. 199–206. Pergamon, Oxford.

Stams, A. J. M. and Hansen, T. A. (1982). Oxygen labile L(+) lactate dehydrogenase activity in *Desulfovibrio desulfuricans*. *FEMS Microbiol. Lett.* **13**, 389–394.

Stephenson, M. P., Jackson, F. A. and Dawes, E. A. (1978). Further observations on carbohydrate metabolism and its regulation in *Azotobacter beijerinckii*. *J. Gen. Microbiol.* **109**, 89–96.

Stewart, V. (1982). Requirement of Fnr and NarL functions for nitrate reductase expression in *Escherichia coli* K-12. *J. Bacteriol.* **151**, 1320–1325.

Stouthamer, A. H. (1976). Biochemistry and genetics of nitrate reductase in bacteria. *Adv. Microb. Physiol.* **14**, 315–375.

Stouthamer, A. H. (1979). The search for correlation between theoretical and experimental growth yields. *Int. Rev. Biochem.* **21**, 1–47.

Takai, Y. and Kamura, T. (1966). The mechanism of reduction in waterlogged paddy soil. *Folia Microbiol. (Prague)* **11**, 304–313.

Tempest, D. W. and Neijssel, O. M. (1981). Comparative aspects of microbial growth yields with special reference to C_1 utilizers. In "Microbial Growth on C_1 Compounds" (Ed. H. Dalton), pp. 325–334. Heyden, London.

Turnbough, C. L. and Switzer, R. L. (1975). Oxygen-dependent inactivation of glutamine phosphoribosylpyrophosphate amidotransferase *in vitro:* model for *in vivo* inactivation. *J. Bacteriol.* **121**, 115–120.

Udvardy, J., Borbely, G., Juhasz, A. and Farkas, G. L. (1984). Thioredoxins and redox modulation of glucose 6-phosphate dehydrogenase in *Anabaena* sp. strain PCC 7120 vegetative cells and heterocysts. *J. Bacteriol.* **157**, 681–683.

Unden, G. and Guest, J. R. (1984) Cyclic AMP and anaerobic gene expression in *Escherichia coli. FEBS Lett.*, submitted for publication.

van Gent-Ruijters, M. L. W., de Meijere, F. A., de Vries, W. and Stouthamer, A. H. (1976). Lactate metabolism in *Propionibacterium pentosaceum*. *Antonie van Leeuwenhoek* **42**, 217–228.

Vollbrecht, D., Schlegel, H. G., Stoschek, G. and Janczikowski, A. (1979). Excretion of metabolites by hydrogen bacteria. IV. Respiration rate-dependent formation of primary metabolites and of poly-3-hydroxybutanoate. *Eur. J. Appl. Microbiol. Biotechnol.* **7**, 267–276.

Wakao, N. and Furusaka, C. (1976). Presence of microaggregates containing sulfate-reducing bacteria in a paddy field soil. *Soil Biol. Biochem.* **8**, 157–159.

Weitzmann, P. D. J. (1981). Unity and diversity in some bacterial citric acid-cycle enzymes. *Adv. Microb. Physiol.* **22**, 185–244.

Wimpenny, J. W. T. (1969a). Oxygen and carbon dioxide as regulators of microbial growth and metabolism. *Symp. Soc. Gen. Microbiol.* **19**, 161–197.

Wimpenny J. W. T. (1969b). Oxygen and microbial metabolism. *Process Biochem.* **4**, 19–22.

Wimpenny, J. W. T.(1981). Spatial order in microbial ecosystems. *Biol. Rev. Cambridge Philos. Soc.* **56**, 295–342.

Wimpenny, J. W. T. and Firth, A. (1972). Levels of nicotinamide adenine dinucleotide and reduced nicotinamide adenine dinucleotide in facultative bacteria and the effect of oxygen. *J. Bacteriol.* **111**, 24–32.

Wimpenny, J. W. T. and Necklen, D. K. (1968). The redox environment and microbial physiology 1. The transition from anaerobiosis to aerobiosis in continuous cultures of facultative anaerobes. *Biochim. Biophys. Acta* **253**, 352–359.

Wimpenny, J. W. T. and Parr, J. A. (1979). Biochemical differentiation in large colonies of *Enterobacter cloacae*. *J. Gen. Microbiol.* **114**, 487–489.

Wimpenny, J. W. T. and Samah, O. A. (1978). Some effects of oxygen on the growth and physiology of *Selenomonas ruminantium*. *J. Gen. Microbiol.* **108**, 329–332.

Yamamoto, I. and Ishimoto, M. (1975). Effect of nitrate reduction on the enzyme levels in carbon metabolism in *Escherichia coli*. *J. Biochem. (Tokyo)* **78**, 307–315.

4

The Utilisation of Light by Microorganisms

C. E. GIBSON

Department of Agriculture for Northern Ireland
Freshwater Biological Investigation Unit
Greenmount, Muckamore, Northern Ireland
and

D. H. JEWSON

Limnology Laboratory
New University of Ulster
Ballyronan, Northern Ireland

Introduction

The effect of light on microorganisms continues to be a very active field of research, and any review is of necessity highly selective. We have covered the areas most familiar to us as ecologists working with the plankton of a turbid lake. This chapter is confined to oxygenic photoautotrophs, and the radiation with which we are concerned therefore lies within the wavelength band 400–700 nm. Since much of the classical work on photosynthesis has been carried out using planktonic organisms, our self-imposed circumscription is not a serious one, but we have tried to broaden this chapter to include organisms other than *Chlorella* and *Anacystis* and to consider life outside the laboratory in intermittent light at moderate temperatures.

Unlike other sources of energy, there are still problems in the measurement of light, and there are important gaps in the understanding of the underwater light field. We have therefore taken space to discuss briefly some of the problems of light measurement and our current understanding of the light climate experienced by planktonic organisms before considering the response of microorganisms to changes in light quality and quantity. Since the preparation of this review (July, 1982), the excellent publication of Kirk (1983) has appeared, which treats more fully some of the topics touched on here. In the main, we have avoided the question of whether such changes are adaptive, in the sense that they better fit the organism to survive, and have preferred the more neutral term "response," which avoids teleology.

Organisms deriving their energy from light face the unique problem that their energy supply is necessarily interrupted. In the final section, the problems this poses are considered and the relationship between photosynthesis and growth are explored.

The Underwater Light Climate

Much of the progress that has been made over the last 2 decades in unravelling the complexities of the underwater light environment has been due to the improved technical and analytical methods employed by physical oceanographers (Tyler and Smith, 1970; Morel, 1973; Jerlov and Steemann-Nielsen, 1973; Jerlov, 1976; Tyler, 1977; Morel and Prieur, 1977; Smith and Baker, 1978a,b; Prieur and Sathyendranath, 1981). However, as pointed out by Weinberg (1976) and Brakel (1979), it is only relatively recently that the task of translating this information into biologically meaningful and readily usable terms has begun to receive attention. One of the reasons for this delay was that measurements of light by aquatic biologists were for many years dominated by the use of Secchi discs and "lux" meters. Although they provided useful preliminary information, their continued use prevented a clear understanding of the response of aquatic organisms, because Secchi discs cannot yield data on any single optical property (Tyler, 1968; Lorenzen, 1980) and lux meters measure light in photometric units (lux, foot-candles), which do not necessarily correlate with the absorption of radiant energy by microorganisms. Although a number of workers have referred to this latter problem (Tyler, 1973a; Arnold, 1975a,b), it is worth repeating here since much of the early laboratory data were reported in lux. Unfortunately, accurate conversion from lux to radiometric units is rarely possible because the ambient light conditions are imprecisely described. Discussions of the correct use of terminology and units can be found in Westlake (1965), Tyler (1973b), Jerlov (1976), Højerslev (1981), and Kirk (1983).

At present, we are in a transitional period in which there is a growing awareness of the complexity of the underwater light climate, but still no consensus on the way in which information on it should be gathered. Some improvement has resulted from a shift to measurements of photon flux density (= photon fluence rate or quantum irradiance), but both methodological and physiological problems remain (Yentsch, 1980).

For instance, differences in the spectral composition of light sources (Golterman et al., 1978), the optical properties of experimental vessels (Loogman and Van Liere, 1978), reflections from the sides of the incubation chamber, and the collecting properties of the measuring instrument (Smith and Wilson, 1972) all contribute to the large errors that may occur when laboratory and field studies are compared (Tyler, 1975). Infrequent calibration of instruments can also be a problem (see Arnold, 1975a,b, for recommendations), and it should perhaps become more widely accepted that all light measurements include details of how, when, and where the last calibration of light meters was carried out.

Talling (1971), in an earlier review, suggested that three general features had hindered the assessment of the underwater light field. "Firstly, it is a complex of variables which cannot be expressed by a single numerical value, as can temperature. Secondly, it is characteristically related to biological processes such as photosynthesis by markedly non-linear responses. Thirdly, its effect on algal cells may be multiple, involving diverse physiological processes." Today, this assessment still applies, although in the intervening period there has been a large upsurge of interest and a greater awareness of the problems (Kirk, 1977, 1980; Westlake *et al.,* 1980; Yentsch, 1980; Dring, 1981b; Talling, 1982). This section summarises some of the most important aspects of the natural light field, concentrating on work published in the last decade as a background to discussions of experimental work that follow.

The factors influencing the irradiance received by a cell can be grouped into the following categories:

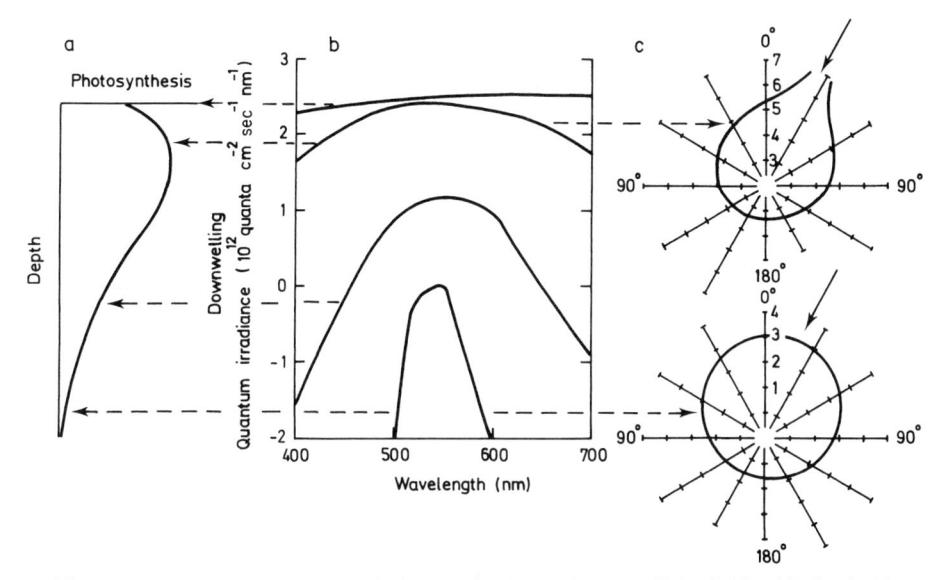

Fig. 1. Summary of the type of changes in the underwater light field with depth (a) photosynthesis–depth profile with (b) the corresponding changes in light quality and quantity and (c) angular distribution. Part (c) is plotted to a logarithmic scale, and the solid arrow indicates the direction of the sun. Data in (b) and (c) adapted from Dubinsky and Berman (1976) and Lythgoe (1975).

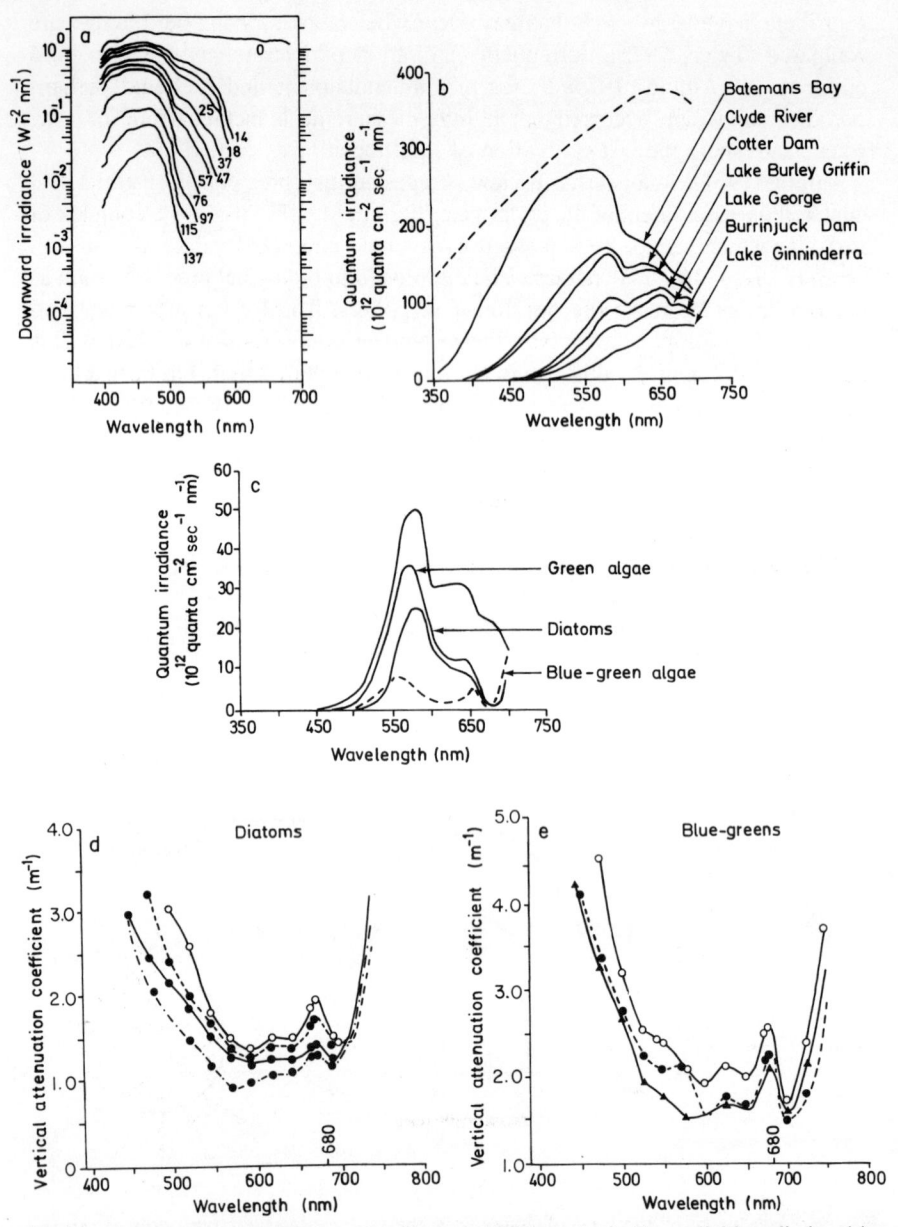

Fig. 2. The spectral distribution of underwater photosynthetically available radiation (a) at selected depths (in metres) in the Sargasso Sea. (b) Calculated for a 2-m depth in southeastern Australian waters. (c) Calculated spectral distribution at 5 m in model suspensions of various algal types when combined with other factors attenuating light. The

1. *Above surface*, such as day length, cloud cover, and solar elevation.
2. *air–water interface*, where the proportion of reflected versus refracted light affects the quantity of light entering the water.
3. *Underwater*, the optical properties of the water body.

The first category has been discussed by Talling (1971), Straskraba and Hammer (1980), and Kirk (1983) with particular reference to variations at different latitudes and times of the year. Other factors, such as day length and fluctuating irradiance, are covered in the later discussion, and problems associated with the entry of light into water are summarised in Jerlov (1976) and Kirk (1977, 1983). The discussion here is mainly concerned with the optical properties of the water body which alter the composition of incident radiation (through scattering and absorption) as it passes down through the water.

Firstly, there is a decrease in quantity; secondly, a change in spectral quality, with a trend towards monochromacity; and thirdly, a change in angular distribution caused by scattering which renders the light field more diffuse (Fig. 1). As a result, the path length of photons increases, and there is a greater chance of their being absorbed.

The wide variation in the relative importance of these factors leads to a correspondingly wide range of underwater light climates. For instance, the depth of the euphotic zone (1% light level) varies from a few centimetres in some African lakes (Talling *et al.*, 1973) to over 100 m in the clearest lakes and oceans (Tyler and Smith, 1967; Tyler, 1977). Light quality also changes. In the clearest waters, there is a shift towards predominance of light between 430 and 470 nm (Tyler, 1977), but in turbid fertile lakes the shift is in the opposite direction (Jewson, 1977; Jewson and Taylor, 1978), and the most penetrating wavelength may be beyond 700 nm (Fig. 2). Such variability emphasises that the underwater light climate of an individual cell or population is not easily defined and cannot be described by a single numerical value.

The main agents of underwater light attenuation may be classified into the following categories:

1. *Gelbstoff*—dissolved and colloidal coloured organic compounds.
2. Water.
3. Nonliving particulate material—detrital and inorganic.
4. Plankton, including phytoplankton and bacteria.

---

uppermost curve is attained in the absence of algae. (d) Spectral variation of the vertical attenuation coefficient in Lough Neagh waters during a growth of diatoms. ●--·●, 19 February 1974; ●——●, 24 February 1974; ●- - -●, 5 March 1974; ○———○, 14 March 1974. (e) Blue-greens. ▲———▲, 4 January 1971; ●- - -●, 26 May 1971; ○———○, 26 April 1972. (a) After Morel and Caloumenos (1974); (b) from Kirk (1976b); (c) from Kirk (1976a); (e) from Jewson (1977).

For present purposes, the first three categories may be considered as background modifiers of the light received by the microorganisms. In the following section, we will discuss the background light attenuation before reviewing the present knowledge of light interception by cells.

Background Light Interception

Gelbstoff. *Gelbstoff* is composed of a variety of dissolved and colloidal organic compounds (Kalle, 1966; Søndergard and Schierup, 1982) which absorb mainly blue light (Kirk, 1976b, 1979; Jewson and Taylor, 1978). In the clearest ocean waters and a few lakes, where *Gelbstoff* concentrations are small, the effect on light attenuation is minor compared to that of water itself (see below), and blue light penetrates farthest (Fig. 2a). However, with increasing concentrations of *Gelbstoff*, the shorter wavelengths are preferentially removed. The series of curves in Fig. 2b illustrates the point. They show a range from coastal marine water, where *Gelbstoff* absorbs 30% of the quanta, to a darkly stained lake water where *Gelbstoff* is responsible for 80% of the quantum absorbance.

Gelbstoff absorbs mainly at the shorter wavelengths, and so competes with the blue-absorbing pigments of photosynthesis, but this is likely to be important only at subsaturating irradiances, and in many cases it is then the reduction in the number of photons rather than their wavelength which is important. In lakes with heavily stained waters, not only is photosynthesis reduced by a decrease in the euphotic zone, but there is also a relative increase in the proportion of the lake that lies in permanent darkness (Jewson and Taylor, 1978; Jones and Ilmavirta, 1978).

Water. Water attenuates light mainly at the red end of the spectrum, and some of the clearest natural waters (e.g., Crater Lake, North Pacific Gyre, Sargasso Sea; see Fig. 2a) have attenuation spectra approaching those of pure water (Tyler and Smith, 1967; Morel, 1974), with broad flat minima between 430 and 470 nm. Molecular scattering becomes increasingly important at the shorter wavelengths (Morel, 1974).

Nonliving Particulates. This is not a natural grouping, as it includes both detrital and inorganic particles. Relatively little is known about their optical properties compared to those of the other groups, partly because it is difficult to separate nonliving from living particulate material. Kirk (1980) has attempted a division for Australian lakes, in some of which nearly 80% of the quanta are absorbed by suspended particles. The contribution of nonliving particulates to light attenuation is very variable, and depends, for example, on the degree of turbulence. In rough weather, resuspension of bottom material may reduce light penetration and delay spring growth (Jewson, 1976, 1977), and in some shallow lakes, continual mixing of the waters can maintain very high concentrations of inorganic material throughout the year (Dokulil, 1979). Scattering of light is particularly important in such cases; this is discussed further in relation to the light climate of planktonic cells.

Planktonic Light Interception

Attenuation of light by phytoplankton involves absorption and scattering, which are both strongly wavelength dependent. In some cases, populations may intercept up to 80% of the incident light, yet only a small proportion of it is ever utilised (Fig. 5 in Talling, 1982). Studies of the interception of light by phytoplankton have followed two lines. Firstly, there is a predominantly physical approach which has investigated how phytoplankton biomass is related to light attenuation, with a view to the optical classification of waters and to remote sensing (Jerlov, 1976; Morel and Prieur, 1977; Smith and Baker, 1978a,b; Prieur and Sathyendranath, 1981). Secondly, biological research has attempted to estimate the amount of light utilised in photosynthesis (see reviews by Westlake *et al.*, 1980; Yentsch, 1980; Talling, 1982). Both lines of enquiry require a knowledge of how much light is intercepted by phytoplankton and how much by "background" factors.

It is possible to reach some general conclusions about the quantities of light intercepted by phytoplankton in the water column. For instance, at low biomass concentrations above 2 µg chlorophyll *a* litre^{-1} algae increasingly dominate light attenuation, but in freshwaters the balance point frequently occurs at higher proportion of the available light and become progressively "self-shading." In marine conditions, Yentsch (1980; see Fig. 3.6) has suggested that at biomass concentrations about 2 µg chlorophyll a litre^{-1} algae increasingly dominate light attenuation, but in freshwaters the balance point frequently occurs at higher population densities, because the amount of background material is normally greater (Jewson and Taylor, 1978; Megard *et al.*, 1979). Biomass concentrations in freshwater also reach much higher values than in the sea, and may even approach the theoretical maximum content of the euphotic zone in some cases (Talling *et al.*, 1973; Bindloss, 1974, 1976; Ganf, 1974).

If more specific questions are asked about the interception of light by an individual cell at a given instant in time, then the limits of our present knowledge are much more clearly shown.

One of the problems has centred on estimates of the specific absorption coefficients of phytoplankton. This has frequently been assumed to be the same as the increment in the attenuation coefficient per unit biomass (usually denoted as either ϵ_s or k_s). It is in reality a complex of variables, but for the sake of modelling partitioning and efficiency measurements, many attempts have been made to "force-fit" it into a single numerical solution. In some cases this may be satisfactory, but difficulties arise if the results are applied more generally and the original assumptions and limitations forgotten. In particular, Prieur and Sathyendranath (1981) have drawn attention to the large variety of ways in which this parameter has been determined and how few of them are truly comparable. Talling (1982) has also discussed some of the problems, with particular emphasis on two points. Firstly, many treatments neglect the "inherent difference in scattering media between attenuation and true absorption"; secondly, ϵ_s is wavelength (and hence depth) dependent. Talling concludes that most estimates are no more than "useful first approximations." Although the pessimism implied by this last remark is well founded (see also Welschmeyer and Lorenzen, 1981; Prieur and Santhyendranath, 1981), there has been some progress in measuring this parameter in three important areas—scattering, wavelength dependence, and cell morphology. These are discussed individually, with an outline of some of the problems that must be solved before our understanding of light interception *in situ* can advance.

Scattering. Most of the light scattered by particulate matter in water, including microorganisms, tends to be forward scattered at a narrow angle (Jerlov, 1976; Kirk, 1981a,b; Morton, 1978), although there may be relatively more backscattering from detrital particles. The importance of scattering in the underwater light field has been recognised by physical oceanographers for many years (Jerlov, 1976; Tyler, 1977; Sugihara and Tsuda, 1979), but because sophisticated equipment is needed to measure it, it has largely been ignored in biological studies (Yentsch, 1980). Attempts have been made to remedy this deficiency by Kirk (1981a,b), who has suggested a simplified measurement technique that can be incorporated into most field studies. If such measurements are more widely made by biologists, the general awareness of the diffuse nature of the underwater light field will be greatly increased (Smith and Wilson, 1972; Højerslev, 1981).

The need to account for scattering in absorption measurements of turbid solutions has long been appreciated in laboratory studies of algal pigments (Shibata, 1958; Lattimer, 1959; Doucha and Kubin, 1976; Roger and Reynaud, 1977; Yentsch, 1980), and it is even possible to identify some species by their scattering signature (Price *et al.*, 1978). In one study linking laboratory and field estimates (Kiefer *et al.*, 1979; Wilson and Kiefer, 1979), there was a wavelength-dependent change in the ratio of scattering to absorption which was correlated with changes in growth rate and cell age. However, this and other work (Mueller, 1976; Prieur and Sathyendranath, 1981) is largely concerned with improving methods of optical classification and remote sensing. What is required now is the combination of such measurements of true absorption with experiments on physiological activity *in situ*. This approach would help reduce the anomalies frequently found when comparing the response to available light at different sites.

Spectral Changes

Aquatic microorganisms have evolved a wide variety of light-harvesting pigments (Jeffrey 1980). These pigments, if present in sufficient quantity, cause phytoplankton to make a considerable modification in the spectral composition of their own light field (Yentsch, 1980; Talling, 1982). In general, photosynthetic pigments shift the most penetrating regions to longer wavelengths (Fig. 3-15 in Yentsch, 1980), even beyond 700 nm if *Gelbstoff* is present.

In an attempt to analyse spectral differences *in situ,* Kirk (1977, Fig. 9), using data from earlier work (Kirk, 1976a), determined how the different pigment contents of diatoms, blue-green and green algae might affect light quality when combinted with other light-attenuating components such as *Gelbstoff* (Fig. 2c). Although he did not account for scattering, the results are similar to those found for dense natural populations of diatoms and blue-greens (Fig. 2d,e) (Jewson, 1977; Jewson and Taylor, 1978). There is also a packaging effect on light quality, which is discussed further in the next section.

In assessing the importance of spectral changes in the ambient light field, it should be remembered that absorption does not necessarily reflect the physiological response. For example, some of the largest modifications to light climate in populations of cyanophytes occur in the blue region of the spectrum, where light is absorbed but relatively poorly utilised (Duysens, 1952; Fay, 1970; D. H. Jewson and A. C. Ley, unpublished). Response changes with irradiance and most published photosynthetic action spectra are relevant only to conditions of light limitation. At light saturation there is a flattening across the spectrum, and photosynthesis becomes progressively independent of wavelength (Smith, 1968; Yentsch, 1980). At even higher irradiances there are deleterious effects of photoinhibition which, by causing rearrangements of chloroplasts (Kiefer, 1973), can also alter absorption properties. The light interception of an individual cell therefore needs to be considered in the context of its position in the water column. Various authors have noted how ϵ_s is likely to change with depth as the light quality changes (Morel, 1978; Atlas and Bannister, 1980; Talling, 1982), but since ϵ_s is so difficult to measure *in situ,* many estimates of quantum efficiency have assumed it to be constant (Tyler, 1975; Dubinsky and Berman, 1976, 1981; Platt and Jassby, 1976; Morel, 1978). This can lead to large inaccuracies (Welschmeyer and Lorenzen, 1981; Talling, 1982), particularly if accessory pigments are more important in deeper water (Shimura and Ichimura, 1973; Shimura and Fujita, 1975) and show changes dependent on growth irradiance (Foy and Gibson, 1982b).

Although simple assumptions about light absorption based on chlorophyll a may have been convenient in the past for giving preliminary information, future work in the field will require more rigorous attention to the other light-harvesting pigments, their arrangement into photosynthetic units (Falkowski and Owens, 1980), and their relation to ambient light conditions.

Cell Morphology. The size and shape of a cell (Kirk, 1975a,b, 1976a), as well as the packaging of the pigment within it (Kiefer, 1973), can make a large difference in the amount of light intercepted. However, although much is known about changes in the pigment content of cells grown at different irradiances (see later), there is relatively little quantitative information on how these pigment changes affect the efficiency of light absorption. Many of the available data are inconclusive, because of the problem of accurately measuring ϵ_s, changes in accessory pigments, and nutrient status (see the discussion of Welschmeyer and Lorenzen, 1981).

Some of the best information is contained in a series of papers by Kirk (1975a,b, 1976a) which analyse the effects of cell size, cell shape, and spectral composition on light interception. In general, the larger the cell or colony, the less light it collects per unit biomass. Further, because this is in part a property of the surface/volume ratio, shape is also important. For a given volume of cell or colony, the more elongated it is, the more light it collects. Shape may also affect differential absorption at different wavelengths. A more elongate shape favours light collection at the most strongly absorbed wavelength (Fig. 3 in Kirk, 1976a) but is relatively less effective at weakly absorbed wavelengths. This implies that a population of straight multicellular filaments is at no disadvantage for light interception compared to a similar population of single cells.

There is evidence from field estimates that the specific attenuation coefficient of microorganisms is related to cell (or colony) size. Harris (Fig. 26 in Harris, 1978) has drawn data from a wide range of published sources which suggest that there is an optimum in light interception at volumes between 100 and 1000 μm^3. The decrease in light interception above 1000 μm^3 probably occurs for the reason discussed above (Talling, 1971), but the decrease with smaller cell sizes (Bindloss, 1974; Jones, 1977) is not readily explained at present. Not all work has confirmed this effect of cell size. Welschmeyer and Lorenzen (1981) found no trend in the rate of light absorption (expressed per unit of chlorophyll a) as a function of cell volume in cultures of six species of marine phytoplankton.

This brings us back to the problem of comparability of data. Future work will almost certainly revise the present estimate of ϵ_s, which lies between 0.005 and 0.03 in square meters per milligram of chlorophyll a, but the essential point is that results should not be averaged (either across the spectrum or across the literature) for inclusion in models. This practice is mathematically incorrect and ignores the adaptive strategies which are available to cells. Droop *et al.* (1982) observed that the value of 0.016, frequently used as an average value in models, lies outside the range of possibility in their experimental material. Furthermore, they pointed out that a small shift in the assumed value meant that a "move from 0.016 to 0.009 was to move from impossibility on one side to extreme improbability on the other." The rare appreciation of the significance of changes in ϵ_s values may result from the tradition of expressing them to the third place of decimals. This practice makes changes seem mathematically small when, in ecological terms, they can be very large.

Responses to the Light Climate

Responses to Light Quantity

It has been known for many years that microorganisms respond to changes in light climate by altering different components of their photosystems, and there has been considerable discussion on whether these changes are truly adaptive in the sense that they optimise cell metabolism to meet changed circumstances. Some confusion arose in the ecological literature because the discussion centred mainly on data gathered from productivity studies which comprised measurements of carbon fixation or oxygen evolution normalised to chlorophyll *a*. Pigment shifts within the cell can obscure the significance of photosynthetic responses for individual cells or populations, and expressing photosynthesis to cell number, cell volume, protein photosynthetic unit, or total pigment may all give quite different pictures from photosynthesis per chlorophyll *a*.

Changes in Photosynthetic Characteristics. The earliest responses noted in ecological studies were changes in photosynthetic characteristics, which were usually examined by means of photosynthesis versus irradiance plots. Talling (1957) defined a light saturation characteristic Ik as the irradiance at which the initial slope of a photosynthesis versus irradiance plot (α) (a photosynthetic efficiency) attains the light-saturated rate of photosynthesis (P max, the photosynthetic capacity). Steeman-Nielsen and Jørgensen (1968a,b) suggested that under low light, cells adapt by reducing Ik. However, Ik is defined by both P max and α; hence, the reduction in Ik may reflect either a reduced photosynthetic capacity or an increased efficiency (Talling, 1966; Yentsch and Lee, 1966). It is probable that Beardall and Morris (1976) are correct in suggesting that a decreased photosynthetic capacity (normalised to chlorophyll *a*) is the predominant response to reduced irradiance. This has been observed widely in field (Platt and Jassby, 1976) and laboratory studies (Falkowski, 1981), although whereas *Oscillatoria* strains respond in this way, some heterocystous cyanophytes (*Anabaena* and *Aphanizomenon*) show relatively little depression of photosynthetic capacity at low irradiance (Foy and Gibson, 1982a).

Much of the decrease in photosynthetic capacity can be caused by an increased cell chlorophyll *a* content and if the same data are expressed per cell or per protein, the apparent response to low irradiance may be abolished (Foy and Gibson, 1982b). In some organisms, however, the photosynthetic capacity per cell is also depressed under low irradiance (Beardall and Morris, 1976). Because chlorophyll is the only unequivocal measure of photosynthetic biomass in the water column, expression of photosynthesis to a cell or protein base is rarely possible in the field, so that there are few critical data on the cells' response to changing irradiance. Prézelin and Sweeney (1978) suggested that photosynthetic capacity was less important than photosynthetic performance (i.e., the rate of photosynthesis at the ambient irradiance), and that the photoadaptive strategy of *Gonyaulax polyedra* was directed towards optimising the latter.

A similar strategy probably also lies behind the commonly observed increase in photosynthetic efficiency (the initial slope of a photosynthesis versus irradiance plot) which occurs in low-light cells (Myers and Kratz, 1955; Platt and Jassby, 1976; Foy and Gibson, 1982b). Low light cells sometimes appear to have a photosynthetic efficiency that is decreased (Perry *et al.*, 1981; Prézelin and Sweeney, 1979) or unaltered (Prézelin, 1976; Vierling and Alberte, 1980), but the differences between organisms result in the main from different responses of the pigment ratios to reducing irradiance. Because photosynthetic efficiency is expressed to chlorophyll *a,* if the low-light response is predominantly to produce more chlorophyll, the efficiency declines, whereas if proportionately more accessory pigments are produced, the more effective light harvesting results in increased efficiency of chlorophyll. Results expressed per photosynthetic unit or per cell consistently show more effective light harvesting in cells adapted to low irradiance, although the quantum efficiency (photosynthesis per light received) remains essentially constant (Welschmeyer and Lorenzen, 1981).

The extent to which differences in photosynthetic characteristics are observed in natural populations depends upon the stability of the water column. In a well-mixed water column, cells probably respond to some integral of the perceived light climate (Harris, 1980), and the significance of intermittent and fluctuating irradiances will be discussed later. Under conditions in which stratification prevents mixing, distinct populations of the same organism can arise which are adapted to higher light near the surface and to low light at depth (Steeman-Nielsen and Hansen, 1959; Talling, 1966, Falkowski, 1980).

Changes in Pigment Content. When irradiance falls below the saturating value for photosynthesis, increases in cell pigment usually occur. In *Chlorella* (Myers, 1946) the chlorophyll *a* content shows an inverse logarithmic relationship to irradiance and a fourfold increase in pigment content takes place at low irradiance. A similar response, but of varying magnitude, occurs in, among others, cyanophytes (Myers and Kratz, 1955; Van Liere and Mur, 1978), *Euglena* (Cook, 1963), and some marine diatoms (Yoder, 1979), although the marine diatoms *Cyclotella* (Jorgensen, 1969) and *Lauderia borealis* (Marra, 1978a) show relatively little response. A low-light pigment response occurs in the cyanophyte *Oscillatoria redekei* grown at high irradiance but in 6:18 light:dark (L:D) cycles (Foy and Gibson, 1982b).

Pigment ratios are also variable (Halldal, 1970), and in general, accessory pigments increase relative to chlorophyll *a* under reduced irradiance. Adaptation is limited to a certain range of irradiances, from the irradiance saturating photosynthesis down to that saturating growth rate. At maximum pigment shift, a considerable proportion of the cell may be composed of photosynthetic pigments, particularly in cyanophytes; in *Anacystis nidulans,* phycocyanin (as a total chromoproteid) can account for 24% of dry weight, 40% of the total cell protein. As a significant proportion of the remaining protein is accounted for by the Calvin cycle enzymes, low-light *A. nidulans* cells invest much of their substance in photosynthetic apparatus. In some organisms, growth at very low irradiances (below 5 μM m^{-2} sec^{-1}) results in a photostress response (Prézelin and Sweeney, 1978), and pigment content declines. A similar response has been shown for cyanophytes grown under small light doses (C. E. Gibson, in preparation), but the "photostress response" was species specific.

The time course of photoadaptation has not been well studied, but as might be expected, adaptation to high irradiance is more rapid than it is to low irradiance. Under a 12:12 L:D cycle, *Glenodinium* cells transferred to a 10-fold greater irradiance (Prézelin and Matlick, 1980) immediately ceased net pigment production, and pigment content was halved by cell divisions. On transfer back to a low irradiance, increase in pigment began within 6 hr but took 3 days to complete. The speed of pigment shift probably depends on the growth rate, and chlorophyll *a* turnover times of a few hours have been reported from *Chlorella* (Grumbach *et al.,* 1978). It is clear that organisms can adapt very rapidly to changes in light climate which in nature are greatly damped by diel and spatial variations. Since chlorophyll *a* synthesis may occur predominantly in the light (Cook, 1966) or in the dark (Foy and Smith, 1980), the mechanism of pigment shift may well vary in different taxonomic groups.

Changes in the Properties of the Photosystems. More fundamental and universal to the metabolism of the cell than gross changes in pigment content are changes in the proportion of pigments performing various photochemical roles. The light reactions of photosynthesis occur in two different photosystems, Photosystem I (PSI) and Photosytem II (PSII), which both consist of antennal (or bulk) pigment molecules that trap light energy and transfer it to catalytic reaction centres, where chemical energy is generated by the oxidation of primary donors named P700 and P680 (from their absorbance maxima) for PSI and PSII, respectively. Since PSI and PSII operate serially, optimum rates for the throughput of electrons will occur only when the two reaction centres receive photons in the optimum proportion (Ley, 1980). It is partly this requirement that dictates the pigment shift occurring at low irradiances. The mechanisms involved range from rapid (millisecond) transitions between photochemical states that are detectable only by sensitive assays to gross pigment shifts visible to the naked eye and taking up to one generation time to complete. Adjustments are possible both to the proportion of light entering PSI and PSII and to the transfer efficiency from PSII to PSI. In low-light-adapted cells, a greater proportion of the received light energy is delivered to PSII, and the transfer efficiency to PSI is decreased (Ghosh and Govindjee, 1966; Ley, 1980; Carthew, 1980).

Increases in pigment content can reflect either an increase in the total number of reaction centres and their associated antennal pigment or an increase in the antennal pigment associated with a constant number of reaction centres. Both types of photoadaptation have been shown to occur by photo-oxidative assay of the P700 number per cell (Falkowski and Owens, 1980; Perry et al., 1981). Although measurements made by this essay do not accord with measurements of the photosynthetic unit (PSU) made with short light flashes (Kawamura et al., 1979), it seems that in most phytoplankton the molar ratio of antenna chlorophyll to P700 chlorophyll rises from a minimum of 350–500 to a maximum of 1200 in low light cells. The increase in the size of the PSUs accounts for the increase in the pigment content, and the number of PSUs per cell remains constant. In the green algae *Chlorella* (Myers and Graham, 1971) and *Dunaliella* (Falkowski and Owens, 1980), however, and possibly also in the dinoflagellate *Peridinium cinctum* (Prézelin and Sweeney, 1979), the number of PSUs increases in response to increased light intensity. The first type of response seems more appropriate to low light environments, whereas the *Chlorella/Dunaliella* type appears better suited to high light environments; the two types have been called shade and sun types, respectively, by analogy to higher plants. The question of optimum rates of light capture will be discussed further in a subsequent section, but it is timely to remark here that there is little advantage in increasing carbon fixation beyond the rate at which it may be assimilated, and even modest irradiances can saturate the carbon requirement for growth, a point that is often neglected in discussions of photoadaptation.

Responses to Light Quality

Chromatic Adaptation. Some cyanophytes have a conspicuous ability to alter their biliprotein pigment composition in response to changes in light quality. Species differ in their response. Some may change either their phycoerythrin content, their phycocyanin content, or both, whereas others show no response at all (Tandeau de Marsac, 1977). The responses are referred to as "complementary chromatic adaptation," (Bogorad, 1975) since the pigment ratio alters in such a way as to optimise capture of the available light. Growth under red (>600 nm) light produces a preponderance of phycocyanin, and the synthesis of phycoerythrin is promoted by light between 500 and 600 nm (Gantt, 1981). Limits to the effect are reached in the far red (>680 nm; Myers et al., 1978) and blue light (450 nm; Pulich and Van Baalen, 1974), where quite different effects may come into play. For a review of blue light effects, see Senger (1980). The alteration in pigment ratios has been utilised to study the role of accessory pigments in photosynthetic energy transfer (Jones and Myers, 1963; Ghosh and Govindjee, 1966; Stevens and Myers, 1976), but the role of biliproteins in cell metabolism is complex since phycocyanin is also used as a nitrogen store. Phycocyanin may be rapidly depleted under nitrogen starvation in high light conditions (Allen and Smith, 1969), but may be maintained under nitrogen starvation when light is also limiting (Zevenboom and Mur, 1978). A balance of factors is involved, the intereation of which has been little studied.

In order to clarify the mechanisms involved, most studies have used highly responsive species under conditions chosen to give a simple change of pigmentation. The response to changed light quality occurs rapidly; in *Fremyella diplosiphon,* phycoerythrin synthesis begins within 90 min of transfer from red to green light (Gendel *et al.,* 1979). Pigment changes may also occur at population level, and stable clones differing widely in pigmentation have been isolated from *Oscillatoria* species (Meffert and Krambeck, 1977; Skulberg, 1978; Kohl and Niklisch, 1981). It has been suggested that the proportion of strains with different pigment ratios may change in response to changes in underwater light climate, so that the strain intercepting the greatest proportion of available light becomes dominant (Kohl and Niklisch, 1981). A general increase in red-pigmented strains has also been noted in some areas (Skulberg, 1978). The most notorious red strain, *Oscillatoria rubescens,* has become a feature of some enriched European lakes.

The adaptive significance of such changes is not clear, and suggested advantages, though plausible, have not been clearly demonstrated in nature. One of the few reported pigment shifts in natural population (Jewson, 1976) may owe as much to nutrient starvation as adaptation to light climate. Chromatic adaptation is widely found in the benthic marine algae (e.g., Levring, 1966), but critical analysis shows that the case is far from proven (Dring, 1981a), and what appears to be chromatic adaptation may be a response to light quantity. Myers *et al.* (1978) considered that the features accompanying growth in far red light represent incompetence rather than adaptation, but Gendel *et al.* (1979) showed that in *F. diplosiphon,* phycoerythrin synthesis was positively induced by red light and positively repressed by green light. There was fundamental chromatic control of the process, probably at the transcriptional level.

Response to Superoptimal Light

Irradiances exceeding approximately 400 μM m^{-2} sec^{-1} are progressively inimical to microorganisms, the more energetic short wavelengths particularly so. Small amounts of dissolved organic matter or suspended particles exert a great effect on the attenuance of short wavelengths in a water column (p. 101), and in general, the plankton of lakes and coastal marine waters are less likely than surface oceanic plankton to be exposed to harmful irradiances. Nevertheless, under stagnant conditions of clear water and in the surface "water blooms" of cyanophytes, very high irradiances may be experienced, and a number of mechanisms occur which enable organisms to mitigate or avoid otherwise lethal circumstances.

Effects of Superoptimal Irradiance. Large doses of ultraviolet (Halldal, 1967) and visible light (Krinsky, 1976) affect virtually every aspect of cell metabolism, and the damage is exacerbated by high oxygen tensions (Halliwell, 1978). In natural populations, even the relatively attenuated short wavelength irradiances below the water surface cause inhibition of photosynthesis (Findenegg, 1966; Harris, 1978), which, due to the low transmission of ultraviolet light by glass, may not be accurately observed in bottle experiments (Ilmavirta, 1977). A quantitative analysis has been published (Smith *et al.*, 1980). Inhibition probably involves both light and dark reactions of photosynthesis (Belay and Fogg, 1978; Belay, 1981), more particularly the flavone-mediated inactivation of ribulose bisphosphate carboxylase (Codd, 1982). Photo-oxidative damage is also caused by the accumulation of H_2O_2 (Abeliovich and Shilo, 1972a; Kaplan, 1981), and increased efficiency of superoxide dismutase confers some resistance to photooxidation in certain cyanophyte strains (Eloff *et al.*, 1976). Incubation under anaerobic conditions lowers resistance to photo-oxidation (Friedberg *et al.*, 1979), and it may be that there are adaptive changes in the stability of superoxide dismutase.

Low CO_2 concentrations under high irradiance in conjunction with high oxygen tensions are equally inimical and favour photorespiration (Tolbert, 1974), as well as causing cell lysis (Meffert, 1973). Where growth is forced by irradiance, depletion of other nutrients may also prove fatal. For example, in the diatom *Asterionella formosa* (Lund, 1950; Hughes and Lund, 1962), depletion of silica precludes new cell formation, and under high irradiances the cells die.

Response to Superoptimal Irradiance. Many of the effects of high irradiance may be alleviated by the diversion of excess light energy into harmless reactions. The principal agents in this process are the carotenoids (Krinsky, 1976), but biliprotein pigments may be involved in cyanophytes (Abeliovich and Shilo, 1972b). Similarly, photorespiration may serve as an overflow mechanism, diverting unneeded photosynthetic energy. Gas vesicles of cyanophytes could have a light-shielding effect, but in practice their presence does not seem to modify the intracellular light climate significantly (Shear and Walsby, 1975).

In general, the capacity of microorganisms to cope with intense irradiation is limited, and avoidance is probably the most widespread strategy. Planktonic diatoms may achieve this by increasing their sinking rate (Moed, 1973; Jaworski *et al.*, 1981), and it is an integral part of the life strategy of some planktonic algae to spend the time of nutrient depletion in darkness on the bottom sediment (Lund, 1954). The gas vesicles of planktonic cyanophytes enable some species to regulate their position in a stable water column, so that superoptimal irradiances are avoided. Metalimnetic populations are frequently observed (Reynolds and Walsby, 1975) where growth is probably rather slow, but the rate of carbon fixation perhaps matches the nutrient supply. Motile organisms are capable of a more flexible response (Foster and Smith, 1980), and may alter their light preference to suit other conditions (Harris *et al.*, 1979).

Photosynthesis and Growth

Obligate phototrophs are unique in that their energy source is transient and subject to very great variation over a time scale ranging from seconds, for planktonic organisms circulating through the water column, to hours over the alternation of day and night. Variations in the quantity of irradiance may also be accompanied by qualitative changes, and in the preceding sections we have attempted to give a brief account of the underwater light climate experienced by planktonic cells and the responses they are able to make.

Until relatively recently, laboratory studies of photosynthesis and growth were generally conducted under continuous irradiance, which is a light climate experienced only rarely in nature. Under continuous irradiance, cells are in a steady state with respect to energy supply, analogous to the supply of nutrients in a chemostat. Although metabolic shifts may occur in response to changes in the rate of irradiance, no accommodation to interruptions in energy supply is necessary. Under these conditions, photosynthesis is closely linked to specific growth rates, and a knowledge of specific net photosynthetic rates, corrected if necessary for the excretion of carbon to the medium, enables an accurate estimate of growth rate to be made. Under intermittent illumination, which is normal in nature, a different situation prevails. Whilst in the light, the cell must fix sufficient carbon to supply contemporary requirements, and must also accumulate sufficient storage material to meet the synthetic and energetic demands of the ensuing darkness, when exogenous energy sources are probably lacking. In the following sections we explore the relationship between photosynthesis and growth, the matching of photosynthetic rate to carbon requirement, and the mechanisms occurring in phototrophs which optimise the growth rate under intermittent irradiance.

Attempts to model growth in phototrophic microorganisms have been made with various motives. One of the most powerful was the search for a system of mass algal culture, which began in the late 1940s. Observations on mass cultures of algae were made with respect to their growth rate under natural conditions

(Burlew, 1953); these studies have continued up to the present time (Soeder and Binsack, 1978; Shelef and Soeder, 1980). Although the prediction of growth in a controlled culture system might seem a modest goal, in practice the problem has proved to be somewhat difficult. In nature, added complications arise, as there may be several possible alternative limitations on growth rate. Attempts to model the interaction of factors have often been arbitrary (Talling, 1979), so that even the basic form of factor interaction, that is, whether it should be threshold or multiplicative, has not been settled. An attempt has been made to model light capture and nutrient uptake by means of bisubstrate kinetics (Falkowski, 1977) for light-dependent nitrate uptake, but because the possible number of factors controlling growth is large, generalisation is hazardous. Other models have been proposed by Rhee and Gotham (1981) and Laws and Bannister (1980).

The resemblance between photosynthesis versus irradiance plots, growth versus irradiance plots, and nutrient uptake kinetic plots makes the treatment of light as a substrate a seductive concept. Because of the complication of time in the treatment of light quantities, the intercovertability of units requires some ingenuity. Relatively recently, the problem has been approached from a new standpoint (Droop et al., 1982), and by expressing all variables in energetic units (joules for quantities, watts for rates), the interaction between light limitation, nutrient limitation, and specific growth rate has been modelled for *Pavlova lutheri*. The experiments were conducted under continuous light at subsaturating irradiances (at which the photosynthesis–irradiance relationship was linear), and the problem of estimating the interception of light energy (p. 103) was solved by an empirical iterative technique. Using concepts formulated earlier for a model of nutrient uptake (Droop, 1974), a cell energy quota could be defined, which was equal to the reciprocal of photosynthetic efficiency. Droop et al. (1982) also defined a "coefficient of luxury" as the ratio of photosynthetic efficiencies under light and nutrient limitation at the same growth rate. The measured efficiency of photosynthesis was significantly higher under light limitation than under nutrient limitation, and in spite of the fact that the respiration rate increased hyperbolically with the growth rate, under light-limiting conditions growth was directly proportional to light absorption.

Van Liere (1979) derived an energy model for growth using dense cultures of *Oscillatoria agardhii* in continuous culture. Growth was varied by dilution rate and a range of irradiances applied. Van Liere concluded that the efficiency of growth, calculated from the yield and the specific light uptake rate, was constant for a given growth rate but declined with increasing irradiance.

Intermittent Irradiance

Simple proportionality between growth and light dose (irradiance × time) is to be expected only if the law of reciprocity is not violated and light is delivered at a rate within the linear portion of the irradiance:growth curve. Put simply, 100 μM

m^{-2} sec^{-1} for 10 sec may have a different effect than 1000 μM m^{-2} sec^{-1} for 1 sec. Regardless of irradiance, microorganisms under short photoperiods are "light limited" in the sense that photosynthesis cannot fix sufficient carbon to fuel the maximum growth rate allowed by other factors, and metabolic shifts within the cell reflect this fact. In *O. redekei*, grown under 6:18 L:D cell pigment contents at 180 μM m^{-2} sec^{-1} were greater than cells grown at 25 μM m^{-2} sec^{-1} under continuous light (Foy and Gibson, 1982b).

Providing that light is delivered at a rate-limiting irradiance, growth is proportional to the light dose under a surprisingly wide range of conditions (Fig. 3). Over the temperature range 5–23°C (Gibson and Foy, 1983) under light-limiting conditions, there was a close relationship between growth and light dose which was maintained in several L:D regimes. This suggests that metabolism in the dark is closely related to the fixation of carbon in the light. The slope of the

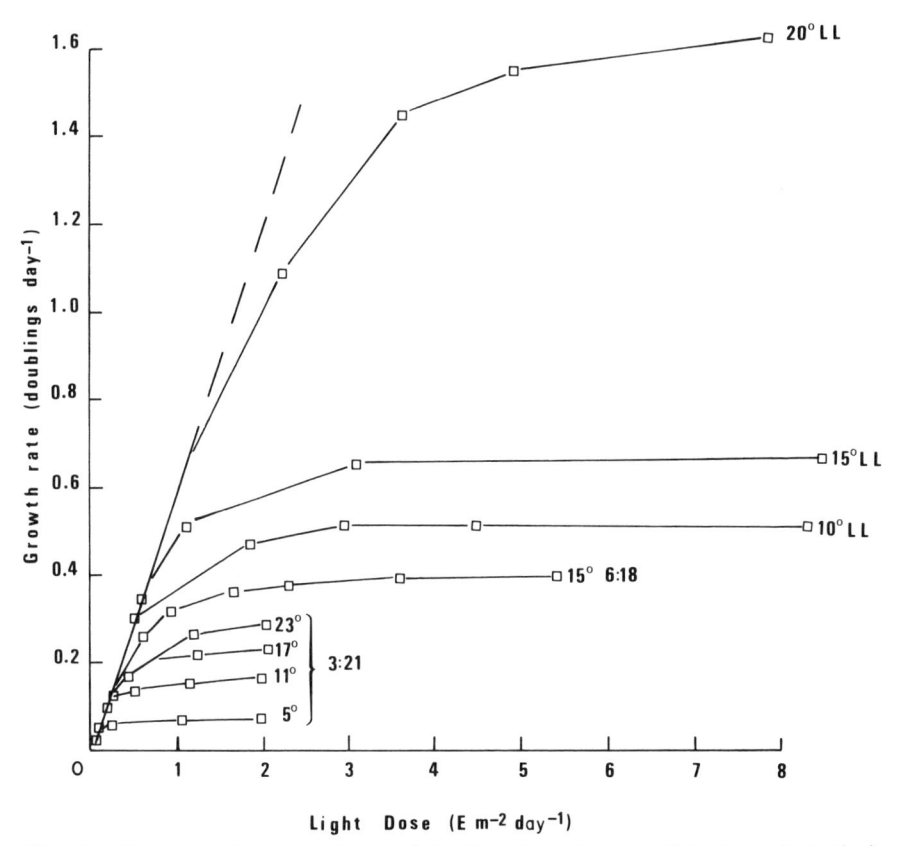

Fig. 3. Response of the growth rate of *Oscillatoria redekei* to a light dose of identical quality but applied under a range of L:D cycles and temperatures. From Gibson and Foy (1983).

initial linear portion of the graph is the maximum growth efficiency, and given a knowledge of the light interception by the cells, efficiency might be calculated in energetic terms. Gibson and Foy (1983) avoided this calculation and used thin layers of dilute cell suspension where self-shading was negligible. The calculated maximum growth efficiency (i.e., the minimum quantum requirement) was 1.7 moles m^{-2} doubling^{-1}. Departure from maximum efficiency occurred at progressively smaller light doses with decreasing temperature and decreasing photoperiod. With decreasing temperature, dark processes dominated growth; under short photoperiods, relatively small light doses saturated the maximum photosynthetic rate.

Oscillatoria redekei is able to make considerable adjustments to its metabolism in response to the form of the light dose. Under continuous light, pigment content is low, and at 15°C photosynthesis saturates at 80–450 μM m^{-2} sec^{-1} depending upon irradiance. Growth rate, however, saturates at 13 μM m^{-2} sec^{-1}, so that only a fraction of the maximum photosynthetic capacity is needed to permit maximum growth. Under 6:18 L:D, on the other hand, light saturation of photosynthesis occurs at much lower irradiances, and maximum growth rate is achieved only in conjunction with maximum photosynthetic rate. The degree of accommodation this allows is shown by the fact that the ratio of growth rates under continuous light and 6:18 L:D was 1.68:1 compared to a ratio of photoperiods of 4:1.

Growth efficiency, expressed as growth per light hour, may decline markedly with increasing photoperiod (Castenholz, 1964; Durbin 1974; Holt and Smayda, 1974; Foy et al., 1976), although in green algae growth often appears to increase directly with photoperiod (Tamiya et al., 1953; Eppley and Coatsworth, 1966).

Growth rate under cycles more rapid than 1 day also shows an inverse relationship to photoperiod. With a photocycle time of 40 sec and saturating irradiance, the growth rate increased linearly with the proportion of light in the cycle but reached a maximum at a light fraction of 0.77 at 37°C (Lee and Pirt, 1981). Planktonic cyanophytes grown under L:D cycles of 60 min show a similar phenomenon. At 15°C, the growth rate of *O. redekei* was saturated at an L:D ratio of 1.7:1, and that of *O. agardhii* achieved a lower maximum growth rate at a light fraction of 0.75:1 (C. E. Gibson, unpublished observations). Considerable pigment shifts took place in these experiments, although the irradiance was constant throughout and the total cycle time was far shorter than in the earlier work (Foy and Gibson, 1982b). This supports the contention (see also Droop et al., 1982) that cells respond to an average light experience, not the highest irradiance encountered.

The significance of these results for natural populations is not yet clear. Cells circulating through a water column probably exhibit a rhythmic pattern of photosynthesis linked to the diel cycle (Eppley et al., 1967; Steeman-Nielsen and Jorgensen, 1968a,b; Kalff, 1969), and when cells are artificially circulated through a natural light gradient (Jewson and Wood, 1975; Marra, 1978b), the net column photosynthesis is greater than when cells are held in stationary bottles. Part of the explanation for this is no doubt to be found in the removal of surface inhibition effects (Harris *et al.*, 1979) and in the relief of the commonly observed afternoon depression of photosynthesis (Marra, 1978a), and it is likely that in natural populations there is an ordering of metabolic priorities to accommodate the whole L:D cycle. The problem of modelling the effect of intermittent illumination on natural populations will not be finally solved until there is a realistic physical model of the movement of the cells through the water column. Even in the shallow, turbulent Lough Neagh, where there is little evidence of temperature stratification, considerable biochemical differentiation of cells from different depths is normal (Gibson, 1975, 1978).

Strategies of Growth in Intermittent Irradiance

As suggested earlier, the optimum strategy of a cell under intermittent irradiance is to so regulate the storage of reserve products during the day that it anticipates the needs of the following dark period. This is possible only in a regular L:D cycle in which the response to one L:D cycle is appropriate to the next. It has been suggested that *Euglena* accumulates paramylum during the day only when carbon is fixed in excess of immediate needs, for example, for protein synthesis (Cook, 1963). However, this straightforward view has been challenged by Cohen and Parnas (1976), who suggested that cells adopt an optimum policy which takes into account cell requirements throughout the L:D cycle. Shuter (1979) proposed a model to test the conclusions of Cohen and Parnas (1976), and divided the cell into four theoretical compartments: photosynthetic, synthetic, structural, and storage. He postulated the existence of a common control mechanism which would regulate the ratio of the four compartments in order to maximise the specific growth rate. However, attempts to apply the model to real data were only partially successful (Schlesinger and Shuter, 1981), and fitting data from our laboratory to the model shows that it does not describe the growth strategy of *O. redekei* very faithfully. One basic problem is that it assumes that growth is regulated by the amount of carbon fixed, that is, growth is directly proportional to photosynthesis. However, carbon fixation need not necessarily increase the synthetic capacity of the cell, and hence growth rates predicted from photosynthesis need not be realised.

More extensive experimental data have been published for the planktonic cyanophytes. *O. redekei* and *O. agardhii* (Foy and Smith, 1980). Both species showed an optimum L:D cycle under which growth per light hour was greatest, which coincided with the cycle in which carbohydrate accumulated in the light period was exhausted just before the end of the dark period. If the dark period was shorter than the optimum, cells entered the light period with the carbohydrate store partially full, so that less of the photosynthetic capacity was usable during the light period. On the other hand, if the dark period was too long, carbohydrate reserves were exhausted before the end of this period, and there was a progressive catabolism of the biosynthetic components of the cell. As a result, there was a delay in the onset of protein synthesis when photosynthesis restarted and more carbohydrate was accumulated. Carbohydrate was exhausted more quickly in *O. redekei* than in *O. agardhii,* which consequently was better able to survive prolonged dark starvation.

In these two organisms, the control mechanisms which match photosynthetic carbon input to synthetic and energetic needs of cells under intermittent-illumination are therefore twofold. Firstly, under large light doses the photosynthetic efficiency is depressed, and carbohydrate accumulation slows accordingly. Secondly, cells respond in a similar manner to long dark periods and to low irradiances, except that under low L:D ratios, depression of protein synthesis produces a greater carbohydrate reserve. To some extent, these results bear out the contention (Cohen and Parnas, 1976) that cells respond to a low energy supply by accumulating larger carbohydrate reserves, but the teleological assertion that cells anticipate the energy needs of the future dark period clearly cannot be true.

Under continuous irradiance, the amount of carbohydrate stored is proportional to the irradiance received (Foy and Gibson, 1982b), and there is no evidence of carbohydrate reserves being accumulated in response to energy

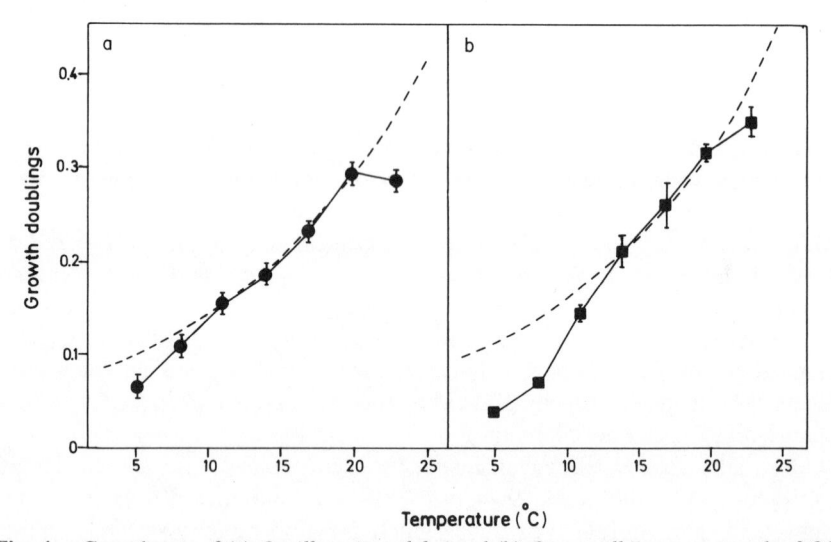

Fig. 4. Growth rate of (a) *Oscillatoria redekei* and (b) *O. agardhii* grown at under 3:21 L:D at a range of temperatures. Dashed lines show a theoretical exponential relationship between growth rate and temperature; vertical bars are 95% confidence limits. Data from Foy (1983).

limitation. As a result of the interaction between temperature and the balance of photosynthesis and protein synthesis, under 3:21 L:D cycles the specific growth rate of *O. redekei* shows a linear, not an exponential, temperature response (Foy, 1983). Under the same irradiance, growth is slowed below the optimum temperature range by the inability of the cells to utilise all the carbohydrate reserves during the dark; above the temperature optimum, growth is slowed by the detrimental effects of dark starvation. A change in temperature under very short days therefore resembles a change in day length. The observed growth rate:temperature relationship is a tangent to a theoretical temperature response, touching the exponential line at the temperature optimum (Fig. 4).

References

Abeliovich, A. and Shilo, M. (1972a). Photooxidative death in blue-green algae. *J. Bacteriol.* **111**, 682–689.
Abeliovich, A. and Shilo, M. (1972b). Photooxidative reactions of c-phycocyanin. *Biochim. Biophys. Acta* **283**, 483–491.
Allen, M. M. and Smith, A. J. (1969). Nitrogen chlorosis in blue-green algae. *Arch. Mikrobiol.* **69**, 114–120.
Arnold, G. P. (1975a). The measurement of irradiance with particular reference to marine biology. *In* "Light as an Ecological Factor II" (Eds. G. C. Evans, R. Bainbridge and O. Rackham), pp. 1–25. Blackwell, Oxford.
Arnold, G. P. (1975b). Standards of spectral power distribution for measuring spectral irradiance. *In* "Light as an Ecological Factor II" (Eds. G. C. Evans, R. Bainbridge and O. Rackham), pp. 573–583. Blackwell, Oxford.
Atlas, D. and Bannister, T. T. (1980). Dependence of mean spectral extinction coefficient of phytoplankton on depth, water colour and species. *Limnol. Oceanogr.* **25**, 157–159.
Beardall, J. and Morris, I. (1976). The concept of light intensity adaptation in marine phytoplankton: some experiments with *Phaeodactylum tricornutum*. *Mar. Biol. (Berlin)* **37**, 377–387.
Belay, A. (1981). An experimental investigation of phytoplankton photosynthesis at lake surfaces. *New Phytol.* **89**, 61–74.
Belay, A. and Fogg, G. E. (1978). Photoinhibition of photosynthesis in *Asterionella formosa* (Bacillariophyceae) *J. Phycol.* **14**, 341–347.
Bindloss, M. E. (1974). Primary productivity of phytoplankton in Loch Leven, Kinross. *Proc. R. Soc. Edinburgh Sect. B: Biol.* **74**, 157–181.
Bindloss, M. E. (1976). The light-climate of Loch Leven, a shallow Scottish lake, in relation to primary production by phytoplankton. *Freshwater Biol.* **6**, 501–518.
Bogorad, L. (1975). Phycobiliproteins and complementary chromatic adaptation. *Annu. Rev. Plant Physiol.* **26**, 369–401.
Brakel, W. H. (1979). Small-scale spatial variation in light available to coral reef benthos: quantum irradiance measurements from a Jamaican reef. *Bull. Mar. Sci.* **29**, 406–413.
Burlew, J. S. (Ed.) (1953). Algal culture from laboratory to pilot plant. *Carnegie Inst. Washington Publ.* **600**.
Carthew, R. W. (1980). The thermodynamics of photosynthetic adaptation to photon fluence rate in the blue-green alga *Oscillatoria limnetica*. *Photosynthetica* **14**, 202–212.

Castenholz, R. W. (1964). The effect of daylength and light intensity on the growth of littoral marine diatoms in culture. *Physiol. Plant.* **17,** 951–963.

Codd, G. A. (1982). Photoinhibition of photosynthesis and photoinactivation of ribulose bisphosphate carboxylase in algae and cyanobacteria. In "Plants and the Daylight Spectrum" (Ed. H. Smith), pp. 315–337. Academic Press, London.

Cohen, D. and Parnas, H. (1976). An optimal policy for the metabolism of storage materials in unicellular algae. *J. Theor. Biol.* **56,** 1–18.

Cook, J. R. (1963). Adaptations in growth and division in *Euglena* affected by energy supply. *J. Protozool.* **10,** 436–444.

Cook, J. R. (1966). Photosynthetic activity during the division cycle in synchronised *Euglena gracilis. Plant Physiol.* **41,** 821–825.

Dokulil, M. (1979). Optical properties, colour and turbidity. In "Neusiedlersee: The Limnology of a Shallow Lake in Central Europe" (Ed. H. Löffler), pp. 151–170. Junk Publ., The Hague.

Doucha, J. and Kubin, S. (1976). Measurement of *in vivo* absorption spectra of microscopic algae using bleached cells as a reference sample. *Arch. Hydrobiol. Suppl.* **49,** *Algol. Stud.* **15,** 199–213.

Dring, M. J. (1981a). Chromatic adaptation of photosynthesis in benthic marine algae: an examination of its ecological significance using a theoretical model. *Limnol. Oceanogr.* **26,** 271–284.

Dring, M. J. (1981b). Photosynthesis and development of marine macrophytes in natural light specta. In "Plants and the Daylight Spectrum" (Ed. H. Smith), pp. 297–314. Academic Press, London.

Droop, M. R. (1974). The nutrient status of algal cells in continuous culture. *J. Mar. Biol. Assoc. U.K.* **54,** 825–855.

Droop, M. R., Mickelson, M. J., Scott, J. M. and Turner, M. F. (1982). Light and nutrient status of algal cells. *J. Mar. Biol. Assoc. U.K.* **62,** 403–434.

Dubinsky, Z. and Berman, T. (1976). Light utilisation efficiencies of phytoplankton in Lake Kinneret. *Limnol. Oceanogr.* **21,** 226–230.

Dubinsky, Z. and Berman, T. (1981). Light utilisation by phytoplankton in Lake Kinneret (Isreel). *Limnol. Oceanogr.* **26,** 660–670.

Durbin, E. G. (1974). Studies on the autecology of the marine diatom *Thalassiosira nordenskioldii* Cleve. I. The influence of daylength, light intensity and temperature on growth. *J. Phycol.* **10,** 220–225.

Duysens, L. N. M. (1952). Transfer of Excitation Energy in Photosynthesis. Ph.D. Thesis, University of Utrecht.

Eloff, J. M., Steinitz, Y. and Shilo, M. (1976). Photooxidation and cyanobacteria in natural conditions. *Appl. Environ. Microbiol.* **31,** 119–126.

Eppley, R. W. and Coatsworth, J. L. (1966). Culture of the marine phytoplankton *Dunaliella tertiolecta* with light dark cycles. *Arch. Mikrobiol.* **55,** 66–80.

Eppley, R. W., Holmes, R. W. and Paasche, E. (1967). Periodicity in cell division and physilogical behaviour of *Ditylum brightwellii*, a marine planktonic diatom, during growth in light-dark cycles. *Arch. Mikrobiol.* **56,** 305–323.

Falkowski, P. G. (1977). A theoretical description of nitrate uptake kinetics based on bisubstrate kinetics. *J. Theor. Biol.* **64,** 375–379.

Falkowski, P. G. (1980). Light-shade adaptation in marine phytoplankton. *Brookhaven Symp. Biol.* **31,** 99–120.

Falkowski, P. G. (1981). Light-shade adaptation and assimilation numbers. *J. Plankton Res.* **3,** 203–216.

Falkowski, P. G. and Owens, T. G. (1980). Light-shade adaptation two strategies in marine phytoplankton. *Plant Physiol.* **66**, 592–595.
Fay, P. (1970). Photostimulation of nitrogen fixation in *Anabaena cylindrica*. *Biochim. Biophys. Acta* **216**, 353–356.
Findenegg, I. (1966). Die Bedeutung kurzwellige Strahlung fur die planktische Primärkproduktion in den Seen. *Verh.—Int. Ver. Theor. Angew. Limnol.* **16**, 314–320.
Foster, K. W. and Smyth, R. D. (1980). Light antennas in phototactic algae. *Microbiol. Rev.* **44**, 572–630.
Foy, R. H. (1983). Interaction of temperature and light on the growth of two planktonic *Oscillatoria* species under a short photoperiod regime. *Br. Phycol. J.* **18**, 267–273.
Foy, R. H. and Gibson, C. E. (1982a). Photosynthetic characteristics of planktonic blue-green algae: the response of 20 strains grown under high and low light. *Br. Phycol. J.* **17**, 169–182.
Foy, R. H. and Gibson, C. E. (1982b). Photosynthetic characteristics of planktonic blue-green algae: changes in photosynthetic capacity and pigmentation of *Oscillatoria redekei* Van Goor under high and low light. *Br. Phycol. J.* **17**, 183–193.
Foy, R. H., Gibson, C. E., and Smith, R. V. (1976). The influence of day length, light intensity and temperature on the growth rates of planktonic blue-green algae. *Br. Phycol. J.* **11**, 151–163.
Foy, R. H. and Smith, R. V. (1980). The role of carbohydrate accumulation in the growth of planktonic *Oscillatoria* species. *Br. Phycol. J.* **15**, 139–150.
Friedberg, D., Fine, M. and Oren, A. (1979). Effect of oxygen on the cyanobacterium *Oscillatoria limnetica*. *Arch. Microbiol.* **123**, 311–313.
Ganf, G. G. (1974). Incident solar irradiance and underwater light penetration as factors controlling the chlorophyll *a* content of a shallow equatorical lake (Lake George, Uganda). *J. Ecol.* **62**, 593–629.
Gantt, E. (1981). Phycobilisomes. *Annu. Rev. Plant Physiol.* **32**, 327–347.
Gendel, S., Ohad, I. and Bogorad, L. (1979). Control of phycoerythrin synthesis during chromatic adaptation. *Plant Physiol.* **64**, 786–790.
Ghosh, A. K. and Govindjee (1966). Transfer of the excitation energy in *Anacystis nidulans* grown to obtain different pigment ratios. *Biophys. J.* **6**, 611–619.
Gibson, C. E. (1975). A field and laboratory study of oxygen uptake by planktonic blue-green algae. *J. Ecol.* **63**, 867–880.
Gibson, C. E. (1978). Field and laboratory observations on the temporal and spatial variation of carbohydrate content in planktonic blue-green algae in Lough Neagh, Northern Ireland. *J. Ecol.* **66**, 97–115.
Gibson, C. E. and Foy, R. H. (1983). The photosynthesis and growth efficiency of a planktonic blue-green alga, *Oscillatoria redekei*. *Br. Phycol. J.* **18**, 39–45.
Golterman, H. L., Clymo, R. S. and Ohnstad, M. A. M. (1978). "Methods for Chemical Analysis of Fresh Waters." Blackwell, Oxford.
Grumbach, K. H., Lichtentaler, H. K. and Erisman, K. H. (1978). Incorporation of carbon-14 labeled carbon di-oxide in the photosynthetic pigments of *Chlorella pyrenoidosa*. *Planta* **140**, 37–43.
Halldal, P. (1967). Ultraviolet action spectra in algology. A review. *Photochem. Photobiol.* **6**, 445–460.
Halldal, P. (1970). The photosynthetic apparatus of microalgae and its adaption to environmental factors. *In* "Photobiology of Microorganisms" (Ed. P. Halldal), pp. 17–56. Wiley, New York.
Halliwell, B. (1978). Biochemical mechanisms accounting for the toxic action of oxygen

on living organisms: the key role of superoxide dismutase. *Cell Biol. Int. Rep.* **2**, 113–128.
Harris, G. P. (1978). Photosynthesis, productivity and growth: the physiological ecology of phytoplankton. *Ergeb. Limnol.* **10**, 1–171.
Harris, G. P. (1980). The measurement of photosynthesis in natural populations of phytoplankton. *In* "The Physiological Ecology of Phytoplankton" (Ed. I. Morris), pp. 129–187. Blackwell, Oxford.
Harris, G. P., Heaney, S. I. and Talling, J. F. (1979). Physiological and environmental constraints in the ecology of the planktonic dinoflagellate *Ceratium hirundinella* O. F. Müller. *Freshwater Biol.* **6**, 531–542.
Højerslev, N. K. (1981). Daylight measurements appropriate for photosynthetic studies in natural seawaters. *J. Cons., Cons. Explor. Mar* **38**, 131–146.
Holt, M. G. and Smayda, T. J. (1974). The effect of daylength and light intensity on the growth rate of the marine diatom *Detonula confervacea* (Cleve) Gran. *J. Phycol.* **10**, 231–237.
Hughes, J. C. and Lund, J. W. G. (1962). The rate of growth of *Asterionella formosa* Hass. in relation to its ecology. *Arch. Microbiol.* **42**, 117–129.
Ilmavirta, V. (1977). Diel periodicity in the phytoplankton community of the oligotrophic lake Paajarvi, Southern Finland. III. The influence of the bottle material on the measurement of production. *Ann. Bot. Fenn.* **14**, 102–111.
Jaworski, G. H. M., Talling, J. F. and Heaney, S. I. (1981). The influence of carbon dioxide-depletion on growth and sinking rate of two planktonic diatoms in culture. **16**, 395–410.
Jeffery, S. W. (1980). Algal pigment systems. *In* "Primary Productivity in the Sea" (Ed. P. Falkowski), pp. 33–58. Plenum, New York.
Jerlov, N. G. (1976) "Marine Optics." Elsevier, Amsterdam.
Jerlov, N. G. and Steeman-Nielsen, E. (1973). "Optical Aspects of Oceanography." Academic Press, New York.
Jewson, D. H. (1976). The interaction of components controlling net phytoplankton photosynthesis in a well mixed lake (Lough Neagh, Northern Ireland). *Freshwater Biol.* **6**, 551–576.
Jewson, D. H. (1977). Light penetration in relation to phytoplankton content of the euphotic zone of Lough Neagh, N Ireland. *Oikos* **28**, 74–83.
Jewson, D. H. and Taylor, J. A. (1978). The influence of turbidity on net phytoplankton photosynthesis in some Irish lakes. *Freshwater Biol.* **8**, 573–584.
Jewson, D. H. and Wood, R. B. (1975). Some effects on integral photosynthesis of artificial circulation of phytoplankton through light gradients. *Verh.—Int. Ver. Theor. Angew. Linmol.* **19**, 1037–1044.
Jones, L. W. and Myers, J. (1963). Pigment variations in *Anacystis nidulans* induced by light of selected wavelengths. *J. Phycol.* **1**, 7–14.
Jones, R. I. (1977). Factors controlling phytoplankton production and succession in a highly eutrophic lake (Kinnego Bay, Lough Neagh). II. Phytoplankton production and its chief determinants. *J. Ecol.* **65**, 561–577.
Jones, R. I. and Ilmavirta, V. (1978). Vertical and seasonal variation of phytoplankton photosynthesis in a brown-water lake with winter ice cover. *Freshwater Biol.* **8**, 561–572.
Jorgensen, E. (1969). The adaptation of plankton algae. IV. Light adaptation in different algal species. *Physiol. Plant.* **22**, 1307–1315.
Kalff, J. (1969). A diel periodicity in the optimum light intensity for maximum photosynthesis in natural phytoplankton populations. *J. Fish. Res. Board Can.* **26**, 463–468.

Kalle, K. (1966). The problem of Gelbstoff in the sea. *Oceanog. Mar. Biol.* **4**, 91–104.

Kaplan, A. (1981). Photoinhibition in *Spirulina platensis:* response of photosynthesis and HCO_3 uptake capability in CO_2-depleted conditions. *J. Exp. Bot.* **32**, 669–677.

Kawamura, M., Mimuro, M. and Fujita, Y. (1979). Quantitative relationship between two reaction centers in the photosynthetic system of blue-green algae. *Plant Cell Physiol.* **20**, 697–705.

Kiefer, D. A. (1973). Chlorophyll *a* fluorescence in marine centric diatoms: response of chloroplasts to light and nutrient stress. *Mar. Biol. (Berlin)* **23**, 39–46.

Kiefer, D. A., Olson, R. J. and Wilson, W. H. (1979). Reflectance spectroscopy of marine phytoplankton. I. Optical properties related to age and growth rate. *Limnol. Oceanogr.* **24**, 664–672.

Kirk, J. T. O. (1975a). A theoretical analysis of the contribution of algal cells to the attenuation of light in natural waters. I. General treatment of suspension of pigmented cells. *New Phytol.* **75**, 11–20.

Kirk, J. T. O. (1975b). A theoretical analysis of the contribution of algal cells to the attenuation of light within natural waters. II. Spherical cells. *New Phytol.* **75**, 21–36.

Kirk, J. T. O. (1976a). A theoretical analysis of the contribution of algal cells to the attenuation of light in natural waters. III. Cylindrical and spheroidal cells. *New Phytol.* **77**, 341–358.

Kirk, J. T. O. (1976b). Yellow substance (gelbstoff) and its contribution to the attenuation of photosynthetically active radiation in some inland and coastal south-eastern Australian waters. *Aust. J. Mar. Freshwater Sci.* **27**, 67–71.

Kirk, J. T. O. (1977). Attenuation of light in natural waters. *Aust. J. Mar. Freshwater Res.* **28**, 497–508.

Kirk, J. T. O. (1979). Spectral distribution of photosynthetically active radiation in some South-eastern Australian waters. *Aust. J. Mar. Freshwater Res.* **30**, 81–91.

Kirk, J. T. O. (1980). Spectral absorption properties of natural waters: contribution of the soluble and particulate fractions to light absorption in some inland waters of south-eastern Australia. *Aust. J. Mar. Freshwater Res.* **31**, 287–296.

Kirk, J. T. O. (1981a). Monte Carlo study of the nature of the underwater light field in, and the relationship between optical properties of turbid yellow waters. *Aust. J. Mar. Freshwater Res.* **32**, 517–532.

Kirk, J. T. O. (1981b). Estimation of the scattering coefficients of natural waters using underwater irradiance meters. *Aust. J. Mar. Freshwater Res.* **32**, 533–539.

Kirk, J. T. O. (1983). "Light and photosynthesis in aquatic ecosystems." Cambridge Univ. Press, London/New York.

Kohl, J. G. and Niklisch, A. (1981). Chromatic adaptation of the planktonic blue-green alga *Oscillatoria redekei* van Goor and its ecological significance *Int. Rev. Gesamten Hydrobiol.* **66**, 83–94.

Krinsky, N. I. (1976). Cellular damage initiated by visible light. *Symp. Soc. Gen. Microbiol.* **26**, 209–239.

Lattimer, P. (1959). Influence of selective light scattering on measurements of the absorption spectra of *Chlorella*. *Plant Physiol.* **34**, 193–199.

Laws, E. A. and Bannister, T. T. (1980). Nutrient- and light-limited growth of *Thalassiosira fluriatilis* in continuous culture, with implications for phytoplankton growth in the ocean. *Limnol. Oceanogr.* **25**, 457–473.

Lee, Y. K. and Pirt, S. J. (1981). Energetics of photosynthetic algal growth: influence of intermittent illumination in short (40S) flashes. *J. Gen. Microbiol.* **124**, 43–52.

Levring, T. (1966). Submarine light and algal shore zonation. *In* "Light as and Ecological Factor" (Eds. R. Bainbridge, A. C. Evans and O. Rackham), pp. 305–319. Blackwell, Oxford.

Ley, A. C. (1980). The distribution of absorbed light energy. *Brookhaven Symp. Biol.* **31**, 59–82.

Loogman, J. G. and Van Liere, L. (1978). An improved method for measuring irradiance in algal culture. *Verh.—Int. Ver. Theor. Angew. Limnol.* **20**, 2322–2328.

Lorenzen, M. W. (1980). Use of chlorophyll-Secchi disc relationships. *Limnol. Oceanogr.* **25**, 371–372.

Lund, J. W. G. (1950). Studies on *Asterionella formosa* Hass. II. Nutrient depletion and the spring maximum. *J. Ecol.* **38**, 1–35.

Lund, J. W. G. (1954). Seasonal cycle of the plankton diatom Melosira italica Ehr) Kutz. subsp. *subarctica* O. Mull. *J. Ecol.* **42**, 151–179.

Lythgoe, J. N. (1975). The ecology, function and phylogeny of iridescent multilayers in fish corneas. *In* "Light as an Ecological Factor II" (Eds. G. C. Evans, R. Bainbridge and O. Rackham), pp. 211–247. Blackwell, Oxford.

Marra, J. (1978a). Effect of short-term variations in light intensity on photosynthesis of a marine phytoplankter: a laboratory simulation study. *Mar. Biol. (Berlin)* **46**, 191–202.

Marra, J. (1978b). Phytoplankton response to vertical movement in a mixed layer. *Mar. Biol. (Berlin)* **46**, 203–208.

Meffert, M.-E. (1973). Kultur und Wachstum von *Oscillatoria redekei* van Goor. Zur Bedeutung der Bakterien. *Verh.—Int. Ver. Theor. Angew. Limnol.* **18**, 1359–1366.

Meffert, M.-E. and Krambeck, H. J. (1977). Planktonic blue-green algae of the *Oscillatoria redekei* group. *Arch. Hydrobiol.* **79**, 149–181.

Megard, R. O., Combs, W. S., Jr., Smith, P. D. and Knoll, A. S. (1979). Attenuation of light and daily integral rates of photosynthesis attained by planktonic algae. *Limnol. Oceanogr.* **24**, 1038–1050.

Moed, J. R. (1973). Effect of combined action of light and silicon depletion on *Asterionella formosa* Hass. *Verh.—Int. Ver. Theor. Angew. Limnol.* **18**, 1367–1374.

Morel, A. (1974). Optical properties of pure water and pure sea water. *In* "Optical Aspects of Oceanography" (Eds. N. G. Jerlov and E. Steeman-Nielsen), pp. 1–24. Academic Press, New York.

Morel, A. (1978). Available, usable and stored radiant energy in relation to marine photosynthesis. *Deep-Sea Res.* **25**, 673–688.

Morel, A. and Caloumenos, L. (1974). Variabilité de la répartition spectrale de l'énergie photosynthétique. *Tethys* **6**, 93–104.

Morel, A. and Prieur, L. (1977). Analysis of variations in ocean color. *Limnol. Oceanogr.* **22**, 709–722.

Morton, R. (1978). A model for light-scattering by algae in water. *Math. Biosci.* **40**, 195–204.

Mueller, J. L. (1976). Ocean colour spectra measured off the Oregon Coast: "Characteristic Vectors." *Appl. Opt.* **15**, 394–402.

Myers, J. (1946). Influence of light intensity of photosynthetic characteristics of *Chlorella*. *J. Gen. Physiol.* **29**, 429–440.

Myers, J. and Graham, J. K. (1971). The photosynthetic unit in *Chlorella* measured by repetitive short flashes. *Plant Physiol.* **48**, 282–286.

Myers, J. and Kratz, W. A. (1955). Relations between pigment content and photosynthetic characteristics in a blue-green alga. *J. Gen. Physiol.* **39**, 11–22.

Myers, J., Graham, J. R. and Wang, R. T. (1978). On spectral control of pigmentation in *Anacystis nidulans* (Cyanophyceae). *J. Phycol.* **14**, 513–518.

Perry, M. J., Talbot, M. C. and Alberte, R. S. (1981). Photoadaptation in marine phytoplankton: response of the photosynthetic unit. *Mar. Biol. (Berlin)* **62**, 91–101.

Platt, T. and Jassby, A. D. (1976). The relationship between photosynthesis and light for natural assemblages of coastal marine phytoplankton. *J. Phycol.* **12**, 421–430.
Prézelin, B. B. (1976). The role of peridinin—chlorophyll*a*—proteins in the photosynthetic light adaption of the marine dinoflagellate *Glenodinium* sp. *Planta* **130**, 225–233.
Prézelin, B. B. and Matlick, H. A. (1980). Time occurs of photoadaptation in the photosynthesis-irradiance relationship of a dinoflagellate exhibiting photosynthetic periodicity. *Mar. Biol. (Berlin)* **58**, 85–96.
Prézelin, B. B. and Sweeney, B. M. (1978). Photoadaptation of photosynthesis in *Gonyaulax polyedra*. *Mar. Biol. (Berlin)* **48**, 27–35.
Prézelin, B. B. and Sweeney, B. M. (1979). Photoadaptation of photosynthesis in two bloomforming dinoflagellates. *In* "Toxic Dinoflagellate Blooms" (Eds. D. L. Taylor and A. N. Seliger), pp. 101–106. Elsevier–North Holland, New York.
Price, B. J., Kollman, V. H. and Salzman, G. C. (1978). Light scatter analysis of microalgae. Correlation of scatter patterns from pure and mixed asynchronous cultures. *Biophys. J.* **22**, 29–36.
Prieur, L. and Sathyendranath (1981). An optical classification of coastal and oceanic waters based on the specific spectral absorption curves of phytoplankton pigments, dissolved organic matter and other particulate materials. *Limnol. Oceanogr.* **26**, 671–689.
Pulich, W. M. and Van Baalen, C. (1974). Growth requirements of blue-green algae under blue-light conditions. *Arch. Mikrobiol.* **97**, 303–312.
Reynolds, C. S. and Walsby, A. E. (1975). Waterblooms. *Biol. Rev. Cambridge Philos. Soc.* **50**, 437–481.
Rhee, G. Y. and Gotham, I. J. (1981). The effect of environmental factors on phytoplankton growth: light and the interactions of light with nitrate limitation. *Limnol. Oceanogr.* **26**, 649–659.
Roger, P. A. and Reynaud, P. A. (1977). Correction de la diffusion pour l'établissement de spectres d'absorption par des cultures d'algues microscopiques. *Cah. ORSTOM, Ser. Biol.* **12**, 129–139.
Schlesinger, D. A. and Shuter, B. J. (1981). Patterns of growth and cell composition of freshwater algae in light-limited continuous cultures. *J. Phycol.* **17**, 250–256.
Senger, H. (Ed.) (1980). "The Blue Light Syndrome." Springer-Verlag, Berlin.
Shear, H. and Walsby, A. E. (1975). An investigation into the possible light-shielding role of gas vacuoles in a planktonic blue-green alga. *Br. Phycol. J.* **10**, 241–251.
Shelef, G. and Soeder, C. J. (Eds.) (1980). "Algae Biomass: Production and Use." Elsevier/North Holland Biomedical Press, Amsterdam.
Shibata, K. (1958). Spectrophotometry of intact biological materials. *J. Biochem. (Tokyo)* **45**, 599–604.
Shimura, S. and Fujita, Y. (1975). Changes in the activity of fucoxanthin—excited photosynthesis in the marine diatom *Phaeodactylum tricornutum* grown under different culture conditions. *Mar. Biol. (Berlin)* **33**, 185–194.
Shimura, S. and Ichimura, S. (1973). Selective transmission of light in the ocean waters and its relation to phytoplankton photosynthesis. *J. Oceanogr. Soc. Jpn.* **29**, 257–266.
Shuter, B. J. (1979). A model of physiological adaptation in unicellular algae. *J. Theor. Biol.* **78**, 519–552.
Skulberg, O. M. (1978). Some observations on red-coloured species of *Oscillatoria* (Cyanophyceae) in nutrient enriched lakes of Southern Norway. *Verh.—Int. Ver. Theor. Angew. Limnol.* **20**, 776–787.
Smith, R. C. (1968). The optical characterization of natural waters by means of an extinction coefficient. *Limnol. Oceanogr.* **13**, 423–429.

Smith, R. C. and Baker, K. S. (1978a). The bio-optical state of ocean waters and remote sensing. *Limnol. Oceanogr.* **23**, 247–259.
Smith, R. C. and Baker, K. S. (1978b). Optical classification of natural waters. *Limnol. Oceanogr.* **23**, 260–267.
Smith, R. C. and Wilson, W. H. (1972). Photon scalar irradiance. *Appl. Opt.* **11**, 934–938.
Smith, R. C., Baker, K. S., Holm-Hansen, O. and Olson, R. (1980). Photoinhibition of photosynthesis in natural waters. *Photochem. Photobiol.* **31**, 585–592.
Soeder, C. J. and Binsack, R. (Eds.) (1978). Microalgae for food and feed. A status analysis. *Ergeb. Limnol.* **11**, 1–300.
Søndergaard, M. and Schierup, H. H. (1982). Dissolved organic carbon during a spring diatom bloom in Lake Moss, Denmark. *Water Res.* **16**, 815–821.
Steeman-Nielsen, E. and Hansen, V. K. (1959). Light adaptations in marine phytoplankton populations and its interrelation with temperature. *Physiol. Plant.* **12**, 353–370.
Steeman-Nielsen, E. and Jorgensen, E. G. (1968a). The adaptation of plankton algae. 1. The general part. *Physiol. Plant.* **21**, 401–413.
Steeman-Nielsen, E. and Jorgensen, E. G. (1968b). The adaptation of plankton algae III with special consideration of the importance in nature. *Physiol. Plant.* **21**, 647–654.
Stevens, C. L. R. and Myers, J. (1976). Characterization of pigment mutants in a blue-green alga *Anacystic nidulans*. *J. Phycol.* **12**, 99–105.
Straskraba, M. and Hammer, U. T. (1980). Solar radiation. In "The Functioning of Freshwater Ecosystems" (Eds. E. D. Le Cren and R. H. Lowe-McConnell), pp. 17–28. Cambridge Univ. Press, London and New York.
Sugihara, S. and Tsuda, R. (1979). Light scattering and size distribution of particles in the surface waters of the North Pacific Ocean. *J. Oceanogr. Soc. Jpn.* **35**, 82–96.
Talling, J. F. (1957). Photosynthetic characteristics of some freshwater plankton diatoms in relation to underwater radiation. *New Phytol.* **56**, 29–50.
Talling, J. F. (1966). Photosynthetic behaviour in stratified and unstratified lake populations of a planktonic diatom. *J. Ecol.* **54**, 99–127.
Talling, J. F. (1971). The underwater light climate as a controlling factor in the production ecology of freshwater phytoplankton. *Mitt.—Int. Ver. Theor. Angew. Limnol.* **19**, 214–243.
Talling, J. F. (1979). Factor interactions and implications for the prediction of lake metabolism. *Ergeb. Limnol.* **13**, 96–109.
Talling, J. F. (1982). Utilisation of solar radiation by phytoplankton. In "Trends in Photobiology" (Eds. C. Helene, M. Charlier, T. Montenay-Garestier and G. Laustriat), pp. 619–631. Plenum, New York.
Talling, J. F., Wood, R. B., Prosser, M. V. and Baxter, R. M. (1973). The upper limit of photosynthetic productivity by phytoplankton: evidence from Ethiopian soda lakes. *Freshwater Biol.* **3**, 53–76.
Tamiya, H., Shibata, K., Sasa, T., Iwamura, T. and Morimura, Y. (1953). Effect of diurnally intermittent illumination on the growth and some cellular characteristics of *Chlorella*. In "Algal Culture from Laboratory to Pilot Plant" (Ed. J. S. Burlew), pp. 63–75. Carnegie Institute of Washington, Washington, D.C.
Tandeau de Marsac, N. (1977). Occurrence and nature of chromatic adaptation in Cyanobacteria. *J. Bacteriol.* **130**, 82–91.
Tolbert, N. E. (1974). Photorespiration. In "Algal Physiology and Biochemistry" (Ed. W. D. P. Stewart), pp. 475–504. Blackwell, Oxford.
Tyler, J. E. (1968). The Secchi disc. *Limnol. Oceanogr.* **13**, 1–6.

Tyler, J. E. (1973a). Lux vs quanta. *Limnol. Oceanogr.* **18**, 810.
Tyler, J. E. (1973b). Applied radiometry. *Oceanogr. Mar. Biol.* **11**, 11–25.
Tyler, J. E. (1975). The *in situ* quantum efficiency of natural phytoplankton populations. *Limnol. Oceanogr.* **20**, 976–980.
Tyler, J. E. (1977). "Light in the Sea." Dowden, Hutchinson & Ross, Inc. Stroudsburg, Pennsylvania.
Tyler, J. E. and Smith, R. C. (1967). Spectroradiometric characteristics of natural light under water. *J. Opt. Soc. Am.* **57**, 595–601.
Tyler, J. E. and Smith, R. C. (1970). "Measurements of Spectral Irradiance Underwater." Gordon & Breach, New York.
Van Liere, L. (1979). An *Oscillatoria agardhii* Gomont Experimental Ecology and Physiology of a Nuisance Bloom-forming Cyanobacterium. Ph.D. Thesis, University of Amsterdam.
Van Liere, L. and Mur, L. R. (1978). Light-limited cultures of the blue-green alga *Oscillatoria agardhii. Mitt.—Int. Ver. Theor. Angew. Limnol.* **21**, 158–167.
Vierling, E. and Alberte, R. S. (1981). Functional organisation and plasticity of the photosynthetic unit of the cyanobacterium *Anacystis nidulans. Physiol. Plant.* **50**, 93–98.
Weinberg, S. (1976). Submarine daylight and ecology. *Mar. Biol. (Berlin)* **37**, 291–304.
Welschmeyer, N. A. and Lorenzen, C. J. (1981). Chlorophyll-specific photosynthesis and quantum efficiency at subsaturating light intensities. *J. Phycol.* **17**, 283–293.
Westlake, D. F. (1965). Some problems in the measurement of radiation underwater: a review. *Photochem. Photobiol.* **4**, 849–868.
Westlake, D. F., Adams, M. S., Bindloss, M. E., Ganf, G. G., Gerloff, G. C., Hammer, U. T., Javornicky, P., Koonce, J. F., Marker, A. F. H., McCracken, M. D., Moss, B., Nauwerck, A., Pyrina, I. L., Steel, J. A. P., Tilzer, M. and Walters, C. J. (1980). Primary production. *In* "The Functioning of Freshwater Ecosystems" (Eds. E. D. Le Cren and R. H. Lowe-McConnell), pp. 141–246. Cambridge Univ. Press, London and New York.
Wilson, W. H. and Kiefer, D. A. (1979). Reflectance spectroscopy of marine phytoplankton. II. A simple model of ocean color. *Limnol. Oceanogr.* **24**, 673–682.
Yentsch, C. S. (1980). Light attenuation and phytoplankton photosynthesis. *In* "The Physiological Ecology of Phytoplankton" (Ed. I. Morris), pp. 95–127. Blackwell, Oxford.
Yentsch, C. S. and Lee, R. W. (1966). A study of photosynthetic light reactions and a new interpretation of sun and shade phytoplankton. *J. Mar. Res.* **24**, 319–337.
Yoder, J. A. (1979). Effect of temperature on light limited growth and chemical composition of *Skeletonema costatum* (Bacillariophyceae). *J. Phycol.* **15**, 362–370.
Zevenboom, W. and Mur, L. R. (1978). N-uptake and pigmentation of N-limited chemostat cultures and natural populations of *Oscillatoria agardhii. Mitt.—Int. Ver. Theor. Angew. Limnol.* **21**, 261–274.

5

Aspects of Carbon Dioxide Assimilation by Autotrophic Prokaryotes

G. A. CODD

Department of Biological Sciences
University of Dundee
Scotland, United Kingdom

Introduction

Autotrophic organisms are capable of growth at the expense of carbon dioxide as the principal carbon source. Their energy needs may be provided by the absorption of light (photoautotrophs) or by the oxidation of inorganic chemical compounds [chemo(litho)autotrophs] (Stanier *et al.*, 1977). The concept of autotrophy has been extended to the methylotrophs, since these organisms can grow at the expense of organic compounds which contain methyl groups, with no carbon–carbon bonds, as the principal source of carbon and the sole source of energy (Whittenbury and Kelly, 1977). The diversity of mechanisms for energy capture and conversion among prokaryotic autotrophs, including anoxygenic and oxygenic photosynthesis and the oxidation of hydrogen and reduced sulphur, nitrogen, and iron compounds, contrasts to a relative uniformity in the mechanism of net carbon dioxide assimilation. Almost all autotrophs share a common pathway for the reduction of carbon dioxide into cell carbon, namely, the reductive pentose phosphate cycle (C_3 or Calvin cycle; Bassham and Calvin, 1957). Exceptions occur among the autotrophic prokaryotes, however, and include the green photosynthetic bacteria and the methanogens.

The details of the Calvin cycle for the photosynthetic incorporation of carbon dioxide into cell material by algae and higher plants have remained essentially correct throughout the three decades since their elucidation (Bassham and Calvin, 1957; Bassham, 1979; Robinson and Walker, 1981). However, it is now clear that the Calvin cycle alone does not fully describe the path of carbon dioxide fixation during plant photosynthesis. Evidence for the C_4-dicarboxylic, or Hatch-Slack pathway, and the C_2-photorespiratory carbon oxidation cycle, which are both complementary to the C_3 cycle, is considerable among plants, and fundamental similarities between the Calvin cycle enzymes of plants and prokaryotes have prompted the study of the presence, or otherwise, of these ancillary reactions among the bacteria.

Two enzymes in particular are unique and essential to the Calvin cycle. They are the carbon dioxide-fixing enzyme D-ribulose 1,5-bisphosphate carboxylase-oxygenase (3-phospho-D-glycerate carboxylase dimerizing EC 4.1.1.39) and phosphoribulokinase (ATP : D-ribulose 5-phosphate 1-phosphotransferase, EC 2.7.1.19), which catalyses the regeneration of the carbon dioxide acceptor. These two enzymes represent key control points in the regulation of autotrophic carbon dioxide fixation and are subjects of considerable current study. Their structures, functions, regulation, and compartmentation in autotrophic bacteria are reviewed here, and comparisons are made to their equivalents in eukaryotes. Aspects of these enzymes and their actions are also considered in terms of the physiological ecology of autotrophic bacteria.

The Utilisation of the Calvin Cycle by Autotrophic Prokaryotes

The Calvin cycle, as summarised in Fig. 1 to show the enzymes exclusive to the cycle, essentially consists of three phases: (1) the carboxylation of ribulose 1,5-bisphosphate (RuBP) by RuBP carboxylase-oxygenase (RuBisCO) to form an unstable 6-carbon intermediate, and thence 3-phosphoglyceric acid (PGA) as the primary stable product, (2) then PGA reduction to the level of cell carbohydrate,

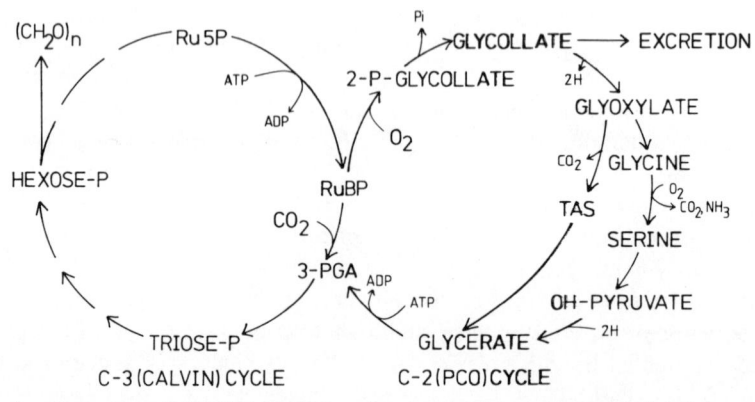

Fig. 1. Summary of the Calvin cycle, showing its integration with the C-2, or photorespiratory carbon oxidation cycle. Glycollate may be excreted by autotrophic microbes or metabolised. Glyoxylate can be further metabolised by glycine and serine, or in some cases via TAS (tartronic semialdehyde). Stoichiometry in the C-2 cycle is not shown, two glyoxylate molecules being required to form one serine or TAS.

and finally (3) the regeneration of the carbon dioxide acceptor, RuBP. The Calvin cycle is integrated with a C-2, or photorespiratory carbon oxidation cycle, the cycles being linked by the bifunctional RuBisCO.

It should be noted that not all of the criteria needed to establish the Calvin cycle as the pathway for net carbon dioxide assimilation, namely, short-term and steady-state labelling kinetics, intramolecular labelling studies, plus the demonstration of all of the enzymes invoked with activities sufficient to account for the rate of carbon fixation by whole cells, have been reported in all of the prokaryotes investigated. However, the evidence that the Calvin cycle is used by almost all autotrophic prokaryotes for net carbon dioxide assimilation is compelling. Carbon-14 labelling kinetics and enzymatic data for the operation of the Calvin cycle have thus been obtained, where sought in the phototrophs, in the purple nonsulphur and purple sulphur bacteria, the cyanobacteria, and among the chemolithoautotrophs, in the colourless sulphur-oxidising bacteria, the ammonia and nitrite-oxidising bacteria, and the hydrogen bacteria. It is not possible to provide extensive coverage of this literature here, and several excellent reviews may be consulted (Kelly, 1971; McFadden, 1973, 1978; Smith, 1973; Smith and Hoare, 1977; Fuller, 1978; Ohmann, 1979; Schlegel, 1975, 1976; Bowien and Schlegel, 1981).

By contrast, data are lacking for the operation of the Calvin cycle in both green sulphur bacteria and green nonsulphur bacteria. RuBisCO has been detected in (Buchanan and Sirevåg, 1976) and successfully purified from (Tabita *et al.*, 1974) strains of the obligately photoautotrophic green sulphur bacterium *Chlorobium thiosulfatophilum*. However, the kinetics of ^{14}C labelling and the results of carbon isotope discrimination studies are consistent with the operation of the reductive tricarboxylic acid in these organisms (Sirevåg, 1974; Sirevåg and Ormerod, 1970; Sirevåg *et al.*, 1977; Takabe and Akazawa, 1977). Early attempts to measure the second Calvin cycle marker enzyme, phosphoribulokinase, in *C. thiosulfatophilum* were unsuccessful, in contrast to parallel assays with extracts of the purple nonsulphur bacterium *Rhodospirillum rubrum* (Sirevåg, 1974). The nutritionally versatile green nonsulphur bacterium *Chloroflexus* is capable of photoheterotrophic, chemoheterotrophic, and possibly photoautotrophic growth (Madigan *et al.*, 1974), and has been found to contain low activities of RuBisCO and phosphoribulokinase in photoheterotrophic cell extracts. Although rapid labelling kinetics need to be performed with phototrophically grown cells, the inhibitory effect of fluoroacetate on CO_2 fixation by photoheterotrophically grown cells, together with the accumulation of citrate due to fluoracetate treatment, indicates the operation of the reductive tricarboxylic acid cycle rather than the Calvin cycle during CO_2 fixation in this organism (Sirevåg and Castenholz, 1979). Further studies with these fascinating organisms are awaited.

The possible autotrophic capability of pure cultures of the sulphide oxidiser

Beggiatoa has been indicated occasionally (Keil, 1912; Kowallik and Pringsheim, 1966), and the mixotrophic capabilities of several isolates are well established (Scotten and Stokes, 1962; Strohl *et al.*, 1981). The possibility of a functional Calvin cycle for carbon dioxide fixation has been examined in chemoheterotrophically and mixotrophically grown cells of *Beggiatoa alba*. However, glutamate and apartate are the principal early stable products of carbon dioxide incorporation. These data, together with the presence of adequate levels of isocitrate dehydrogenase (nicotinamide adenine dinucleotide phosphate, reversed) and malate dehydrogenase (nicotinamide adenine dinucleotide phosphate, decarboxylating), the suggested carbon dioxide-fixing enzymes in extracts of chemoheterotrophically grown cells, indicate that carbon dioxide fixation occurs into the tricarboxylic acid, though not via typical autotrophic pathways (Strohl *et al.*, 1981). Neither RuBisCO nor phosphoenolpyruvate carboxylase activities were detected in extracts from chemoheterotrophic or mixotrophic *B. alba* cells (Strohl *et al.*, 1981). Strohl and Larkin (1978) have been unable to obtain (chemolitho) autotrophic growth of their *Beggiatoa* strains, and the likelihood of a Calvin cycle in *Beggiatoa* appears to be remote at present.

Evidence for the chemolithoautotrophic growth of the iron bacterium *Gallionella ferruginea* has been found, however, almost a century after Winogradsky's hypothesis on chemolithotrophy. Phosphoribulokinase and RuBisCO have been found in *G. ferruginea* (Hanert, 1982).

Several bacteria capable of growing on 1-carbon compounds, namely, carbon monoxide, formate, methanol, formamide, and methylamine, first oxidise them to carbon dioxide and then presumably assimilate the carbon via the Calvin cycle (Colby *et al.*, 1979). These organisms include bacteria normally viewed as carbon dioxide-utilising phototrophs or chemolithoautotrophs. For example, *Rhodopseudomonas palustris* and *Rps. acidophila* can grow on formate, and *Rps. acidophila* can also grow on methanol. Among the chemolithoautotrophs, *Hydrogenomonas eutropha* (*Alcaligenes eutrophus*), *Alcaligenes* FOR$_1$, *Microcylus aquaticus, Thiobacillus novellus,* and *Paracoccus denitrificans* can use formate as a growth substrate, the last three organisms additionally being able to utilise methanol (Colby *et al.*, 1979; Manian and O'Gara, 1982). Several carbon monoxide-oxidising bacteria have been isolated which oxidise carbon monoxide to carbon dioxide. The carbon dioxide may be then assimilated via the Calvin cycle (Zavarzin and Nozhevnikova, 1976).

The enzymes of the Calvin cycle are present in the methane oxidizer *Methylococcus capsulatus* (Bath strain) (Taylor, 1977), which assimilates formaldehyde via the ribulose monophosphate cycle. *In vivo* evidence for the operation of RuBisCO during carbon dioxide fixation by methane-grown cells was obtained in labelling studies, although attempts to obtain carbon dioxide-dependent growth of *M. capsulatus* (Bath) have been unsuccessful (Taylor *et al.*, 1981). The operation of the Calvin cycle for carbon assimilation by *Pseudomonas oxalaticus* when grown on formate as the sole source of carbon and energy is well documented (Quayle, 1961). This versatile organism can also grow on oxalate when the Calvin cycle is inoperative, and carbon assimilation occurs via the glycerate pathway (Quayle, 1961).

Characteristics of Enzymes of the Calvin Cycle from Autotrophic Prokaryotes

The structure, functions, regulation, and compartmentation of the key Calvin cycle enzymes, phosphoribulokinase and RuBisCO, are being studied in autotrophic bacteria. In the case of RuBisCO, a start has been made on the molecular genetics of the enzyme. These investigations can aid the understanding of autotrophic carbon dioxide assimilation *in vivo* and have implications for the ecology of autotrophic prokaryotes.

Phosphoribulokinase

In contrast to the much studied RuBisCO, relatively little is known about phosphoribulokinase, although this enzyme is unique to the Calvin cycle, essential for its operation, and thus is an important control point (McFadden, 1973, 1978; Bassham, 1979; Lilley and Walker, 1979; Latzko and Kelly, 1979).

Purification and Molecular Properties of Bacterial Phosphoribulokinases. The enzyme has been purified from few prokaryotic sources and from fewer eukaryotes (Table 1). Similarities in the molecular weights of the prokaryotic enzymes can be seen. Dissociation of the homogeneous *Rhodopseudomonas capsulata* (Tabita, 1980) and *A. eutrophus* (Siebert *et al.*, 1981) enzymes has shown only one type of subunit in each case, and hexameric and octameric

Table 1. *Molecular properties of phosphoribulokinases*

Source	Native enzyme (mol wt)	Subunit (SU) (mol wt)	Suggested quaternary structure	Reference
Chromatium D	240,000	nr	—	Hart and Gibson (1971)
Rhodopseudomonas capsulata	220,000	36,000	6 SU	Tabita (1980)
Alcaligenes eutrophus	256,000	33,000	8 SU	Siebert *et al.* (1981)
Spinach	nr[a]	46,000	—	Lavergne and Bismuth (1973)

[a] nr, Not reported.

quaternary structures, respectively, have been suggested. The subunits of these bacterial enzymes appear to be smaller than those reported for the spinach enzyme (Table 1). It will be of obvious interest to purify phosphoribulokinase from other prokaryotes, especially cyanobacteria, and compare their properties with those of the chloroplast enzyme.

Catalytic and Regulatory Properties. Apart from the studies on the enzymes from *Chromatium* D (Hart and Gibson, 1971), *Rps. capsulata* (Tabita, 1980, 1981), and *A. eutrophus* H 16 (Abdelal and Schlegel, 1974a; Siebert *et al.*, 1981; Bowien, 1984), the catalytic and regulatory properties of phosphoribulokinases have been investigated only in crude or partially purified extracts. Nevertheless, a range of allosteric effects has been noted with the enzymes from chemolithoautotrophic and some phototrophic bacteria. Inhibition of the *Thiobacillus thioparus* enzyme by AMP was found in early studies by MacElroy *et al.* (1968). Allosteric inhibition by PEP and AMP of the *Chromatium, Thiobacillus neapolitanus,* and *Pseudomonas facilis* enzymes is also well established (Hart and Gibson, 1971; MacElroy *et al.*, 1969). Rindt and Ohmann (1969) and MacElroy *et al.* (1969) provided the first evidence for the activation of the *Rps. sphaeroides* and *P. facilis* phosphoribulokinases by NADH. Findings on the activation of *A. eutrophus* H 16 enzyme by NADH and inhibition by AMP and PEP (Abdelal and Schlegel, 1974a; Siebert *et al.*, 1981) have extended to the phosphoribulokinases from the following hydrogen bacteria: *Alcaligenes* type strain and strains H 1 and H 20, *A. paradoxus, Pseudomonas pseudoflava, Paracoccus denitrificans, Xanthobacter autotrophicus, Nocardia opaca* 1 b, and *Arthrobacter* 11/x (Bowien, 1984). These data, together with the activation of the *Nitrobacter winogradsky* enzyme by NADH (Kiesow *et al.*, 1977), clearly indicate that regulation of phosphoribulokinase in chemolithoautrophic bacteria is under regulation by energy charge and NADH. Activation of the *Alcaligenes eutrophus* enzyme by NADH affects the affinity of the enzyme for both substrates, ribulose 5-phosphate, and ATP (Abdelal and Schlegel, 1974a; Siebert *et al.*, 1981). Furthermore, the apparent activation constant, K_a, for NADH (0.19 mM) lies well below the intracellular NADH pool concentration (about 0.8 mM) in autotrophic *A. eutrophus* cells (Cook and Schlegel, 1978).

Requirement for NADH also appears to be a general feature of the phosphoribulokinase of purple nonsulphur bacteria. Besides the *Rps. sphaeroides* enzyme (Rindt and Ohmann, 1969), the kinases of *Rps. capsulata, Rps. palustris, Rhodospirillum rubrum, R. tenue,* and *R. molischianum* show an absolute or partial requirement for NADH (Joint *et al.*, 1972; Tabita, 1981). The sole exception found among the purple nonsulphur bacterial enzymes so far is from *Rhodomicrobium vannielli* (Tabita, 1981). The rationale for allosteric regulation by energy charge and NADH in chemolithoautotrophs and phototrophic bacteria appears straightforward. Phosphoribulokinase is specifically activated by NADH, which in turn is the product of chemolithoautotrophic and phototrophic energy conversion and is needed to provide the reducing power for the Calvin cycle. The inhibition of phosphoribulokinase by PEP can be viewed as a feedback inhibition, since PEP can be derived from the Calvin cycle (Bassham, 1979).

Few data are available on the regulation of cyanobacterial phosphoribulokinase. Tabita (1981) found that the enzymes from *Agmenellum quadruplicatum* and *Anabaena* CA were not influenced by NADH or NADPH. In this respect, the cyanobacterial enzymes resemble the kinases of *Chlorella* (Tabita, 1981) and spinach chloroplasts (Kiesow *et al.*, 1977). It is noteworthy that chloroplast phosphoribulokinase is not influenced by AMP, ADP, or PEP (MacElroy *et al.*, 1972; Anderson, 1973), but the enzyme is very rapidly inactivated in *Chlorella pyrenoidosa* cells and spinach chloroplasts on transfer from light to darkness (Bassham, 1979; Flügge *et al.*, 1982). In addition to suggested regulation by stromal metabolite levels (Flügge *et al.*, 1982), chloroplast phosphoribulokinase is light activated via a membrane-bound vicinal-dithiol-containing factor or via thioredoxin (Anderson, 1979; Flügge *et al.*, 1982). The activity of the cyanobacterial enzyme may be modulated in a similar manner, since Duggan and Anderson (1975) found a light activation of *Anacystis nidulans* phosphoribulokinase, which could be mimicked in the dark by the addition of dithiothreitol.

The Compartmentation of Phosphoribulokinase in Prokaryotes. At the subcellular level, phosphoribulokinase has been isolated in a complex with fructose bisphosphatase from *Rhodospirillum rubrum* (Joint *et al.*, 1972). However, the complex was unstable, and subsequent studies with phosphoribulokinase from *Rps. capsulata* and *Alcaligenes eutrophus* have not revealed any association of phosphoribulokinase with other Calvin cycle enzymes during purification (Abdelal and Schlegel, 1974; Tabita, 1980; Siebert *et al.*, 1981). The concept of complexes of Calvin cycle enzymes in such organisms which lack inclusion bodies is attractive, and the possibility remains that such associations may exist *in vivo* (Tabita, 1980, 1981; Siebert *et al.*, 1981; Bowien, 1983).

Many autotrophic prokaryotes possess polyhedral inclusion bodies (carboxysomes), and the presence of RuBisCO in these organelles is well established (Shively, 1974; Codd and Stewart, 1976; Lanaras and Codd, 1981a). Details of these organelles are considered later in this chapter, but it may be noted here that Beudeker and Kuenen (1981) have reported the association of phosphoribulokinase with carboxysomes isolated from *Thiobacillus neapolitanus*. However, other studies on carboxysomes isolated from this organism, and judged to be homogeneous by a variety of criteria, have not shown carboxysomal phosphoribulokinase activity (Cannon and Shively, 1982); this question needs to be resolved. Although a small fraction of the cellular pool of phosphoribulokinase in the cyanobacteria *Chlorogloeopsis fritschii* and *Nostoc* 6720 is particulate, sedimentation through density gradients of Percoll plus sucrose has so far indicated that the particulate activity is associated with the thylakoids of these organisms rather than with their carboxysomes.

Intercellular compartmentation of phosphoribulokinase activity occurs in the filamentous, heterocystous, nitrogen-fixing cyanobacterium *Anabaena cylindrica* (Codd et al., 1980). Photosynthetic carbon dioxide fixation occurs only in the vegetative cells of this organism, whereas during aerobic growth in the absence of combined nitrogen, the heterocysts are the site of nitrogen fixation (Stewart, 1973, 1980). To the absence from heterocysts of Photosystem II and RuBisCO may be added the lack of heterocystous phosphoribulokinase activity, in accounting for the inability of these specialized cells to perform the photofixation of carbon dioxide (Codd et al., 1980).

Ribulose Bisphosphate Carboxylase-Oxygenase (RuBisCO)

This bifunctional enzyme, the most abundant protein in the biosphere (Ellis, 1979), has received considerable attention in recent years. Its structural, catalytic, regulatory, and genetic features are of interest in prokaryotes and eukaryotes (McFadden, 1973, 1978, 1980; McFadden and Tabita, 1974; Jensen and Bahr, 1977; Akazawa, 1979; Lorimer and Andrews, 1981; Lorimer, 1981a).

Purification and Molecular Properties of Bacterial RuBisCOs. Most RuBisCOs are of large molecular weight ($\simeq$ 500,000–550,000). The enzymes are also typically among the most abundant proteins in autotrophic cells. This feature, plus the high molecular weight and stability of most RuBisCOs, has permitted the straightforward purification of the enzyme from many sources. The usual purification method involves sedimentation through linear sucrose density gradients as the most effective step (Goldthwaite and Bogorad, 1971; Tabita and McFadden, 1974a; Bowien et al., 1976; Codd and Stewart, 1977a). All eukaryotic RuBisCOs examined are of high molecular weight (500,000–550,000) and consist of large (L) and small (S) subunits with molecular weights of about 50,000–55,000 and 11,000–15,000, respectively, arranged in an 8L8S quaternary structure (Akazawa, 1979). The L subunit bears the site of catalysis; the role of the S subunit remains unclear.

A diversity in molecular weight and subunit composition has been found, however, among prokaryotic RuBisCOs (Table 2). The *R. rubrum* enzyme is well established as a low molecular weight form consisting of two L subunits and no S subunits (Tabita and McFadden, 1974b; Schloss et al., 1979). Enzymes of intermediate molecular weight, lacking S subunits, have also been purified from various photosynthetic bacteria, including 6L enzymes from *Chlorobium thiosulfatophilum* (Tabita et al., 1974), *Rps. sphaeroides* (Gibson and Tabita, 1977a), and *Rps. capsulata* (Gibson and Tabita, 1977b). The *Rhodopseudomonas* strains studied by Gibson and Tabita (1977a,b) are particularly

Table 2. Molecular properties of prokaryotic RuBisCOs

Source	Molecular weight	L[a]	S[b]	Quaternary structure[c]	Reference
Purple sulphur bacteria					
Chromatium D	520,000	+	+	8L8S	Takabe and Akazawa (1975a)
Ectothiorhodospira halophila	600,000	+	+	8L8S	Tabita and McFadden (1976)
Thiocapsa roseopersicina	~500,000	+	+	8L8S	Purohit et al. (1979)
Purple nonsulphur bacteria					
Rhodospirillum rubrum	114,000	+	−	2L	Tabita and McFadden (1974b); Schoss et al. (1979)
Rhodopseudomonas sphaeroides	~500,000	+	+		Akazawa et al. (1972)
Form I	360,000	+	−	6L	Gibson and Tabita (1977a)
Form II	550,000	+	+	8L8S	Gibson and Tabita (1977a)
Rhodopseudomonas capsulata					
Form I	360,000	+	−	6L	Gibson and Tabita (1977b)
Form II	550,000	+	+	8L8S	Gibson and Tabita (1977b)
Rhodomicrobium vannielii	430,000	+	+	6L6S	Taylor and Dow (1980)
Green sulphur bacteria					
Chlorobium thiosulfatophilum	360,000	+	−	6L	Tabita et al. (1974)
Nitrifying bacteria					
Nitrobacter agilis (Delwiche)	500,000	+	+	8(LL′)8S	Harrison et al. (1979)
Hydrogen bacteria					
Alcaligenes eutrophus PHB-4	505,000	+	+	8L8S	Bowien et al. (1976)
A. eutrophus H16	534,000	+	+	8L8S	Bowien and Mayer (1977)
A. eutrophus	516,000	+	+	8L8S	McFadden and Purohit (1978)
Nocardia opaca 1b	541,000	+	+	8L8S	Bowien and Mayer (1977)
Arthrobacter 11x	571,000	+	+	8L8S	Bowien and Mayer (1977)
Paracoccus denitrificans	510,000	+	+	8L8S	Bowien (1977)
	525,000	+	+	8L8S	Shively et al. (1978)

(continued)

Table 2. *Continued*

Source	Molecular weight	Subunits L[a]	Subunits S[b]	Quaternary structure[c]	Reference
Colourless sulphur bacteria					
Thiobacillus denitrificans	350,000	nr[d]	nr		McFadden and Denend (1972)
T. novellus	496,000	+	+	8L8S	McCarthy and Charles (1975)
Thiobacillus A2	521,000	+	+	8L8S	Charles and White (1976)
T. neapolitanus	500,000	+	+	8L8S	Snead and Shively (1978)
T. intermedius	550,000	+	+	8L8S	Bowman and Chollet (1980)
Legume nodule bacteria					
Rhizobium japonicum SR	x[e]	+	+	8(LL')8S	Purohit *et al.* (1982)
Methylotrophic bacteria					
Methylococcus capsulatus (Bath)	360,000	+	+	6L6S	Taylor *et al.* (1981)
Cyanobacteria					
Anabaena cylindrica	500,000	+	+	8L8S	Takabe (1977); Okabe and Codd (1980)
A. variabilis	18 S	+	+	8L8S	Takabe *et al.* (1976)
Plectonema boryanum	18 S	+	+	8L8S	Takabe *et al.* (1976)
Anabaena CA	nd[f]	+	+	8L8S	Gibson and Tabita (1979)
Aphanocapsa 6308	525,000	+	+	8L8S	Codd and Stewart (1977a)

Microcystis aeruginosa	518,000	+	8L8S	Stewart et al. (1977)
Nostoc commune	~500,000	+	8L8S	Cook (1980)
N. canina	~500,000	+	8L8S	Cook (1980)
Nostoc 1453/28	~500,000	+	8L8S	Cook (1980)
Oscillatoria limnetica	~500,000	+	8L8S	Cook (1980)
Chlorogloeopsis fritschii	520,000	+	8L8S	Lanaras and Codd (1981b)
Nostoc 6720	~500,000	+	8L8S	Leadbeater (1981)
Gloeobacter violaceus	~500,000	+	8L8S	G. A. Codd and K. Beattie (unpublished)
Synechococcus RRIMP/N1	530,000	+	8L8S	Andrews et al. (1981)
Aphanothece halophytica	y^g	+		Asami et al. (1983)
Prochlorophytes				
Prochloron from Lissoclinum patella	x^e	+		G. A. Codd and K. Beattie (unpublished)
Cyanelles				
Cyanophora paradoxa	520,000	+	8L8S	Codd and Stewart (1977b)
Glaucosphaera vacuolata	>500,000	+	8L8S	G. A. Codd (unpublished)

[a] Large subunit.
[b] Small subunit.
[c] Mostly derived from molecular weight determinations of constituent subunits.
[d] nr, Not reported.
[e] Not determined, but the position of the enzyme after density gradient centrifugation is consistent with a high molecular weight enzyme (~500,000 daltons).
[f] nd, No data.
[g] Molecular weight of holoenzyme unknown because the enzyme readily dissociates into large and small subunits.

interesting, since both organisms also contain a high molecular weight (8L8S) RuBisCO which is kinetically and immunologically distinct from the 6L enzyme. The reason for two RuBisCOs in the same cell is unclear, but such a system has provided a useful means of studying relations between enzyme structure and function (see later). RuBisCOs containing L and S subunits have been purified from *Rhodomicrobium vannielii* (Taylor and Dow, 1980) and *Methylococcus capsulatus* (Taylor et al., 1981), with respective molecular weights of 430,000 and 360,000 derived from gel filtration, and 6L6S quanternary structures inferred. Similar conclusions were reached after equilibrium sedimentation studies (C. S. Dow, personal communication). However, gel filtration and pore gradient electrophoresis of RuBisCO from a cyanobacterium *Synechococcus* sp. yield a molecular weight of 430,000 (6L6S; Andrews and Abel, 1981), but a value of 530,000 (L8S8) has been found by equilibrium sedimentation (Andrews et al., 1981). The anomalous behaviour of the *Synechococcus* enzyme during pore gradient electrophoresis and gel filtration may be due to the asymmetric shape of the molecules (Andrews et al., 1981).

Although many prokaryotic RuBisCOs are stable *in vitro*, several examples of enzyme breakdown during purification have occurred. Rapid dissociation of the enzyme from the purple sulphur bacterium *Thiocapsa roseopersicina* occurs, to give a spectrum of intermediate breakdown products and a catalytically active 53,000-dalton polypeptide, the presumed L subunit (McFadden and Purohit, 1978). The native enzyme, however, is of the 8L8S type (Table 2). We have purified the enzyme from the obligately halophilic cyanobacterium *Aphanothece halophytica* (Codd et al., 1979). No evidence for S subunits was found after purification from frozen or fresh cells, in the presence of the protease inhibitor phenylmethylsulphomyl fluoride, or after enzyme immunoprecipitation from extracts of freshly broken cells. The molecular weight of the purified L subunit oligomer, determined by equilibrium centrifugation, was 237,000, suggesting a 4L structure. However, we have found the association of a 14,000-dalton polypeptide with the *Aph. halophytica* RuBisCO (Asami et al., 1983). This polypeptide appears to be the S subunit of a readily dissociable high molecular weight multimeric *Aph. halophytica* RuBisCO. The ready dissociation of the *Aph. halophytica* and *Synechococcus* sp. enzymes has provided a further means of studying relations between the structure and function of the cyanobacterial enzyme (Andrews and Abel, 1981; Asami et al., 1983; see later).

As shown in Table 2, the remaining prokaryotic RuBisCOs purified from the photosynthetic bacteria, chemolithoautotrophs, cyanobacteria, *Prochloron,* and cyanelles are all of the 8L8S type. In virtually all cases, only one class each of L and S subunits appears after sodium dodecyl sulphate (SDS) polyacrylamide gel electrophoresis of an SDS-dissociated enzyme. However, L subunit heterogeneity has been observed with the enzymes from *Rho. vannielii* (Taylor and

Dow, 1980), *Nitrobacter agilis* (Harrison *et al.*. 1979), *Rhizobium japonicum* (Purohit *et al.*, 1982), *Alcaligenes eutrophus* (Purohit and McFadden, 1976, 1977), and in 2 out of 10 purifications of the enzyme from *Chlorogloeopsis fritschii* carboxysomes (Lanaras and Codd, 1981b). The significance of possible L subunit heterogeneity of these prokaryotic enzymes is not understood. Although such heterogeneity may persist despite the presence of protease inhibitors during enzyme purification (Purohit and McFadden, 1976, 1977), the possibility of its being due to proteolysis *in vitro* has not been discounted (Harrison *et al.*, 1979; Lanaras and Codd, 1981b).

A structural model of the *A. eutrophus* 8L8S RuBisCO has been constructed on the basis of molecular weight measurements, electron microscopy, and the use of antibodies against the complete enzyme and the L and S subunits (Bowien *et al.*, 1976; Bowien and Mayer, 1978). The model is a four-layered structure with 4:2:2 symmetry, consisting of two central eclipsed layers, each of four U-shaped L subunits arranged perpendicular to the fourfold axis of symmetry, with the arms of the subunits pointing outwards. The two outer layers of the molecule each consist of four S subunits in eclipsed positions to the L subunits, so that each S subunit is in contact with one L subunit. A central hole runs through the axis of symmetry, and the dimensions of the structure are about $13 \times 13 \times 10.5$ nm (Bowien *et al.*, 1976). X-ray analysis of crystalline *A. eutrophus* has supported this model (Bowien *et al.*, 1980). Fourfold symmetry of RuBisCO from *Synechococcus* sp. (Andrews *et al.*, 1981) and *Anabaena cylindrica* (this laboratory, unpublished) has also been observed. These cyanobacterial enzymes also show a central hole, U-shaped arms which may be part of the L subunits, and have a diameter of about 10 to 13 nm. The location of the S subunits on the cyanobacterial enzyme is unknown. The Eisenberg model for tobacco RuBisCO is, by contrast, a two-layer structure, but also with 4:2:2 symmetry. The arrangement of the L subunits resembles that of the Bowien *Alcaligenes* model, but the tobacco L subunits do not appear to possess the V- or U-shaped arms (Baker *et al.*, 1977a,b). Again, the arrangement of the S subunits on the higher plant enzyme is unknown (Eisenberg *et al.*, 1978).

Comparative studies on the quaternary structure and amino acid composition of prokaryotic and eukaryotic RuBisCOs have been performed with a view to providing information on the origin(s) and evolution of RuBisCO and possibly the molecular evolution of autotrophy (McFadden, 1973, 1978, 1980). Considerable homology of the amino acid composition of L subunits from diverse prokaryotic and eukaryotic sources exists, although, as pointed out by Takabe and Akazawa (1975b), that of the 2L *Rhodospirillum rubrum* enzyme is markedly different.

Wide differences in amino acid composition occur between the S subunits of different RuBisCOs (Takabe and Akazawa, 1975b). These studies have provided preliminary evidence for a possible common ancestor for the catalytically functional L subunit, which has been conserved through evolution, in contrast to the functionally enigmatic S subunits.

Further structural comparisons have been sought by immunological means. We have tested antisera raised against the 8L8S *Ana. cylindrica* enzyme and a L subunit oligomer from the *Aph. halophytica* enzyme (Table 3). Both antisera give single bands in Ouchterlony double immunodiffusion gels against the homologous antigens. Cross-reactions with full or partial identity occur between both antisera and RuBisCO-containing extracts from all of 17 other cyanobacterial strains, 2 cyanelles, and all eukaryotic enzymes tested. However, no cross-reactions occur between the cyanobacterial antisera and RuBisCO-containing extracts from any of the other prokaryotic sources tested, which included representatives of the photosynthetic bacteria, colourless sulphur oxiders, hydrogen bacteria, and nitrifiers. Effects of the antisera on enzyme activity confirm the immunodiffusion results (Table 3).

The immunological relatedness between the cyanobacterial, cyanelle, and chloroplast enzymes concurs with the considerable evidence for an endosymbiotic origin for plastids from an ancestral oxygenic prokaryotic phototroph (Gray and Doolittle, 1982). The absence of immunological relatedness between the two cyanobacterial enzyme preparations and RuBisCOs from any of the other groups of prokaryotes tested suggests considerable diversity in the molecular topology between the cyanobacterial and other prokaryotic enzymes. Ouchterlony double immunodiffusion tests using antisera to native enzymes and isolated L and S subunits have revealed full and partial identities between the enzymes of several hydrogen bacteria (Bowien and Schlegel, 1981), but not between the *A. eutrophus* enzyme and antiserum versus RuBisCO from *Euglena gracilis* (B. Bowien, personal communication). Further immunological differences between prokaryotic and chloroplast RuBisCOs were reported by Purohit *et al.* (1982), who found no cross-reactions between *Rhizobium japonicum* RuBisCO antiserum and the enzymes from wheat, spinach, or soybean, or between the tobacco enzyme antiserum and the *Rh. japonicum* enzyme. In contrast to the absence of immunological relatedness between diverse bacterial RuBisCOs, other than cyanobacterial enzymes, and chloroplast enzymes (Table 3, Purohit *et al.*, 1982) the partial immunological identity between the 8L8S enzyme from the purple bacterium *Chromatium vinosum* and the spinach enzyme has long been known (Akazawa *et al.*, 1972, 1978).

The nature of the diversity, if any, between RuBisCOs may be better explored via primary structure and genetic studies. The traditional approach of amino acid sequence determination has been greatly aided by the ability to clone and sequence the genes for the L and S subunits. Partial, and in some cases complete, primary structures are available for the L subunit from maize, tobacco, spinach, and barley (McIntosh *et al.*, 1980; Hartman *et al.*, 1978; Lorimer, 1981a; Poul-

Table 3. Immunological cross-reactions between RuBisCOs from diverse sources and antisera to cyanobacterial enzyme preparations[a]

Source of enzyme[b]	Ouchterlony double immunodiffusion		Effect on enzyme activity	
	Anti-Anabaena cylindrica[c]	Anti-Aphanothece halophytica[d]	Anti-Anabaena cylindrica	Anti-Aphanothece halophytica
Prokaryotes				
Cyanobacteria				
17 strains tested[e]	Yes	Yes	Inhibition	Inhibition
Phototrophic bacteria				
Rhodospirillum rubrum	No	No	No inhibition	No inhibition
Thiocapsa roseopersicina	No	No	No inhibition	No inhibition
Rhodopseudomonas acidophila	No	No	No inhibition	No inhibition
Rhodopseudomonas palustris	No	No	No inhibition	No inhibition
Rhodomicrobium vannielii	No	No	No inhibition	No inhibition
Chemolithoautotrophic bacteria				
Alcaligenes eutrophus H16	No	No	No inhibition	No inhibition
Thiobacillus neapolitanus	No	No	No inhibition	No inhibition
Nitrobacter agilis	No	No	No inhibition	No inhibition
Cyanelles				
From Cyanophora paradoxa	Yes	Yes	nt[f]	nt
From Glaucosphaera vacuolta	Yes	Yes	nt	nt
Eukaryotes				
Chlorella fusca	Yes	Yes	inhibition	inhibition
Scenedesmus braunii	Yes	Yes	nt	nt
Spinacia oleracea	Yes	Yes	inhibition	inhibition

[a] With acknowledgements to Cook (1980) and K. Beattie.
[b] High-speed supernatants of cell-free extracts were used throughout (Cook, 1980).
[c,d] Antisera produced by Okabe and Codd (1980) and Cook (1980), respectively; each produced a single precipitin band versus homologous purified RuBisCOs.
[e] Cyanobacterial extracts tested from Anacystis nidulans, Aphanocapsa 6308, Aphanothece halophytica, Microcystis aeruginosa, Gloeobacter violaceus, Plectonema boryanum, Anabaena cylindrica, A. flos-aquae, A. variabilis, Nostoc 6720, Nostoc canina, N. commune, N. muscorum, Nostoc 1453/28, Cylindrospermum majus, Scytonema javanicum, and Chlorogloeopsis fritschii.
[f] nt,

sen et al., 1979; Zurawski et al., 1981; Shinozaki and Sugiura, 1982) and for the S subunit from spinach and pea (Martin, 1979; Bedbrook et al , 1980). A high degree of primary structure homology occurs between the different higher plant L subunits, for example, 90% between tobacco, maize, and spinach, deduced from DNA sequencing (Shinozaki and Sugiura, 1982).

Reichelt and Delaney (1982) have used a cloned L subunit gene from spinach to hybridize with DNA from the cyanobacterium *Synechococcus* 6301. Nucleotide sequencing of the *Synechococcus* L subunit DNA thus obtained yielded an amino acid sequence which shows about 80% homology to the amino acid sequences of the maize and spinach L subunits. Further examples of homology between cyanobacterial and chloroplast L subunits have been provided by DNA hybridization studies with cloned plant subunit probes. Haselkorn et al. (1983) have found hybridization between *Anabaena* 7120 DNA and probes containing the complete maize L subunit gene and a fragment of the corresponding gene from the green alga *Chlamydomonas reinhardtii*. We have obtained similar results using DNA from the cyanobacteria *Microcystis aeruginosa, Nostoc* 6720, *Aphanothece halophytica*, and *Gloeobacter violaceus* versus the *C. reinhardtii* L subunit probe (S. Heinhorst, G. C. Cannon, and G. A. Codd, unpublished observations). As predicted from amino acid composition differences (Takabe and Akazawa, 1975b), little homology has been found between the amino acid sequences of higher plant L subunits and the 71% of the *Rhodospirillum rubrum* subunit residues so far placed in sequence (Hartman et al., 1982). It is of considerable interest that amino acid sequence homology does exist, however, between plant and cyanobacterial L subunits and that of *R. rubrum* in the region around lysine-175 (Fig. 2). These findings are consistent with the location of lysine-175 in the domain of the catalytic site and suggest that such functionally important regions of the L subunits have been conserved in evolution.

Activation and Catalysis

The enhancement of RuBisCO activity by preincubation with CO_2 and Mg^{2+} (Pon et al., 1963) has been studied by kinetic, physical, and chemical analogue methods, and it is now well established that the enzyme can exist in both an inactive and an active form (Miziorko and Mildvan, 1974; Lorimer et al., 1976, 1977; Badger and Lorimer, 1976; Laing and Christeller, 1976). The principles of RuBisCO activation, though mainly studied with the higher plant and *R. rubrum* enzymes (Lorimer, 1981a), appear to be characteristic of all RuBisCOs. Activation is an ordered, reversible process involving the slow binding of an activating CO_2 molecule, $^ACO_2$, followed by the rapid addition of Mg^{2+}:

$$\text{RuBisCO} + {}^A CO_2 \underset{}{\overset{\text{slow}}{\leftrightarrows}} \text{RuBisCo} - {}^A CO_2 + Mg^{2+} \underset{}{\overset{\text{fast}}{\leftrightarrows}} \text{RuBisCO} - {}^A CO_2 - Mg^{2+} \quad (1)$$
$$\text{inactive} \hspace{4cm} \text{active}$$

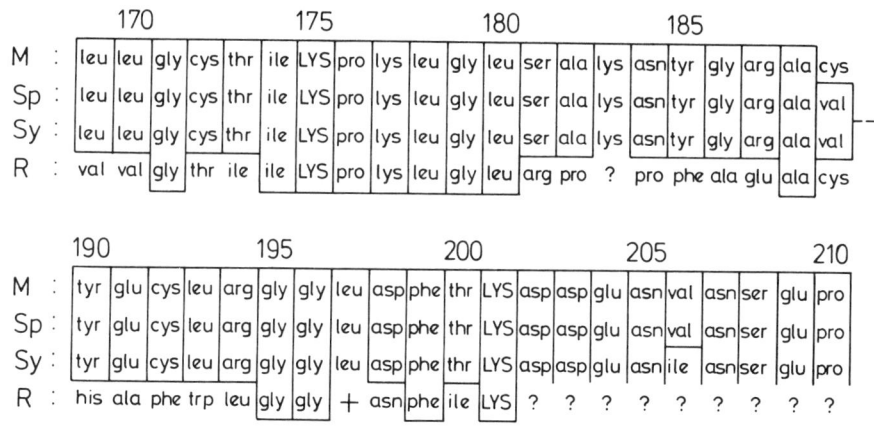

5. CARBON DIOXIDE ASSIMILATION BY AUTOTROPHIC PROKARYOTES

	170				175					180					185				
M :	leu	leu	gly	cys	thr	ile	LYS	pro	lys	leu	gly	leu	ser	ala	lys	asn	tyr	gly	arg	ala	cys
Sp :	leu	leu	gly	cys	thr	ile	LYS	pro	lys	leu	gly	leu	ser	ala	lys	asn	tyr	gly	arg	ala	val
Sy :	leu	leu	gly	cys	thr	ile	LYS	pro	lys	leu	gly	leu	ser	ala	lys	asn	tyr	gly	arg	ala	val
R :	val	val	gly	thr	ile	ile	LYS	pro	lys	leu	gly	leu	arg	pro	?	pro	phe	ala	glu	ala	cys

	190				195					200					205					210
M :	tyr	glu	cys	leu	arg	gly	gly	leu	asp	phe	thr	LYS	asp	asp	glu	asn	val	asn	ser	glu	pro
Sp :	tyr	glu	cys	leu	arg	gly	gly	leu	asp	phe	thr	LYS	asp	asp	glu	asn	val	asn	ser	glu	pro
Sy :	tyr	glu	cys	leu	arg	gly	gly	leu	asp	phe	thr	LYS	asp	asp	glu	asn	ile	asn	ser	glu	pro
R :	his	ala	phe	trp	leu	gly	gly	+	asn	phe	ile	LYS	?	?	?	?	?	?	?	?

Fig. 2. Amino acid sequences surrounding the catalytic site (LYS 175) and the CO_2-binding (activation) site (LYS 201) in the L subunits of RuBisCO from eukaryotic and prokaryotic sources. M, maize (McIntosh et al., 1980); Sp, spinach (Zurawski et al., 1981); Sy, *Synechococcus* 6301 (poster demonstration, Reichelt and Delaney, 1982); R, *Rhodospirillum rubrum* (Hartman et al., 1982). ?, Regions not yet sequenced; +, gap attributed to deletion or insertion. Amino acid sequences deduced from DNA base sequences.

In the case of amino acids 173, 185, 186, 194, and 198, a single base change in the spinach gene sequence could generate the amino acid found at that position in the *R. rubrum* enzyme (Hartman et al., 1982).

Reaction with $^ACO_2$ is the rate-determining step. Activation occurs on the L subunit (Whitman et al., 1979; Lorimer, 1981a,b) and applies equally to the carboxylase and oxygenase reactions. As in catalysis, CO_2 rather than HCO_3^- is involved. The activation mechanism involves the formation of a carbamate residue on the ϵ-amino group of a lysyl residue (Lorimer et al., 1976):

$$\sim lys - NH_3^+ \underset{\pm H^+}{\rightleftarrows} \sim lys - NH_2 + {}^ACO_2 \underset{\pm H^+}{\rightleftarrows} \sim lys - NH - {}^ACOO^- \qquad (2)$$

The equilibrium for the reaction shown in Equation 2 is displaced to the right by Mg^{2+}.

Elegant studies by Lorimer and Miziorko have enabled the activation site to be located (Lorimer, 1981b). These studies involved 2-carboxyarabinitol 1,5-bisphosphate (CABP), which is a "transition state analogue" of the carboxylation intermediate, 3-carboxy-3-ketoarabinitol 1,5-biphosphate (Pierce et al., 1980; Schloss and Lorimer, 1982). CABP binds irreversibly at the catalytic site, locks $^ACO_2$ and Mg^{2+} onto the activation site, and thereby stabilises the otherwise labile carbamate. The use of CABP plus $^{14}CO_2$ as $^ACO_2$, followed by tryptic digestion of the L subunit, sequencing, and location of radioactivity, has identified lysine-201 as the activation site on the spinach enzyme L subunit (Lorimer, 1981b, 1983). The amino acid sequences of the chloroplast and *Synechococcus* L subunits are homologous, and comparison with the completed *R. rubrum* sequence data will be of considerable interest (Fig. 2; Hartman *et al.*, 1982). Studies on the binding of CABP to the RuBisCOs of *Chromatium vinosum* (Brown and Chollet, 1982) and *Thiobacillus neapolitanus* (Cannon and Shively, 1982) suggest that the activation mechanism outlined above may indeed apply to the bacterial enzymes.

Activation may result in a conformational change in the enzyme affecting molecular volume and/or shape. Bowien and Gottschalk (1982) have observed reversible changes in the sedimentation coefficient $S_{20,w}$, of *Alcaligenes eutrophus* RuBisCO from 17.5 S (inactive form) to 14.3 S (fully activated form).

Few bifunctional enzymes are known, and the existence of one such catalyst, RuBisCO, at the centre of autotrophic carbon metabolism, has placed physiological and ecological constraints on this mode of nutrition. Besides catalyzing CO_2 fixation, RuBisCO acts as an oxygenase in a reaction involving the cleavage of RuBP to form one molecule each of 3-phosphoglycerate and phosphoglycollate (Bowes *et al.*, 1971). Phosphoglycollate is then hydrolyzed to glycollate, which in autotrophic microbes can then be metabolized via a C-2 or photorespiratory carbon oxidation cycle, or excreted (Fig. 1). The oxygenase reaction of RuBisCO appears to be an unavoidable feature of the enzyme (Andrews and Lorimer, 1978; Lorimer, 1981a), and all RuBisCOs examined, from the 2L *R. rubrum* enzyme to the higher plant enzymes, act as carboxylases and oxygenases. O_2 acts as a linear competitive inhibitor of the carboxylation reaction, and CO_2 competitively and linearly inhibits oxygenation. Both reactions occur at the same site, in the domain of lysine-176 on the L subunit (see Lorimer and Andrews, 1981). As seen in Fig. 2, the amino acid sequences in this region of the higher plant, *Synechococcus*, and *R. rubrum* L subunits show homology.

CO_2, not HCO_3^-, is the reactive species carboxylated. Besides a requirement in activation, there is some evidence for a role of divalent metal ions in carboxylase and oxygenase catalysis. Mg^{2+} can be replaced by Mn^{2+} and Co^{2+}, and interesting reports have appeared on the different effects of these cations on carboxylation and oxygenation by the higher plant (Christeller and Laing, 1979; Jordan and Ogren, 1981) and *R. rubrum* enzymes (Robison *et al.*, 1979). The main response to Mg^{2+} substitution was to increase enzyme affinity for O_2. It is not yet clear, however, whether the same metal ion plays a role in activation and catalysis. It is beyond the scope of this chapter to consider further the mechanisms of catalysis, but an atomic audit of the carboxylation and oxygenation reactions is available (Lorimer, 1981a).

Although a detailed structural picture of the eukaryotic RuBisCO S subunit is emerging, the function of both the plant and bacterial S subunits is essentially unknown. The S subunits are not essential for activation, or catalysis, since they are absent from *R. rubrum* RuBisCO, which requires CO_2 and Mg^{2+} for the activation of carboxylation and oxygenation (Whitman *et al.*, 1979). Antisera to the S subunits of the enzymes from *Ch. vinosum* and *A. eutrophus* cross-react with the respective native RuBisCOs but do not inhibit activities (Akazawa *et al.*, 1978; Bowien and Mayer, 1978). These data concur with the location of the activation and catalytic sites on the L subunit, but the possibility remains that the antibodies raised against the S subunits are not against antigenic determinants in functional regions.

Comparisons of the activation and catalytic characteristics of the 8L8S enzyme before and after S subunit depletion and comparative studies of the 8L8S and 6L enzymes present in *Rhodopseudomonas sphaeroides* have provided some indications of the roles of the S subunit. A possible role of the S subunit in the Mg^{2+} effect on RuBisCO was indicated: The shift in the 8L8S carboxylase pH optimum from 9.0 to about 7.5 observed when Mg^{2+} levels in the assay were increased from 0 to 10 mM did not occur when tested with the L8 oligomer or when the native enzyme was incubated with S subunit antiserum (Nishimura and Akazawa, 1973; Akazawa, 1979). Gibson and Tabita (1979) have reported that (1) the *Rps. sphaeroides* S subunit enables Ni^{2+} and Co^{2+} ions to substitute for Mg^{2+} or Mn^{2+} in supporting activation and/or catalysis, and that the presence of S subunits (2) accelerates the rate of Mg^{2+} plus high HCO_3^--induced activation, (3) inhibits this activation if RuBP is present, and (4) causes higher affinity for CO_2. These changes may be caused by a conformational change provided by the association of S subunits with the L subunit oligomer (Gibson and Tabita, 1979). Small increases in carboxylase activity have been found after adding S subunits to spinach RuBisCO L subunit oligomers (Nishimura and Akazawa, 1974; Kobayashi *et al.*, 1979), two- to threefold increases have been measured after the addition of S subunits to the S subunit-depleted *Synechococcus* enzyme (Andrews and Abel, 1981), and large increases in carboxylation rates occur when a 14,000-dalton peptide (the presumed S subunit) is added to the *Aphanothece halophytica* L subunit catalytic core (Asami *et al.*, 1983). This peptide, which readily associates with the *Aph. halophytica* L subunit core, markedly enhances carboxylation V_{max}, but does not affect the K_m values for CO_2 or RuBP (Table 4). The effect of S subunits on the activation of the cyanobacterial enzyme is unknown but the data so far suggest that the cyanobacterial S subunit, though incapable of catalysis per se, may have a role in catalysis (Andrews and Abel, 1981; Asami *et al.*, 1983). The presence of S subunits in almost all bacterial RuBisCOs examined and their ubiquitous presence in chloroplast enzymes certainly implies function(s) for these peptides. Recognition of these functions may eventually be aided by the considerable efforts directed at their genetic analysis in higher plants (Ellis, 1981; Smith and Ellis, 1981).

Table 4. *Effect of increasing additions of the 14,000-dalton peptide on catalytic reactions of the large subunit core of Aphanothece halophytica RuBisCO*[a]

14,000-dalton peptide (μM)	Experiment I K_m (NaHCO$_3$) (μM)	V_{max} (nmol · mg^{-1} · min^{-1})	Experiment II K_m (RuBP) (μM)	V_{max} (nmol · mg^{-1} · min^{-1})
0.15	21	98	25	41
0.40	21	151	25	103
0.80	21	222	25	137

[a] Increasing amounts of the 14,000-dalton peptide were incubated with 27.3 μg ($\simeq$0.62 μM) of the *A. halophytica* RuBisCO L subunit catalytic core, and activation was performed with HCO$_3^-$ and Mg^{2+}. NaHCO$_3^-$ and RuBP concentrations for catalysis were varied, the carboxylation reaction being started by the addition of RuBP. For full details, see Asami *et al.* (1983).

Regulatory Properties of RuBisCO

The discovery that RuBisCO can exist *in vitro* in both an inactive and an active form has helped to clarify a long-standing paradox in plant photosynthesis: that the K_m (CO$_2$) for carboxylation *in vitro* (ca. 10–30 mM CO$_2$ + HCO$_3^-$) was much higher than the estimated leaf CO$_2$ concentration, which is about equal to that in air (0.03% CO$_2$ $\simeq$ 10 μM), and that *in vitro* RuBisCO specific activities were insufficient to account for *in vivo* photosynthesis (Lilley and Walker, 1979; Akazawa, 1979). It is now clear that RuBisCO studies before the mid-1970s often involved assay of the inactive enzyme, and that when activated *in vitro,* the affinities and velocities of carboxylation can adequately account for *in vivo* photosynthesis. However, whether the plant enzyme is fully activated *in vivo* is unclear. Evidence for the presence of active RuBisCO *in vivo* was suggested by the finding of an unstable, high-activity, low-K_m (CO$_2$) form of the enzyme in freshly broken chloroplasts (Jensen and Bahr, 1977). Nevertheless, as argued by Lorimer (1981a), if plant RuBisCO is incubated *in vitro* under conditions thought to exist *in vivo* (10 μM CO$_2$, 5–10 mM Mg^{2+}, pH ~8.0), then the enzyme is essentially inactive. This paradox is even more obvious in the autotrophic prokaryotes. The *in vitro* K_m (CO$_2$) values of several bacterial RuBisCOs are at least one order of magnitude greater that the average plant enzyme values, for example, *Rhodospirillum rubrum,* 1400 μM (Schloss *et al.,* 1979); *Anabaena variabilis,* 293 μM (Badger, 1980); and *Synechococcus* sp., 241 μM (Andrews and Abel, 1981).

Several *in vitro* observations have been suggested to contribute to the maintenance of activated RuBisCO *in vivo* and the subsequent regulation of CO$_2$ fixation. Effects of sugar phosphates, including Calvin cycle intermediates and nucleotides, have been studied for years. Inconsistent reports of stimulation and inhibition of higher plant RuBisCO activity *in vitro* exist, and whether the metabolites exert their effect on activation and/or catalysis is unclear (Lorimer et al., 1978a; Lorimer, 1981a). Inhibition of the carboxylase reaction of several 8L8S prokaryotic enzymes by 6-phosphogluconate has been found when tested against the HCO_3^-, Mg^{2+}-activated form (e.g., Tabita and McFadden, 1972; Codd and Stewart, 1977a; Gibson and Tabita, 1977a; Snead and Shively, 1978; Shively et al., 1978; Purohit and McFadden, 1979; Tabita and Colletti, 1979). On the other hand, 6-phosphogluconate addition to inactive RuBisCO (in the presence of 1 mM bicarbonate) causes activation of the enzyme from several sources, for example, *Pseudomonas oxalaticus* (Lawlis et al., 1978) and three cyanobacteria (Tabita and Colletti, 1979). Similar findings for the 2L *R. rubrum* enzyme *in vitro* with 6-phosphogluconate, fructose 1,6-biphosphate, 2-phosphoglycollate, and NADPH have established the L subunit as the location of such effector binding sites (Whitman et al., 1979).

Studies with higher plant RuBisCO have indicated that 6-phosphogluconate, NADPH, and other effectors interact at a single site on the L subunit, the catalytic binding site for RuBP, and promote their response by stabilizing the inactive or active form of the enzyme (Lorimer, 1981a). However, if phosphogluconate or NADPH actually occupies the RuBP-binding site (Badger et al., 1980), it is difficult to visualise how such effectors would stabilize the active enzyme *and* permit catalysis *in vivo*. Nevertheless, there is indirect evidence for the regulation of bacterial RuBisCO *in vivo* by such effectors. The immediate cessation of CO_2 fixation via the Calvin cycle in unicellular cyanobacteria on transfer from light to dark conditions is accompanied by the appearance of 6-phosphogluconate among the soluble metabolites in ^{14}C and ^{32}P steady state labelling experiments (Pelroy and Bassham, 1972; Pelroy et al., 1976). The 6-phosphogluconate, produced via the oxidative pentose phosphate pathway and active in cyanobacteria in the dark (Stanier and Cohen-Bazire, 1977), could inhibit the activated form of RuBisCO *in vivo*. Tabita and Colletti (1979) have measured *in situ* RuBisCO in three cyanobacteria after permeabilising the cells with toluene to permit entry of RuBP and effectors. If RuBP-dependent $^{14}CO_2$ incorporation by permeabilised cells into acid-stable material is taken as an index of RuBisCO activity, then inhibition of the activated cyanobacterial enzyme by added 6-phosphogluconate, NADPH, fructose 6-phosphate, fructose 1,6-biphosphate, and ATP, and activation of the inactive enzyme, particularly by 6-phosphogluconate, NADPH, fructose phosphates, and adenine nucleotides, can occur *in situ*. Interestingly, the degrees of activation of the inactive RuBisCO *in situ* by 6-phosphogluconate and NADPH were considerably greater than when tested with crude or homogeneous enzyme preparations *in vitro* (Tabita and Colletti, 1979).

As already stated, O_2 is a competitive inhibitor of CO_2 fixation by RuBisCO, and vice versa. As with the enzyme in plants, these features of bacterial RuBisCOs (Bowien et al., 1976; Codd and Stewart, 1977a; McFadden and

Purohit, 1978; Badger, 1980) can perhaps be most clearly seen as having significance *in vivo*. Further consideration of the regulation of bacterial CO_2 fixation by O_2 and of photorespiratory C-2 production and metabolism is presented later in an ecophysiological context.

The Compartmentation of RuBisCO in Prokaryotes. The subcellular environment of RuBisCO in prokaryotes may prove important in influencing the regulation of the enzyme *in vivo*. The discovery of Shively and co-workers (1973) that the polyhedral bodies of *Thiobacillus neapolitanus* contain RuBisCO has provided an impetus for studying the structure and function of these long known but functionally enigmatic bacterial inclusion bodies (Jensen and Bowen, 1961) and the regulation of RuBisCOs in organisms which contain these organelles. The term "carboxysome" has been introduced for such RuBisCO-containing organelles (Shively, 1974). Though few isolations have been made, RuBisCO has been found in the carboxysomes isolated from all of the other sources examined so far, namely, *Anabaena cylindrica* (Codd and Stewart, 1976), *Nitrobacter* spp. (Shively *et al.*, 1977; Bock *et al.*, 1979), *Nitrosomonas* sp. (Harms *et al.*, 1981) and *Chlorogloeopsis fritschii* (Lanaras and Codd, 1981b).

Polyhedral bodies have been found in several, but not all, groups of autotrophic bacteria. Among the chemolithoautotrophs, they are present in some colourless sulphur oxidisers, for example, *Thiobacillus intermedius, T. thioparus, T. thiooxidans*, and, of course, *T. neapolitanus,* but not in others, for example, *Thiobacillus* A2 and *T. novellus* (Shively *et al.*, 1973; Holt *et al.*, 1974). Some nitrite- and ammonia-oxidising bacteria contain carboxysomes, including *Nitrobacter winogradsky, N. agilis,* and a marine *Nitrosomonas* sp. (Pope *et al.*, 1969; Bock *et al.*, 1974; Wullenweber *et al.*, 1977), although others appear to lack the organelles, for example, *Nitrospina gracilis, Nitrosovibrio tenuis,* and *Nitrosococcus mobilis* (Watson and Waterbury, 1971; Harms *et al.*, 1976; Koops *et al.*, 1976). Until relatively recently, no carboxysomes had been reported in hydrogen-oxidising bacteria. However, Kostrikina *et al.* (1981) have found polyhedral, membrane-bound particles, apparently carboxysomes, in *Pseudomonas thermophila* K2, grown on CO_2 and H_2.

Among the phototrophic bacteria, carboxysomes have not been reported in the green sulphur, green nonsulphur, and purple sulphur bacteria, although they may be present in the purple nonsulphur bacterium *Rhodomicrobium vannielii* (Whittenbury and Dow, 1977). Cyanobacteria, in contrast, characteristically contain polyhedral bodies (Fogg *et al.*, 1973; Shively, 1974), although in the filamentous, heterocystous N_2-fixing species, the organelles occur only in the vegetative cells and spores (akinetes), not in the heterocysts (Stewart and Codd, 1975). Polyhedral bodies also occur in the chlorophyll b-containing oxygenic prokaryote *Prochloron,* living in association with colonial didemnid ascidians (Lewin, 1977; Whatley, 1977; Fisher and Trench, 1980; Cox and Dwarte, 1981). Finally, polyhedral inclusions, possibly carboxysomes, occur in the cyanobacteria-like cyanelles present in the apoplastidic unicellular algae *Cyanophora, Glaucosphaera, Glaucocystis,* and *Gloeochaete* (Richardson and Brown, 1970; Kies, 1976; Trench *et al.*, 1980).

Can any indication of carboxysome function, particularly in the regulation of RuBisCO and CO_2 fixation, be obtained from this pattern of carboxysome, or presumed carboxysome, distribution among the autotrophic bacteria? Carboxysomes occur in autotrophic organisms and are not apparent in chemoheterotrophs. They are prokaryotic inclusions and are not seen in eukaryotic phototrophs, although the RuBisCO-containing pyrenoids of some algal chloroplasts (Griffiths, 1980) may be analogous structures. Among the autotrophic bacteria, carboxysomes are apparently confined to groups which fix CO_2 via the Calvin cycle, although, as seen, not all Calvin cycle bacteria contain the inclusions. It is thus clear that some organisms, such as purple sulphur bacteria, fix CO_2 via the Calvin cycle without any involvement of carboxysomes. Other groups, including *Thiobacillus* spp. and the cyanobacteria, which contain carboxysomes, may involve these organelles in CO_2 fixation and phototrophic growth.

The only example of intercellular RuBisCO compartmentation known in prokaryotes is found in the filamentous heterocystous cyanobacteria, where, as with carboxysome distribution (Stewart and Codd, 1975), RuBisCO is present in the photosynthetically competent vegetative cells but is undetectable immunologically in the N_2-fixing heterocysts, which are incapable of photosynthetic CO_2 fixation (Codd and Stewart, 1977c; Stewart, 1980).

Ecophysiological Aspects of Prokaryotic Carbon Dioxide Assimilation

Aspects of CO_2 assimilation and the growth of autotrophic bacteria can be found in several reviews (Kelly, 1971; Smith, 1973; McFadden, 1973, 1978; Schlegel, 1975, 1976; Smith and Hoare, 1977; Tabita, 1981; Bowien and Schlegel, 1981). These reviews encompass a large body of data on CO_2 assimilation by chemolithoautotrophic, photoheterotrophic, and photoautotrophic bacteria. The present account is restricted to recent findings in three "ecophysiological" areas of CO_2 assimilation by these organisms, with particular reference to the roles and regulation of phosphoribulokinase and RuBisCO in the whole cell.

Changes in Calvin Cycle Enzyme Levels and Operation During the Growth of Autotrophic Bacteria

A range of nutritional capabilities has long been recognised among the autotrophic bacteria (Schlegel, 1975; Smith and Hoare, 1977). On the one hand, there are specialist (obligate) autotrophs which utilise CO_2 as the principal carbon source and derive energy from light or by oxidising inorganic compounds.

These organisms are unable to utilise or tolerate organic carbon compounds as the principal carbon and energy sources. On the other hand, there are versatile (facultative) autotrophs, able to grow on organic compounds, which can provide the bulk of cell carbon and the energy for growth (Smith and Hoare, 1977). Other nutritional modes also occur, including photoheterotrophy, in which exogenous organic compounds are used to support growth in the light in place of CO_2 (Stanier and Cohen-Bazire, 1977), and mixotrophy, the ability to use light/inorganic or organic compounds concomitantly as energy and/or principal carbon sources (Bowien and Schlegel, 1981). With the exception of the green photosynthetic bacteria, which fix CO_2 via the reverse tricarboxylic acid cycle (Fuller, 1978), all autotrophs require a functioning Calvin cycle to permit photoautotrophic or chemo(litho)-autotrophic growth. A permanent requirement for Calvin cycle enzymes is thus apparent for growth of specialist autotrophs.

Several factors have been invoked to account for the metabolic basis of specialist autotrophy, including toxicity of organic compounds to some specialist strains, an incomplete tricarboxylic cycle, and a lack of transport mechanisms for potential organic substrates (Smith and Hoare, 1977; Smith, 1981). Besides these, an inability to control enzyme synthesis at the transcriptional level has been proposed (Carr, 1973). This concept was reasonably suggested from earlier findings that the activities of enzymes of amino acid and intermediary carbon metabolism in specialist cyanobacteria were unaffected by changes in nutrients and energy supply. By extrapolation, levels of CO_2-fixing enzymes in specialist autotrophs may be expected to remain constant, regardless of environmental conditions. If permanently expressed in specialist photoautotrophs, for example, phosphoribulokinase and RuBisCO would be metabolically redundant whilst the cells were in a dark environment, but such cells would be able to commence photoautotrophic metabolism without delay upon illumination. The cost of maintaining this state of readiness for photosynthesis over the diurnal cycle is unclear, though some ecological advantage may be conferred. Most cyanobacteria are specialist photoautotrophs (Stanier and Cohen-Bazire, 1977).

However, it is now apparent that levels of CO_2 assimilation enzymes do vary in specialist autotrophs in response to environmental changes. Karagouni and Slater (1979) reported a 15-fold variation in the specific activity of RuBisCO in extracts of the specialist cyanobacterium *Anacystis nidulans* (*Synechococcus*) during chemostat culture under CO_2-limited conditions, enzyme activity *in vitro* increasing with decreasing dilution rate. RuBisCO activity was lower in extracts of light-limited (excess CO_2) cells, and no variation in activity was found with changes in the dilution rate of these cultures. No variation in phosphoribulokinase activity was apparent during light-limited growth, though a doubling in activity occurred with increasing dilution rate during CO_2 limitation. The higher level of RuBisCO activity during CO-limited rather than light-limited growth of *A. nidulans* was paralleled by differences in cellular CO_2 photofixation rates

(Karagouni and Slater, 1979). The chemostat approach has also demonstrated variations in RuBisCO levels and CO_2-fixing capacities in the specialist *Thiobacillus neapolitanus*. As with *A. nidulans*, *T. neapolitanus* RuBisCO activities were maximal in extracts of CO_2-limited cells, being about 3.5–5.5 times higher than in extracts of thiosulphate-limited cells grown at the same dilution rate (Beudeker et al., 1980). Total RuBisCO protein levels, determined immunologically, paralleled the variations in activity. At a fixed dilution rate of $D = 0.07$ hr^{-1}, the enzyme constituted 17% of total protein during CO_2-limited growth, but only 4–5% of total protein during thiosulphate- or nitrogen-limited growth (Beudeker et al., 1981a). However, the maximal CO_2-fixing capacity of *T. neapolitanus* cells was greater during thiosulphate limitation than during CO_2 limitation. CO_2 assimilation by CO_2-limited cells nevertheless occurred via the Calvin cycle (Beudeker et al., 1981a), and the data together indicate that RuBisCO activity per se was not rate-limiting for maximal CO_2 fixation by the CO_2-limited cultures. Besides variations in total RuBisCO levels, changes in enzyme distribution between cytoplasm and carboxysomes were found, with more of the enzyme being carboxysomal during CO_2-limited growth. Thus, though specialist autotrophs apparently retain the enzymes of CO_2 assimilation during wide fluctuations in environmental conditions, it is seen that changes in the complement of these enzymes do occur. CO_2 availability during growth can clearly influence RuBisCO levels in *A. nidulans* and *T. neapolitanus*, and the synthesis of more RuBisCO by these organisms may contribute to their ability to scavenge CO_2 for growth in CO_2-depleted environments. No significant difference in RuBisCO activity was found in extracts of another specialist cyanobacterium, *Anabaena variabilis*, after growth on low CO_2 (air) compared to high CO_2 (5% CO_2 in air) (Kaplan et al., 1980), though this organism was grown in batch culture, and whether CO_2 was the limiting factor throughout growth is not apparent. It is clear, nevertheless, that the CO_2 concentration in the batch growth medium influences the apparent photosynthetic affinity of *Ana. variabilis* for CO_2, although this is primarily ascribed to a CO_2-concentrating mechanism (Kaplan et al., 1980).

Changes in the capacity for CO_2 assimilation, and of the levels of enzymes and other energy conversion components with varying environmental conditions, are documented for several versatile prokaryotes (McFadden, 1973; Schlegel, 1975; Smith and Hoare, 1977; Bowien and Schlegel, 1981), although the mechanisms involved are still little understood. RuBisCO can account for 8% of total protein in CO_2/H_2-grown *Alicaligenes eutrophus* (Leadbeater et al., 1982), and the ability of versatile strains such as *A. eutrophus* to decrease RuBisCO synthesis when usable organic carbon compounds are available is seen as an important strategy in the protein economy of the cell. The degree of repression of RuBisCO formation in individual versatile autotrophs depends on the particular organic compounds provided, and is almost complete in *A. eutrophus* when supplied with pyruvate, acetate, or lactate. Intermediate RuBisCO levels have been found in cells grown in the presence of fructose, alanine, citrate, or gluconate, whereas high activities persist in cells grown with glycerol (Bowien and Schlegel, 1981; Friedrich *et al.*, 1981; Leadbeater *et al.*, 1982). The regulation of RuBisCO in hydrogen bacteria is considered further by Schlegel, Chapter 7, this volume. No RuBisCO activity was detected in extracts of *Thiobacillus intermedius* when the CO_2-thiosulphate medium was supplemented with yeast extract, though utilisation of exogenous glutamate did not reduce RuBisCO activities from levels found in CO_2-thiosulphate-supported cultures (Purohit *et al.*, 1976).

Earlier studies on CO_2-assimilating enzymes in versatile cyanobacteria have indicated little if any repression and derepression of RuBisCO formation. No significant differences in RuBisCO activity were apparent in soluble extracts of chemoheterotrophic, photoheterotrophic, or photoautotrophic cultures of *Chlorogloeopsis fritschii* or *Aphanocapsa* 6714 (Evans and Carr, 1975; Joset-Espardellier *et al.*, 1978). Indeed, relatively recent studies of growth rates, growth yields, pigment levels, and respiratory and photosynthetic O_2 evolution capacities in *Aphanocapsa* have indicated that modulation of the photoautotrophic capacity in this organism is achieved in part through changes in the photosynthetic apparatus rather than in the enzymes of autotrophy (Der-Vartanian *et al.*, 1981). However, Raboy *et al.* (1976) found that RuBisCO activities in photoautotrophic cell extracts of *Plectonema boryanum* were double those measured in dark-grown cells and correlated with changes in the *in vivo* CO_2 photoassimilation capacity.

The compartmentation of RuBisCO between the cytoplasm and carboxysomes of cyanobacteria (Codd and Stewart, 1976; Lanaras and Codd, 1981a,b) requires consideration of particulate RuBisCO in estimating cellular RuBisCO levels *in vitro*. We have measured RuBisCO activities and enzyme protein concentrations in extracts of *Nostoc* 6720, which grows readily on glucose in the dark (Table 5). Similar levels of RuBisCO protein were present in midexponential phase cultures of photoautotrophic and chemoheterotrophic cells cultured under N_2-fixing conditions, but growth in excess nitrate resulted in over three times greater RuBisCO levels in the light than in the dark. Though little or no changes in total RuBisCO protein per cell have been found, we have noted a marked difference in the subcellular distribution of the enzyme, which is mainly soluble in the light-grown log-phase extracts, with a shift to the carboxysomes in dark-grown cell extracts (Table 5). Light-dark transition experiments with *Nostoc* 6720 have also indicated a redistribution of RuBisCO from cytoplasm to carboxysomes rather than changes in the overall RiBisCO cell complement (Leadbeater and Codd, 1984). Clearly, the subcellular compartmentation of RuBisCO in carboxysome-containing prokaryotes introduces additional factors in the regulation of autotrophy, and perhaps ultrastructural observations on the cellular abundance of car-

Table 5. *RuBisCO protein levels in extracts of photoautotrophic and chemoheterotrophic cultures of Nostoc 6720*[a]

Growth conditions[b]	Total protein mg · dm^{-3} culture	Total RuBisCO protein mg · dm^{-3} culture[c]	Total RuBisCO protein as % of total protein	Enzyme location[d]	Subcellular distribution of RuBisCO protein (%)
Light (N$_2$)	133.6	0.75	0.56	Soluble	93
				Particulate	7
Dark (N$_2$)	34.51	0.23	0.66	Soluble	51
				Particulate	49
Light (NO$_3$)	162.35	1.40	0.86	Soluble	70
				Particulate	30
Dark (NO$_3$)	30.54	0.08	0.26	Soluble	31
				Particulate	69

[a] With acknowledgements to Leadbeater (1981).
[b] Batch culture, midexponential phase cultures harvested. Dark cultures grown on 0.5% (w/v) glucose.
[c] RuBisCO protein measured by rocket immunoelectrophoresis (Lanaras and Codd, 1981a).
[d] Soluble enzyme assumed to be cytoplasmic; particulate enzyme assumed to be carboxysomal (Lanaras and Codd, 1981b).

boxysomes in natural populations and any changes with varying environmental conditions would be useful.

Multiple control of synthesis of the key enzymes of CO$_2$ assimilation would be expected to benefit versatile autotrophs in natural environments. The repression and derepression of RuBisCO formation in response to competent external organic carbon compounds and CO$_2$ may be exerted via changes in the intracellular pool sizes of signal carbon compounds whose identity is yet unknown (Dijkhuizen *et al.*, 1978; Friedrich, 1982). Indeed, Sarles and Tabita (1983) have found that RuBisCO may constitute up to 50% of the total protein in *Rhodospirillum rubrum* when cultured in mineral medium under low CO$_2$. Autotrophic growth in the absence of organic carbon sources cannot occur unless energy can be obtained from light or inorganic compounds, and *Alicaligenes eutrophus*, though carbon limited, does not synthesise RuBisCO unless excess reducing equivalents are available for energy generation (Friedrich, 1982).

Phosphoribulokinase and RuBisCO are contiguous enzymes in the Calvin cycle (Fig. 1), and there is evidence for the coordinated synthesis of the enzymes in versatile purple nonsulphur bacteria (Tabita, 1981). Phosphoribulokinase-specific activities are 2.3 and 18.8 times higher in N$_2$-fixing and nitrate-utilising, exponential phase photoautotrophic *Nostoc* 6720 extracts, respectively, than in corresponding extracts from dark (glucose)-grown cells (Leadbeater and Codd, 1984), though whether these differences are accompanied by changes in enzyme protein is not known. Leadbeater et al. (1982) found that changes in *A. eutrophus* RuBisCO protein levels with the mode of nutrition are accompanied by changes in phosphoribulokinase protein. It will be of interest to know whether or not the structural genes for phosphoribulokinase and RuBisCO in such organisms are associated in one operon.

RuBisCO and Glycollate Production by Autotrophic Prokaryotes

The Warburg effect, the inhibition of photosynthesis by O_2, as first studied in *Chlorella* (Warburg, 1920), can now be partly accounted for by the competitive inhibition of the carboxylase reaction of RuBisCO by O_2 (Lorimer, 1981a). The oxygenation of RuBP to form glycollate via phosphoglycollate (Fig. 1), and the subsequent metabolism of glycollate through the C-2 carbon oxidation cycle, further account for the photorespiration and loss of net productivity observed in terrestrial C-3 plants and several green algae.

The inhibition of photosynthetic CO_2 assimilation by O_2 has been found in laboratory studies on purple sulphur and nonsulphur bacteria (Codd and Smith, 1974; Takabe and Akazawa, 1977; Khanna et al., 1981) and on cyanobacteria (Lex et al., 1972; Glover and Morris, 1981). Glycollate production and excretion have been found with a wide range of autotrophic bacteria, including the O_2-sensitive purple bacteria *Rhodospirillum rubrum* (Codd and Smith, 1974; Takabe et al., 1979; Størro and McFadden, 1981), *Rho. vannielii* (Codd and Turnbull, 1975), *Rhodopseudomonas palustris* and *Rps. acidophila* (Codd et al., 1976b), and *Chromatium vinosum* (Takabe and Akazawa, 1977). Glycollate excretion has also been found using the cyanobacteria. *A. nidulans* and *Plectonema boryanum* (Döhler and Braun, 1971; Han and Eley, 1973), *Oscillatoria* and *Anabaena flos-aquae* (Cheng et al., 1972), and *A. cylindrica, Microcystis aeruginosa,* and *Aphanocapsa* 6308 (Codd et al., 1976b). In all cases with the phototrophs, glycollate formation and excretion are light dependent and enhanced by high O_2. Chemolithoautotrophic glycollate production and excretion has been found using *Alcaligenes eutrophus* (Codd et al., 1976a; King and Andersen, 1980) and *Thiobacillus neapolitanus* (Cohen et al., 1979). Under low O_2 and/or elevated CO_2, glycollate production and excretion by these organisms account for 1% or less of total CO_2 fixation, whereas under limiting or low CO_2 and elevated O_2, between one-quarter and one-half of the CO_2 fixed can be excreted as glycollate (Cohen et al., 1979; King and Andersen, 1980).

Abundant circumstantial evidence indicates that, as in plants and algae, the photorespiratory, or C-2 chemorespiratory, production of glycollate by autotrophic prokaryotes occurs via RuBP oxygenation by RuBisCO, namely, the bifunctional nature of all RuBisCOs (Lorimer, 1981a); glycollate production by all autotrophic Calvin cycle bacteria examined and the similarity of response to

5. CARBON DIOXIDE ASSIMILATION BY AUTOTROPHIC PROKARYOTES

relative CO_2 and O_2 levels; and glycollate production by autotrophic but not heterotrophic cultures of the versatile *A. eutrophus* (Codd et al., 1976a). The enrichment of the glycollate excreted by *Chromatium* with ^{18}O, after supplying photosynthetic cells with $^{18}O_2$, indicated that almost all of the glycollate created by this organism, as with *Chlorella* (Lorimer et al., 1978b), was produced by a mechanism involving the incorporation of an atom of molecular oxygen. King and Andersen (1980) have elegantly shown that glycollate production and CO_2 fixation are competitive processes in *A. eutrophus*. It is clear that several mechanisms are involved in the inhibition of photosynthesis by O_2 in photosynthetic bacteria and cyanobacteria. Photooxidative damage, involving oxygen radicals and hydrogen peroxide, can occur to the photosynthetic electron transport chain, to pigments, and to RuBisCO itself (Codd, 1982). Indeed, photosynthesis by *Chlorobium thiosulfatophilum*, which does not fix CO_2 via the Calvin cycle, is also sensitive to inhibition by O_2, but in this case, no glycollate production was detected under photoinhibitory conditions (Takabe and Akazawa, 1977). However, we may generalise that Calvin cycle prokaryotes, if challenged in the laboratory with high O_2/low CO_2 conditions in the presence of excess light or inorganic energy sources, can show decreased CO_2 assimilation and increased glycollate production.

The glycollate formed can undergo alternative changes in autotrophic bacteria. Glycollate oxidation to glyoxylate, and thence via glycine–serine or tartronic semialdehyde to glycerate, and 3-phosphoglycerate, or via acetylation to malate, returns organic carbon to central pools, although CO_2 is lost during glycine–serine conversion and in tartronic semialdehyde production from glyoxylate (Asami et al., 1977; Codd and Stewart, 1973; Beudeker et al., 1981b; Bowien and Schlegel, 1981). If prokaryotic glycollate production rates exceed the capacity for oxidation by glycollate dehydrogenase, if the latter is inhibited or enzyme levels are repressed, then, as in algae, glycollate excretion occurs (Codd and Smith, 1974; Codd and Turnbull, 1975; Codd et al., 1976a,b; King and Andersen, 1980). Although exogenous glycollate can support the growth of *Alcaligenes eutrophus* as a carbon and energy source (Friedrich et al., 1979; King and Andersen, 1980), it is not a significant electron donor or carbon source for purple nonsulphur bacteria (Trüper and Pfennig, 1978). Exogenous glycollate can be assimilated by cyanobacteria (Codd and Stewart, 1973), but the compound can support neither chemoheterotrophic nor photoheterotrophic growth of these organisms (Codd and Stewart, 1974; Stanier and Cohen-Bazire, 1977). It thus seems likely that apart from some versatile chemolithoautotrophs, glycollate, once excreted, represents an essentially irreversible loss of organic carbon from autotrophic prokaryotes.

Glycollate is often a major constituent of the extracellular products of photosynthesis released by phytoplankton into natural waters, and marine and freshwaters have been found to contain glycollate in field studies which have ranged from the tropics to the Antarctic (Fogg, 1975, 1983). Eukaryotic microalgae largely have been implicated in this release of glycollate, although cyanobacteria are involved, and it is possible that glycollate excretion by other groups of autotrophic prokaryotes may occur in natural waters, if the cells are challenged with conditions favouring RuBP oxygenation and glycollate production. Extracellular products may comprise as little as 5% of total carbon fixation in eutrophic waters, but can account for 40% in oligotrophic waters. As pointed out by Fogg (1983), since nutrient-poor waters make up about 90% of the oceans, gross photosynthesis in these environments is seriously underestimated unless the extracellular products (including glycollate) are taken into account. The utilisation of glycollate by marine microorganisms has been studied in several locations. Rapid utilisation of undefined phytoplankton extracellular products by chemoheterotrophic bacteria can occur, and the biomass produced can approach about one-quarter of primary production. Glycollate itself, though often a principal component, does not appear to serve as an adequate carbon source for growth, but is used to provide energy to support the uptake and assimilation of other organic compounds (Wright and Shah, 1975, 1977; Fogg, 1983).

Photorespiration and Intracellular CO_2 Concentration in Cyanobacteria

Varying estimations of cyanobacterial photorespiration, the O_2-sensitive loss of CO_2 during photosynthesis, have been made over recent years. Measurements of CO_2 exchange by infrared gas analysis (IRGA) using cyanobacteria and several microalgae on filter paper gave CO_2 compensation points of <10 µlitre litre^{-1}, characteristic of C_4 plants which do not exhibit photorespiration (Lloyd et al., 1977). The low CO_2 compensation points and lack of effect of increasing O_2 on apparent photosynthesis, whether cells were previously grown in low CO_2 (air) or high CO_2 (1.5 or 5% CO_2 in air), led Lloyd et al. (1979) to infer that these organisms did not exhibit photorespiration. The IRGA method requires rapid equilibration of gaseous and dissolved CO_2, and is suitable for use at acid pH levels (4–5) but less so under the alkaline conditions necessary for cyanobacterial growth. Colman and colleagues have applied a method involving total inorganic carbon acidification to CO_2, conversion to methane, and measurement by gas chromatography. Initial studies using this approach again generally showed low cyanobacterial CO_2 compensation points at pH 5.0–6.0 and indicated that the compensation point decreased further at pH 7.5–8.0 (Birmingham and Colman, 1979). Although low (<1 µlitre litre^{-1}), the alkaline CO_2 compensation points for *Ana. flos-aquae* and *Anacystis nidulans* showed a 50% increase with a change in O_2 level from 2 to 21% (Birmingham et al., 1982). Two other cyanobacteria, *Coccochloris peniocystis* and *Phormidium molle,* did not show this response. However, ^{14}C counting, combined with gas chromatographic total dissolved inorganic carbon (DIC) measurements, after cells were supplied with $NaH^{14}CO_3$, provided estimations of true and apparent photosynthesis and photorespiration. The true photosynthetic rate was calculated from the initial rate of decrease of ^{14}C activity in the medium. The rate of apparent photosynthesis was estimated from the rate of DIC depletion over the same initial period of photosynthesis (2–4 min), and the rate of photorespiration (CO_2 evolution) was calculated from the difference between true and apparent rates. Photosynthetic rates were low due to the use of subsaturating initial DIC concentrations, but under 21% O_2 and alkaline conditions, rates of photorespiration amounting to 18% and 10% of true photosynthesis were displayed by *Ana. flosaquae* and *A. nidulans* (Birmingham et al., 1982). These rates decreased to 7 and 5%, respectively, under 2% O_2. Confirming the lack of response of the lower CO_2 compensation points to O_2, rates of photorespiration by *C. peniocystis* and *P. molle* were lower (2–6% of true photosynthesis) and were unaffected by O_2 changes.

The possession of low CO_2-insensitive compensation points by *C. peniocystis* and *P. molle* is characteristic of terrestrial C_4 plants, although $NaH^{14}CO_3$-labelling kinetics studies with *C. peniocystis* (Coleman and Colman, 1980), as with other cyanobacteria (Pelroy and Bassham, 1972; Ihlenfeldt and Gibson, 1975; Créach et al., 1981), have demonstrated that the main pathway of cyanobacterial photosynthesis is via the Calvin cycle. The concept of an intracellular, inorganic carbon transport mechanism in cyanobacteria (and green algae) offers an explanation of this problem (Badger et al., 1977; Kaplan et al., 1980; Miller and Colman, 1980a,b). This transport mechanism results in an increase in intracellular CO_2 concentration in *Anabaena variabilis* (Kaplan et al., 1980) and *C. peniocystis* (Miller and Colman, 1980a,b), giving an increase in apparent photosynthetic affinity for inorganic carbon. The apparent affinity of photosynthesis for inorganic carbon by *Ana. variabilis* and *C. peniocystis* is higher in cells grown in the presence of low CO_2 than in high CO_2-grown cells (0.03 and 5% CO_2 in air, respectively), and the induction of an HCO_3^- transport system by low CO_2 has been proposed to account for this (Marcus et al., 1982; Miller and Coleman, 1980a,b).

An accumulation ratio of about 500 (internal/external inorganic carbon concentration) has been found in low CO_2-grown *Ana. variabilis* (Kaplan et al., 1980), and, as proposed earlier for algae (Raven, 1970), evidence is accumulating for HCO_3^- uptake by *Ana. variabilis* via an active process, probably involving a primary electrogenic pump (Kaplan et al., 1980). Estimations of the kinetics of inorganic carbon transport by high and low CO_2-grown cells have indicated a similar affinity (K_m about 150 μM), although the V_{max} for inorganic carbon transport is about 10 times higher in the low CO_2 cells (Kaplan et al.,

1980). This difference in *Ana. variabilis* may be due to a larger number of active transport sites for bicarbonate in low CO_2-grown cells. Different sensitivities between low and high CO_2-grown *Ana. variabilis* cell walls to lysozyme attack; in the resulting sphaeroplasts to osmotic rupture, and ultrastructural differences as seen by electron microscopy, have indicated that considerable changes can occur to the cell boundaries of this cyanobacterium with varying CO_2 supply. These changes may be related to an increase in the number of active bicarbonate transport sites. Marcus *et al.* (1982) showed that the increased ability of *Ana. variabilis* to accumulate bicarbonate upon adaptation to growth under low CO_2, plus the accompanying changes in lysozyme sensitivity, are prevented by the protein synthesis inhibitors spectinomycin and chloramphenicol, and by the RNA synthesis inhibitor rifampicin. The possiblity of an adaptive inorganic carbon-concentrating mechanism under transcriptional control in cyanobacteria is thus raised, and it will be of interest to see whether peptide mapping of the cell membrane of organisms such as *Ana. variabilis* and *C. peniocystis,* after growth on high and low CO_2, will reveal particular components of the concentrating mechanism.

The operation of the cyanobacterial inorganic CO_2-concentrating mechanism in natural aquatic environments has not been investigated. However, such a mechanism could account for the ability of *Microcystis aeruginosa* populations to perform photosynthesis at rates greater than could be supported by spontaneous bicarbonate dehydration under alkaline conditions (Talling, 1976). An adaptive CO_2-concentrating mechanism, sensitive to external inorganic carbon levels, or a related subcellular signal, would imply that photorespiration and perhaps glycollate release by cyanobacteria containing the mechanism were transient events in natural waters. Photorespiration and glycollate excretion by such populations would tend to occur with a rapid change to photorespiratory conditions until the CO_2-concentrating mechanism was induced/derepressed.

Finally, the possibility that carboxysomes may function in a CO_2-concentrating mechanism in autotrophs has been raised (Beudeker *et al.*, 1981a). If so, then it is unlikely that they do so via a C_4-dicarboxylic acid pathway, as in C_4 plants (Beudeker *et al.*, 1981a). Further studies of the cellular abundance and composition of carboxysomes in different autotrophic prokaryotes are needed to understand the role(s) of these inclusions in bacterial autotrophy.

Acknowledgements

The contributions of several colleagues and former students to these studies on autotrophic prokaryotes are gratefully acknowledged; in particular, Drs. R. F. Beudeker, B. Bowien, C. M. Cook, T. Lanaras, and L. Leadbeater, and Professors J. G. Kuenen and W. D. P. Stewart.

References

Abdelal, A. T. H. and Schlegel, H. G. (1974a). Purification and regulatory properties of phosphoribulokinase from *Hydrogenomonas eutropha* H16. *Biochem. J.* **139**, 481–489.
Abdelal, A. T. H. and Schlegel, H. G. (1974b). Separation of phosphoribulokinase from enzymes of the Calvin cycle in *Hydrogenomonas eutropha* H16. *Arch. Mikrobiol.* **95**, 139–143.
Akazawa, T. (1979). Ribulose-1,5-bisphosphate carboxylase. *Encycl. Plant Physiol., New Ser.* **6**, 208–229.
Akazawa, T., Kondo, H., Shimazue, T., Nishimura, M. and Sugiyama, T. (1972). Further studies on ribulose 1,5-diphosphate carboxylase from *Chromatium* strain D. *Biochemistry* **11**, 1298–1303.
Akazawa, T., Takabe, T., Asami, S. and Kobayashi, H. (1978). Ribulose bisphosphate carboxylases from *Chromatium vinosum* and *Rhodospirillum rubrum* and their role in photosynthetic carbon assimilation. In "Photosynthetic Carbon Assimilation" (Eds. H. W. Siegelman and G. Hind), pp. 209–226. Plenum, New York.
Anderson, L. E. (1973). Dithiothreitol activation of some chloroplast enzymes in extracts of etiolated pea seedlings. *Plant Sci. Lett.* **1**, 331–334.
Anderson, L. E. (1979). Interaction between photochemistry and activity of enzymes. *Encycl. Plant Physiol., New Ser.* **6**, 271–281.
Andrews, T. J. and Abel, K. M. (1981). Kinetics and subunit interactions of ribulose bisphosphate carboxylase-oxygenase from the cyanobacterium *Synechococcus* sp. *J. Biol. Chem.* **256**, 8445–8451.
Andrews, T. J. and Lorimer, G. H. (1978). Photorespiration—still unavoidable? *FEBS Lett.* **90**, 1–9.
Andrews, T. J., Abel, K. M., Menzel, E. and Badger, M. R. (1981). Molecular weight and quaternary structure of ribulose bisphosphate carboxylase from the cyanobacterium, *Synechococcus* sp. *Arch. Microbiol.* **130**, 344–348.
Asami, S.. Takabe, T. and Akazawa, T. (1977). Biosynthetic mechanism of glycolate in *Chromatium*. IV. Glycolate-glycine transformation. *Plant Cell Physiol.* **18**, 149–159.
Asami, S., Takabe, T., Akazawa, T. and Codd, G. A. (1983). Ribulose 1,5-bisphosphate carboxylase from the cyanobacterium *Aphanothece halophytica*. *Arch. Biochem. Biophys.* (in press).
Badger, M. R. (1980). Kinetic properties of ribulose 1,5-bisphosphate carboxylase/oxygenase from *Anabaena variabilis*. *Arch. Biochem. Biophys.* **201**. 247–254.
Badger, M. R. and Lorimer, G. H. (1976). Activation of ribulose-1,5-bisphosphate oxygenase. The role of Mg^{2+}, CO_2 and pH. *Arch. Biochem. Biophys.* **175**, 723–729.
Badger, M. R., Kaplan, A. and Berry, J. A. (1977). A mechanism for concentrating CO_2 in *Chlamydomonas reinhardtii* and *Anabaena variabilis* and its role in photosynthetic CO_2 fixation. *Year Book—Carnegie Inst. Washington* **77**, 251–261.
Badger, M. R., Andrews, T. J., Canvin, D. T. and Lorimer, G. H. (1980). Interactions of hydrogen peroxide with ribulose bisphosphate carboxylase/oxygenase. *J. Biol. Chem.* **255**, 7870–7875.
Baker, T. S., Eisenberg, D. and Eiserling, F. (1977a). Ribulose bisphosphate carboxylase: a two-layered square-shaped molecule of symmetry 422. *Science* **196**, 293–295.
Baker, T. S., Suh, S. W. and Eisenberg, D. (1977b). Structure of ribulose-1,5-bisphosphate carboxylase-oxygenase: form III crystals. *Proc. Natl. Acad. Sci. U.S.A.* **74**, 1037–1041.

Bassham, J. A. (1979). The reductive pentose phosphate cycle. *Encycl. Plant Physiol. New Ser.* **6,** 9–30.

Bassham, J. A. and Calvin, M. (1957). "The Path of Carbon in Photosynthesis." Prentice-Hall, Englewood Cliffs, New Jersey.

Bedbrook, J. R., Smith, S. M. and Ellis, J. R. (1980). Molecular cloning and sequencing of cDNA encoding the precursor to the small subunit of chloroplast ribulose-1,5-bisphosphate carboxylase. *Nature (London)* **237,** 692–697.

Beudeker, R. F. and Kuenen, J. G. (1981). Carboxysomes: "Calvinosomes"? *FEBS Lett.* **131,** 269–274.

Beudeker, R. F., Cannon, G. C., Kuenen, J. G. and Shively, J. M. (1980). Relations between D-ribulose-1,5-bisphosphate carboxylase, carboxysomes and CO_2 fixing capacity in the obligate chemolithotroph *Thiobacillus neapolitanus* grown under different limitations in the chemostat. *Arch. Microbiol.* **124,** 185–189.

Beudeker, R. F., Codd, G. A. and Kuenen, J. G. (1981a). Quantification and intracellular distribution of ribulose-1,5-bisphosphate carboxylase in *Thiobacillus neapolitanus*, as related to possible functions of carboxysomes. *Arch. Microbiol.* **129,** 361–367.

Beudeker, R. F., Kuenen, J. G. and Codd. G. A. (1981b). Glycollate metabolism in the obligate chemolithotroph *Thiobacillus neapolitanus* grown in continuous culture. *J. Gen. Microbiol.* **126,** 337–346.

Birmingham, B. C. and Colman, B. (1979). Measurement of carbon dioxide compensation points of freshwater algae. *Plant Physiol.* **64,** 892–895.

Birmingham, B. C., Coleman, J. R. and Colman, B. (1982). Measurement of photorespiration in algae. *Plant Physiol.* **69,** 259–262.

Bock, E., Düvel, D. and Peters, K. R. (1974). Charakterisierung eines phagenähnlichen Partikels aus Zellen von *Nitrobacter*. *Arch. Microbiol.* **97,** 115–127.

Bock, E., Cannon, G. C. and Shively, J. M. (1979). Comparison of carboxysomes from various *Nitrobacter* strains. *Abstr. Annu. Meet. Am. Soc. Microbiol.* p. 95.

Bowes, G., Ogren, W. L. and Hageman, R. H. (1971). Phosphoglycolate production catalysed by ribulose diphosphate carboxylase. *Biochem. Biophys. Res. Commun.* **45,** 716–722.

Bowien, B. (1977). D-ribulose 1,5-bisphosphate carboxylase from *Paracoccus denitrificans*. *FEMS Microbiol. Lett.* **2,** 263–266.

Bowien, B. (1984). Structural and regulatory studies on the key enzymes of the Calvin cycle from hydrogen bacteria. *Proc. Int. Symp. Mol. Cell. Regul. Enzyme Act.* (Ed. H. Barth), pp. 255–272. Universität Halle–Wittenberg, DDR.

Bowien, B. and Gottschalk, E. M. (1982). Influence of the activation state on the sedimentation properties of ribulose bisphosphate carboxylase from *Alcaligenes eutrophus*. *J. Biol. Chem.* **257,** 11845–11847.

Bowien, B. and Mayer, F. (1977). Ribulose 1,5-bisphosphate carboxylase from several hydrogen bacteria. *In* "Proceedings of the Second International Symposium on Microbial Growth on C, Compounds," pp. 103–105. USSR Acad. Sci. Pushchino.

Bowien, B. and Mayer, F. (1978). Further studies on the quaternary structure of D-ribulose-1,5-bisphosphate carboxylase from *Alcaligenes eutrophus*. *Eur. J. Biochem.* **88,** 97–107.

Bowien, B. and Schlegel, H. G. (1981). Physiology and biochemistry of aerobic hydrogen-oxidising bacteria. *Annu. Rev. Microbiol.* **35,** 405–452.

Bowien, B., Mayer, F., Codd, G. A. and Schlegel, H. G. (1976). Purification, some properties and quaternary structure of the D-ribulose, 1,5-diphosphate carboxylase of *Alcaligenes eutrophus*. *Arch. Microbiol.* **110,** 157–166.

Bowien, B., Mayer, F., Spiess, E., Pähler, A., Englisch, U. and Saenger, W. (1980). On the structure of crystalline ribulosebisphosphate carboxylase from *Alcaligenes eutrophus. Eur. J. Biochem.* **106**, 405–410.
Bowman, L. H. and Chollet, R. (1980). Presence of two subunit types in ribulose 1,5-bisphosphate carboxylase from *Thiobacillus intermedius. J. Bacteriol.* **141**, 652–657.
Brown, H. M. and Chollet, R. (1982). Isolation of a stable enzyme $^{14}CO_2 \cdot Mg^{2+} \cdot$ carboxyarabinitol bisphosphate complex with ribulose bisphosphate carboxylase from *Chromatium vinosum. J. Bacteriol.* **149**, 1159–1161.
Buchanan, B. B. and Sirevåg, R. (1976). Ribulose 1,5-diphosphate carboxylase and *Chlorobium thiosulfatophilum. Arch. Microbiol.* **109**, 15–19.
Cannon, G. C. and Shively, J. M. (1982). The carboxysomes of *Thiobacillus neapolitanus*. *In* "Abstracts of the IVth International Symposium on Photosynthetic Prokaryotes" (Ed. G. Stanier), C28. Institut Pasteur, Paris.
Carr, N. G. (1973). Metabolic control and autotrophic physiology. *In* "The Biology of Blue-Green Algae" (Eds. N. G. Carr and B. A. Whitton), pp. 39–66. Blackwell, Oxford.
Charles, A. M. and White, B. (1976). Ribulose bisphosphate carboxylase from *Thiobacillus* A2. Its purification and properties. *Arch. Microbiol.* **108**, 195–202.
Cheng, K. H., Miller, A. G. and Colman, B. (1972). An investigation of glycolate excretion by two species of blue-green algae. *Planta* **103**, 110–116.
Christeller, J. T. and Laing, W. A. (1979). Effects of manganese ions and magnesium ions on the activity of soya-bean ribulose bisphosphate carboxylase/oxygenase. *Biochem. J.* **183**, 747–750.
Codd, G. A. (1982). Photoinhibition of photosynthesis and photoinactivation of ribulose bisphosphate carboxylase in algae and cyanobacteria. *In* "Plants and the Daylight Spectrum" (Ed. H. Smith), pp. 315–337. Academic Press, New York.
Codd, G. A. and Smith, B. M. (1974). Glycollate production and excretion by the purple photosynthetic bacterium *Rhodospirillum rubrum. FEBS Lett.* **48**, 105–108.
Codd, G. A. and Stewart, W. D. P. (1973). Pathways of glycollate metabolism in the blue-green alga *Anabaena cylindrica. Arch. Mikrobiol.* **94**, 11–28.
Codd, G. A. and Stewart, W. D. P. (1974). Glycollate oxidation and utilization by *Anabaena cylindrica. Plant Sci. Lett.* **3**, 199–205.
Codd, G. A. and Stewart, W. D. P. (1976). Polyhedral bodies and ribulose 1,5-diphosphate carboxylase of the blue-green alga *Anabaena cylindrica. Planta* **130**, 323–326.
Codd, G. A. and Stewart, W. D. P. (1977a). D-ribulose 1,5-diphosphate carboxylase from the blue-green alga *Aphanocapsa* 6308. *Arch. Microbiol.* **113**, 105–110.
Codd, G. A. and Stewart, W. D. P. (1977b). Quaternary structure of the D-ribulose 1,5-diphosphate carboxylase from the cyanelles of *Cyanophora paradoxa. FEMS Microbiol. Lett.* **1**, 35–38.
Codd, G. A. and Stewart, W. D. P. (1977c). Ribulose 1,5-diphosphate carboxylase in heterocysts and vegatative cells of *Anabaena cylindrica. FEMS Microbiol. Lett.* **2**, 247–249.
Codd, G. A. and Turnbull, F. (1975). Enzymes of glycollate formation and oxidation in two members of the *Rhodospirillaceae* (Purple Non-Sulphur Bacteria). *Arch. Microbiol.* **104**, 155–158.
Codd, G. A., Bowien, B. and Schlegel, H. G. (1976a). Glycollate production and excretion by *Alcaligenes eutrophus. Arch. Microbiol.* **110**, 167–171.
Codd, G. A., Sallal, A-K.J. and Stewart, R. (1976b). Glycollate production, excretion and metabolism by photosynthetic prokaryotes. *In* "Proceedings of the Second Interna-

tional Symposium on Photosynthetic Prokaryotes'' (Eds. G. A. Codd and W. D. P. Stewart), pp. 193–195. University of Dundee.

Codd, G. A., Cook, C. M. and Stewart, W. D. P. (1979). Purification and subunit structure of D-ribulose 1,5-bisphosphate carboxylase from the cyanobacterium *Aphanothece halophytica*. *FEMS Microbiol. Lett.* **6,** 81–86.

Codd, G. A., Okabe, K. and Stewart, W. D. P. (1980). Cellular compartmentation of photosynthetic and photorespiratory enzymes in the heterocystous cyanobacterium *Anabaena cylindrica*. *Arch. Microbiol.* **124,** 149–154.

Cohen, Y., de Jonge, I. and Kuenen, J. G. (1979). Excretion of glycollate by *Thiobacillus neapolitanus* grown in continuous culture. *Arch. Microbiol.* **122,** 189–194.

Colby, J., Dalton, H. and Whittenbury, R. (1979). Biological and biochemical aspects of microbial growth on C_1 compounds. *Annu. Rev. Microbiol.* **33,** 481–517.

Coleman, J. R. and Colman, B. (1980). Demonstration of C_3-photosynthesis in a blue-green alga, *Coccochloris peniocystis*. *Planta* **149,** 318–320.

Cook, A. M. and Schlegel, H. G. (1978). Metabolite concentrations in *Alcaligenes eutrophus* H16 and a mutant defective in poly-β-hydroxybutyrate synthesis. *Arch. Microbiol.* **119,** 231–235.

Cook, C. M. (1980). The Characterization of Ribulose Bisphosphate Carboxylase from Cyanobacteria. Ph.D. Thesis, University of Dundee.

Cox, G. and Dwarte, D. M. (1981). Freeze-etch ultrastructure of a *Prochloron* species—the symbiont of *Didemnum molle*. *New Phytol.* **88,** 427–438.

Créach, E., Codd, G. A. and Stewart, W. D. P. (1981). Primary products of photosynthesis and studies on carboxylating enzymes in the filamentous cyanobacterium *Anabaena cylindrica*. *In* "Photosynthesis IV. Regulation of Carbon Metabolism'' (Ed. G. Akoyunoglou), pp. 49–56. Balaban Internation Science Services, Philadelphia, Pennsylvania.

Der-Vartanian, M., Joset-Espardellier, F. and Astier, C. (1981). Contributions of respiratory and photosynthetic pathways during growth of a facultative photoautotrophic cyanobacterium, *Aphanocapsa* 6714. *Plant Physiol.* **68,** 974–978.

Dijkhuizen, L., Knight, M. and Harder, W. (1978). Metabolic regulation in *Psuedomonas oxalaticus* OX1. *Arch. Microbiol.* **116,** 77–83.

Döhler, G. and Braun, F. (1971). Untersuchung der Beziehung zwischen extracellular Glykolsäure-Ausscheidung und der photosynthetischen CO_2-Aufnahme bei der Blaualge *Anacystis nidulans*. *Planta* **98,** 357–361.

Duggan, J. X. and Anderson, L. E. (1975). Light-regulation of enzyme activity in *Anacystis nidulans* (Richt). *Planta* **122,** 293–297.

Eisenberg, D., Baker, T. S., Suh, S. W. and Smith, W. W. (1978). Structural studies of ribulose 1,5-bisphosphate carboxylase/oxygenase. *In* "Photosynthetic Carbon Assimilation'' (Eds. H. W. Siegelman and G. Hind), pp. 271–281. Plenum, New York.

Ellis, J. R. (1979). The most abundant protein in the world. *Trends Biochem. Sci.* **4,** 241–244.

Ellis, J. R. (1981). Chloroplast protein:synthesis, transport and assembly. *Annu. Rev. Plant Physiol.* **32,** 111–137.

Evans, E. H. and Carr, N. G. (1975). Dark-light transitions with a heterotrophic culture of a blue-green alga. *Biochem. Soc. Trans.* **3,** 373–376.

Fisher, C. R. and Trench, R. K. (1980). *In vitro* carbon fixation by *Prochloron* sp. isolated from *Diplosoma virens*. *Biol. Bull.* (*Woods Hale, Mass.*) **159,** 639–648.

Flügge, U. I., Stitt, M., Freisl, M. and Heldt, H. W. (1982). On the participation of phosphoribulokinase in the light regulation of CO_2 fixation. *Plant Physiol.* **69,** 263–267.

Fogg, G. E. (1975). Biochemical pathways in unicellular plants. *In* "Photosynthesis and

Productivity in Different Environments" (Ed. J. P. Cooper), pp. 437–457. Cambridge Univ. Press, London and New York.
Fogg, G. E. (1983). The ecological significance of extracellular products of phytoplankton photosynthesis. *Bot. Mar.* **26**, 3–14.
Fogg, G. E., Stewart, W. D. P., Fay, P. and Walsby, A. E. (1973). "The Green-blue Algae." Academic Press, New York.
Friedrich, C. G. (1982). Derepression of hydrogenase during limitation of electron donors and derepression of ribulosebisphosphate carboxylase during carbon limitation of *Alcaligenes eutrophus*. *J. Bacteriol.* **149**, 203–210.
Friedrich, C. G., Bowien, B. and Friedrich, B. (1979). Formate and oxalate metabolism in *Alcaligenes eutrophus*. *J. Microbiol.* **115**, 185–192.
Friedrich, C. G., Friedrich, B. and Bowien, B. (1981). Formation of enzymes of autotrophic metabolism during heterotrophic growth of *Alcaligenes eutrophus*. *J. Gen. Microbiol.* **122**, 69–78.
Fuller, R. C. (1978). Photosynthetic carbon metabolism in the green and purple bacteria. *In* "The Photosynthetic Bacteria" (Eds. R. K. Clayton and W. R. Sistrom), pp. 691–705. Plenum, New York.
Gibson, J. L. and Tabita, F. R. (1977a). Different molecular forms of D-ribulose-1,5-bisphosphate carboxylase from *Rhodopseudomas sphaeroides*. *J. Biol. Chem.* **252**, 943–949.
Gibson, J. L. and Tabita, F. R. (1977b). Isolation and preliminary characterization of two forms of ribulose 1,5-bisphosphate carboxylase from *Rhodopseudomonas capsulata*. *J. Bacteriol.* **132**, 818–823.
Gibson, J. L. and Tabita, F. R. (1979). Activation of ribulose 1,5-bisphosphate carboxylase from *Rhodopseudomonas sphaeroides:* probable role of the small subunit. *J. Bacteriol.* **140**, 1023–1027.
Glover, H. E. and Morris, I. (1981). Photosynthetic characteristics of coccoid marine cyanobacteria. *Arch. Microbiol.* **129**, 42–46.
Goldthwaite, J. J. and Bogorad, L. (1971). A one-step method for the isolation and determination of leaf ribulose 1,5-diphosphate carboxylase. *Anal. Biochem.* **41**, 57–66.
Gray, M. W. and Doolittle, W. F. (1982). Has the endosymbiont hypothesis been proven? *Microbiol. Rev.* **46**, 1–42.
Griffiths, D. J. (1980). The pyreniod and its role in algal metabolism. *Sci. Prog. (Oxford)* **66**, 537–553.
Han, T. W. and Eley, J. H. (1973). Glycolate excretion by *Anacystis nidulans:* effect of HCO_3^- concentration, O_2 concentration and light intensity. *Plant Cell Physiol.* **14**, 285–291.
Hanert, H. H. (1982). On the evidence of chemolithotrophic life of *Gallionella ferruginea*. *Abstr. Int. Congr. Microbiol., 13th 1982* p. 61.
Harms, H., Koops, H.-P. and Wehrmann, H. (1976). An ammonia-oxidising bacterium, *Nitrosovibrio tenuis* nov. gen. nov. sp. *Arch. Microbiol.* **108**, 105–111.
Harms, H., Koops, H.-P., Martiny, H. and Wullenweber, M. (1981). D-ribulose 1,5-bisphosphate carboxylase and polyhedral inclusion bodies in *Nitrosomonas spec*. *Arch. Microbiol.* **128**, 280–281.
Harrison, D., Rogers, L. J. and Smith, A. J. (1979). D-ribulose 1,5-bisphosphate carboxylase of the nitrifying bacterium, *Nitrobacter agilis*. *FEMS Microbiol. Lett.* **6**, 47–51.
Hart, B. A. and Gibson, J. (1971). Ribulose 5-phosphate kinase from *Chromatium* sp. strain D. *Arch. Biochem. Biophys.* **144**, 308–321.
Hartman, F. C., Norton, I. L., Stringer, C. D. and Schloss, J. V. (1978). Attempts to

apply affinity labelling techniques to ribulose bisphosphate carboxylase/oxygen. *In* "Photosynthetic Carbon Assimilation" (Eds. H. W. Siegelman and G. Hind), pp. 245–269. Plenum, New York.

Hartman, F. C., Stringer, C. D., Omnaas, J., Donnelly, M. I. and Fraij, B. (1982). Purification and sequencing of cyanogen bromide fragments from ribulosebisphosphate carboxylase/oxygenase from *Rhodospirillum rubrum*. *Arch. Biochem. Biophys.* **219**, 422–437.

Haselkorn, R., Curtis, S. E., Fisher, R., Mazur, B. J., Mevarech, M., Rice, D., Nagaraja, R., Robinson, S. J. and Tuli, R. (1983). Cloning and physical characterization of *Anabaena* genes that code for important functions in heterocyst differentiation: nitrogenase, glutamine synthetase and RmBP carboxylase. *In* "Photosynthetic Prokaryotes: Cell Differentiation and Function" (Eds. G. Papageorgiou and L. Packer), pp. 315–331. Elseview, Amsterdam.

Holt, S. C., Shively, J. M. and Greenwalt, J. W. (1974). Fine structure of selected species of the genus *Thiobacillus* as revealed by chemical fixation and freeze-etching. *Can. J. Microbiol.* **20**, 1347–1351.

Ihlenfeldt, M. J. A. and Gibson, J. (1975). Phosphate utilization and alkaline phosphatase activity in *Anacystis nidulans* (*Synechococcus*). *Arch. Microbiol.* **102**, 23–28.

Jensen, R. G. and Bahr, J. T. (1977). Ribulose 1,2-bisphosphate carboxylase-oxygenase. *Annu. Rev. Plant Physiol.* **28**, 379–400.

Jensen, T. E. and Bowen, C. C. (1961). Organization of the centroplasm in *Nostoc pruniforme*. *Proc. Iowa Acad. Sci.* **68**, 86–89.

Joint, I. R., Morris, I. and Fuller, R. C. (1972). Purification of a complex of alkaline fructose 1,6 bisphosphatase and phosphoribulokinase from *Rhodospirillum rubrum*. *J. Biol. Chem.* **247**, 4833–4838.

Jordan, D. B. and Ogren, W. L. (1981). A sensitive assay procedure for simultaneous determination of ribulose 1,5-bisphosphate carboxylase and oxygenase activities. *Plant Physiol.* **67**, 237–245.

Joset-Espardellier, F., Astier, C., Evans, E. H. and Carr, N. G. (1978). Cyanobacteria grown under photoautotrophic, photoheterotrophic and chemoheterotrophic regimes: sugar metabolism and carbon dioxide fixation. *FEMS Microbiol. Lett.* **4**, 261–264.

Kaplan, A., Badger, M. R. and Berry, J. A. (1980). Photosynthesis and the intracellular inorganic pool in the bluegreen alga *Anabaena variabilis*: response to external CO_2 concentration *Planta* **149**, 219–226.

Karagouni, A. D. and Slater, J. H. (1979). Enzymes of the Calvin cycle and intermediary metabolism in the cyanobacterium *Anacystis nidulans* grown in chemostat culture. *J. Gen. Microbiol.* **115**, 369–376.

Keil, F. (1912). Beiträge zur physiologie der Farblosen Schwefelbakterien *Beitr. Biol. Pflan.* **11**, 335–372.

Kelly, D. P. (1971). Autotrophy: concepts of lithotrophic bacteria and their organic metabolism. *Annu. Rev. Microbiol.* **25**, 117–210.

Khanna, S., Kelley, B. C. and Nicholas, D. J. D. (1981). Oxygen inhibition of the photoassimilation of CO_2 in *Rhodopseudomonas capsulata*. *Arch. Microbiol.* **128**, 421–423.

Kies, L. (1976). Untersuchungen zur Feinstruktur and taxonomischen Einordnung von *Gloeochaete Wittrockiana*, Einer apoplastidalen capsalan Alge mit blaugrünen Endosymbionten (Cyanellen). *Protoplasma* **87**, 419–446.

Kiesow, L. A., Lindsley, B. F. and Bless, J. W. (1977). Phosphoribulkinase from *Nitrobacter winogradskyi* activated by reduced nicotinamide adenine dinucleotide and inhibition by pyridoxal phosphate. *J. Bacteriol.* **130**, 20.

King, W. R. and Andersen, K. (1980). Efficiency of CO_2 fixation in a glycolate oxidore-

ductase mutant of *Alcaligenes eutrophus* which exports fixed carbon as glycolate. *Arch. Microbiol.* **128**, 84–90.
Kobayashi, H., Takabe, T., Nishimura, M. and Akazawa, T. (1979). Roles of large and small subunits of ribulose 1,5-bisphosphate carboxylase in the activation by CO_2 and $MgCl_2$. *J. Biochem. (Tokyo)* **85**, 923–930.
Koops, H.-P., Harms, H. and Wehrmann, H. (1976). Isolation of a moderate halophilic ammonia-oxidising bacterium, *Nitrosococcus mobilis* nov. sp. *Arch. Microbiol.* **107**, 277–282.
Kostrikina, N. A., Emnova, E. E., Biryuzova, V. I. and Romanova, A. K. (1981). Ultrastructural organization of the thermophilic hydrogen bacterium *Pseudomonas thermophila*. *Microbiology (Engl. Transl.)* **50**, 187–191.
Kowallik, U. and Pringsheim, E. G. (1966). The oxidation of hydrogen sulfide by *Beggiatoa*. *Am. J. Bot.* **53**, 801–806.
Laing, W. A. and Christeller, J. T. (1976). A model for the kinetics of activation and catalysis of ribulose-1,5-bisphosphate carboxylase. *Biochem. J.* **159**, 563–570.
Lanaras, T. and Codd, G. A. (1981a). Ribulose 1,5-bisphosphate carboxylase and polyhedral bodies of *Chlorogloeopsis fritschii*. *Planta* **153**, 279–285.
Lanaras, T. and Codd, G. A. (1981b). Structural and immunoelectrophoretic comparison of soluble and particulate ribulose bisphosphate carboxylases from the cyanobacterium *Chlorogloeopsis fritschii*. *Arch. Microbiol.* **130**, 213–217.
Latzko, E. and Kelly, G. J. (1979). Enzymes of the reductive pentose phosphate cycle. *Encycl. Plant Physiol., New Ser.* **6**, 239–250.
Lavergne, D. and Bismuth, E. (1973). Simultaneous purification of two kinases from spinach leaves: ribulose-5-phosphate kinase and phosphoglycerate kinase. *Plant Sci. Lett.* **1**, 229–236.
Lawlis, V. B., Gordon, G. L. R., and McFadden, B. A. (1978). Regulation of activation of ribulose bisphosphate carboxylase from *Pseudomonas oxalaticus*. *Biochem. Biophys. Res. Commun.* **84**, 699–705.
Leadbeater, L. (1981). Pathways and Enzymes of CO_2 Fixation in the cyanobacterium *Nostoc* 6720. Ph.D. Thesis, University of Dundee.
Leadbeater, L., and Codd, G. A. (1984). In preparation.
Leadbeater, L., Siebert, K., Schobert, P. and Bowien, B. (1982). Relationship between activities and protein levels ribulosebisphosphate carboxylase and phosphoribulokinase in *Alcaligenes eutrophus*. *FEMS Microbiol. Lett.* **14**, 263–266.
Lewin, R. A. (1977). *Prochloron*, type genus of the Prochlorophyta. *Phycologia* **16**, 217.
Lex, M., Silvester, W. B. and Stewart, W. D. P. (1972). Photorespiration and nitrogenase activity in the blue-green alga *Anabaena cylindrica*. *Proc. R. Soc. London, Ser. B* **180**, 87–102.
Lilley, R. McC. and Walker, D. A. (1979). Studies with the reconstituted chloroplast system. *Encycl. Plant Physiol., New Ser.* **6**, 41–53.
Lloyd, N. D. H., Canvin, D. T. and Culver, D. A. (1977). Photosynthesis and photorespiration in algae. *Plant Physiol.* **59**, 936–940.
Lorimer, G. H. (1981a). The carboxylation and oxygenation of ribulose 1,5-bisphosphate: the primary events in photosynthesis and photorespiration. *Annu. Rev. Plant Physiol.* **32**, 349–383.
Lorimer, G. H. (1981b). Ribulose bisphosphate carboxylase: amino acid sequence of a peptide bearing the activator for carbon dioxide. *Biochemistry* **20**, 1236–1240.
Lorimer, G. H. (1983). Carbon dioxide and carbamate formation: the makings of a biochemical control system. *Trends Biochem. Sci.* **8**, 65–68.
Lorimer, G. H. and Andrews, T. J. (1981). The C-2 photo- and chemo- respiratory

carbon oxidation cycle. *In* "The Biochemistry of Plants" (Eds. M. D. Hatch and N. K. Boardman), Vol. 8, pp. 329–374. Academic Press, New York.

Lorimer, G. H., Badger, M. R. and Andrews, T. J. (1976). The activation of ribulose bisphosphate carboxylase by carbon dioxide and magnesium ions. Equilibria, kinetics, a suggested mechanism and physiological implications. *Biochemistry* **15,** 529–536.

Lorimer, G. H., Badger, M. R. and Andrews, T. J. (1977). D-ribulose-1,5-bisphosphate carboxylase-oxygenase. Improved methods for activation and assay of catalytic activities. *Anal. Biochem.* **78,** 66–75.

Lorimer, G. H., Badger, M. R. and Heldt, H. W. (1978a). The activation of ribulose 1,5-bisphosphate carboxylase/oxygenase. *In* "Photosynthetic Carbon Assimilation" (Eds. H. W. Siegelman and G. Hind), pp. 283–306. Plenum, New York.

Lorimer, G. H., Osmond, C. B., Akazawa, T. and Asami, S. (1978b). On the mechanism of glycolate synthesis by *Chromatium*. *Arch. Biochem. Biophys.* **185,** 49–56.

MacElroy, R. D., Johnson, E. J. and Johnson, M. K. (1968). Allosteric regulation of phosphoribulokinase activity. *Biochem. Biophys. Res. Commun.* **30,** 678–682.

MacElroy, R. O., Johnson, E. J. and Johnson, M. K. (1969). Control of ATP-dependent CO_2 fixation in extracts of *Hydrogenomonas facilis:* NADH regulation of phosphoribulokinase. *Arch. Biochem. Biophys.* **131,** 272–275.

MacElroy, R. D., Mack, H. M. and Johnson, E. J. (1979). Properties of phosphoribulokinase from *Thiobacillus neapolitanus. J. Bacteriol.* **112,** 532–538.

McCarthy, J. T. and Charles, A. M. (1975). Properties and regulation of ribulose diphosphate carboxylase from *Thiobacillus novellus. Arch. Microbiol.* **105,** 51–59.

McFadden, B. A. (1973). Autotrophic CO_2 assimilation and the evolution of ribulose diphosphate carboxylase. *Bacteriol. Rev.* **37,** 289–319.

McFadden, B. A. (1978). Assimilation of one-carbon compounds. *In* "The Bacteria" (Eds. L. N. Ornstein and J. R. Sokatch), vol. 6, pp. 219–304. Academic Press, New York.

McFadden, B. A. (1980). A perspective of ribulose bisphosphate carboxylase/oxygenase, the key catalyst in photosynthesis and photorespiration. *Acc. Chem. Res.* **13,** 394–399.

McFadden, B. A. and Denend, A. R. (1972). Ribulose diphosphate carboxylase from autotrophic microorganisms. *J. Bacteriol.* **110,** 633–642.

McFadden, B. A. and Purohit, K. (1978). Chemosynthetic, photosynthetic and cyanobacterial ribulose bisphosphate carboxylase. *In* "Photosynthetic Carbon Assimilation" (Eds. H. W. Siegelman and G. Hind), pp. 179–207. Plenum, New York.

McFadden, B. A. and Tabita, F. R. (1974). D-ribulose-1,5-diphosphate carboxylase and the evolution of autotrophy. *BioSystems* **6,** 93–112.

McIntosh, L., Pouslen, C. and Bogorod, L. (1980). The DNA sequence of the gene encoding the large subunit of maize ribulose bisphosphate carboxylase. *Nature (London)* **288,** 556–560.

Madigan, M. T., Peterson, S. P. and Brock, T. D. (1974). Nutritional studies on *Chloroflexus aurantiacus,* a filamentous, photosynthetic gliding bacterium. *Arch. Microbiol.* **100,** 97–103.

Manian, S. S. and O'Gara, F. (1982). Induction and regulation of ribulose bisphosphate carboxylase activity in *Rhizobium japonium* during formate dependent growth. *Arch. Microbiol.* **131,** 51–54.

Marcus, Y., Zenvirth, D., Harel, E. and Kaplan, A. (1982). Induction of HCO_3^- transporting capability and high photosynthetic affinity to inorganic carbon by low concentrations of CO_2 in *Anabaena variabilis. Plant Physiol.* **69,** 1008–1012.

Martin, P. G. (1979). Amino acid sequence of the small subunit of ribulose 1,5-bisphosphate carboxylase from spinach. *Aust. J. Plant Physiol.* **6,** 401–408.

Miller, A. G. and Colman, B. (1980a). Evidence for HCO_3^- transport by the blue-green alga (cyanobacterium) *Coccochloris peniocystis*. *Plant Physiol.* **65**, 397–402.

Miller, A. G. and Colman, B. (1980b). Active transport and accumulation of bicarbonate in a unicellular cyanobacterium. *J. Bacteriol.* **143**, 1253–1259.

Miziorko, H. M. and Mildvan, A. S. (1974). Electron paramagnetic resonance, H and ^{13}C nuclear magnetic resonance studies on the interaction of magnanese and bicarbonate with ribulose 1,5-bisphosphate carboxylase. *J. Biol. Chem.* **249**, 2743–2750.

Nishimura, M. and Akazawa, T. (1973). Further proof for the catalytic role of the larger subunit in the spinach leaf ribulose-1,5-diphosphate carboxylase. *Biochem. Biophys. Res. Commun.* **54**, 842–848.

Nishimura, M. and Akazawa, T. (1974). Structure and function of chloroplast proteins. XXII. Dissociation and reconstruction of spinach leaf ribulose-1,5-diphosphate carboxylase. *J. Biochem. (Tokyo)* **76**, 169–179.

Ohmann, E. (1979). Autotrophic carbon dioxide assimilation in prokaryotic microorganisms. *Encycl. Plant Physiol., New Ser.* **6**, 54–67.

Okabe, K. and Codd, G. A. (1980). Structural and regulatory studies on the ribulose bisphosphate carboxylase-oxygenase of *Anabaena cylindrica*. *Plant Cell Physiol.* **21**, 1117–1127.

Pelroy. R. A. and Bassham, J. A. (1972). Photosynthetic and dark carbon metabolism in unicellular blue-green algae. *Arch. Mikrobiol.* **86**, 25–38.

Pelroy, R. A., Levine, G. A. and Bassham, J. A. (1976). Kinetics of light-dark CO_2 fixation and glucose assimilation by *Aphanocapsa* 6714. *J. Bacteriol.* **128**, 633–643.

Pierce, J., Tolbert, N. E. and Barker, R. (1980). Interaction of ribulose bisphosphate carboxylase/oxygenase with transition state analogues. *Biochemistry* **19**, 934–942.

Pon, N. G., Rabin, B. R. and Calvin, M. (1963). Mechanism of the carboxydismutase reaction. I. The effect of preliminary incubation of substrates, metal ion and enzyme on activity. *Biochem. Z.* **338**, 7–19.

Pope, L. M., Hoare, D. S. and Smith, A. J. (1969). Ultrastructure of *Nitrobacter agilis* grown under autotrophic and heterotrophic conditions. *J. Bacteriol.* **97**, 936–939.

Poulsen, C., Martin, B. and Svendsen, I. (1979). Partial amino acid sequence of the large subunit of ribulosebisphosphate carboxylase from barley. *Carlsberg Res. Commun.* **44**, 191–199.

Purohit, K. and McFadden, B. A. (1976). Heterogeneity of large subunits of ribulose-1,5-bisphosphate carboxylase from *Hydrogenomonas eutropha*. *Biochem. Biophys. Res. Commun.* **71**, 1220–1227.

Purohit, K. and McFadden, B. A. (1977). Quaternary structure and oxygenase activity of D-ribulose-1,5-bisphosphate carboxylase from *Hydrogenomonas eutropha*. *J. Bacteriol.* **129**, 415–421.

Purohit, K. and McFadden, B. A. (1979). Ribulose 1,5-bisphosphate carboxylase and oxygenase from *Thiocapsa roseopersicina*: activation and catalysis. *Arch. Biochem. Biophys.* **194**, 101–106.

Purohit, K., McFadden, B. A. and Shaykh, M. M. (1976). D-ribulose-1,5-bisphosphate carboxylase and polyhedral inclusion bodies in *Thiobacillus intermedius*. *J. Bacteriol.* **127**, 516–522.

Purohit, K., McFadden, B. A. and Lawlis, V. B. (1979). Ribulose bisphosphate carboxylase/oxygenase from *Thiocapsa roseopersicina*. *Arch. Microbiol.* **121**, 79–82.

Purohit, K., Becker, R. R. and Evans, H. J. (1982). D-ribulose-1,5-bisphosphate carboxylase/oxygenase from chemolithotrophically-grown *Rhizobium japonicum*. *Biochim. Biophys. Acta* **715**, 230–239.

Quayle, J. R. (1961). Metabolism of C1 compounds in autotrophic and heterotrophic microorganisms. *Annu. Rev. Microbiol.* **15**, 119–152.

Raboy, B., Padan, E. and Shilo, M. (1976). Heterotrophic capacities of *Plectonema boryanum*. *Arch. Microbiol.* **110**, 77–85.
Raven, J. A. (1970). Exogenous inorganic carbon sources in plant photosynthesis. *Biol. Rev. Cambridge Philos. Soc.* **45**, 167–221.
Reichelt, B. Y. and Delaney, S. F. (1982). Structure of the ribulose 1,5-bisphosphate carboxylase large subunit gene from *Synechococcus* PCC6301 compared to large subunit genes from spinach and maize. *In* "Abstracts of the IV International Symposium on Photosynthetic Prokaryotes" (Ed. G. Stanier), D6. Institut Pasteur, Paris.
Richardson, F. L. and Brown, T. E. (1970). *Glaucosphaera vacuolata*, its ultrastructure and physiology. *J. Phycol.* **6**, 165–171.
Rindt, K.-P. and Ohmann, E. (1969). NADH and AMP as allosteric effectors of ribulose-5-phosphate kinase in *Rhodopseudomonas sphaeroides*. *Biochem. Biophys. Res. Commun.* **36**, 357–364.
Robinson, S. P. and Walker, D. A. (1981). Photosynthetic carbon reduction cycle. *In* "The Biochemistry of Plants" (Eds. M. D. Hatch and N. K. Boardman), Vol. 8, pp. 193–236. Academic Press, London.
Robison, P. D., Martini, M. N. and Tabita, F. R. (1979). Differential effects of metal ions on *Rhodospirillum rubrum* ribulose bisphosphate carboxylase/oxygenase and stoichiometric incorporation of HCO_3 into a cobalt (III)-enzyme complex. *Biochemistry* **18**, 4453–4458.
Sarles, L. S. and Tabita, F. R. (1983). Derepression of synthesis of D-ribulose 1,5-bisphosphate carboxylase/oxygenase from *Rhodospirillum rubrum*. *J. Bacteriol.* **153**, 458–464.
Schlegel, H. G. (1975). Mechanisms of chemo-autotrophy. *In* "Marine Ecology" (Ed. O. Kinne), Vol. 2, Part I, pp. 9–60. Wiley, London.
Schlegel, H. G. (1976). The physiology of hydrogen bacteria. *Antonie van Leeuwenhoek* **42**, 181–201.
Schloss, J. V. and Lorimer, G. H. (1982). The stereochemical course of ribulose bisphosphate carboxylase. *J. Biol. Chem.* **257**, 4691–4694.
Schloss, J. V., Phares, E. F., Long, M. V., Norton, I. L., Stringer, C. D. and Hartman, F. C. (1979). Isolation, characterization and crystallization of ribulose bisphosphate carboxylase from autotrophically grown *Rhodospirillum rubrum*. *J. Bacteriol.* **137**, 490–501.
Scotten, H. L. and Stokes, J. L. (1962). Isolation and properties of *Beggiatoa*. *Arch. Mikrobiol.* **42**, 535–368.
Shinozaki, K. and Sugiura, M. (1982). The nucleotide sequences of the tobacco chloroplast gene for the large subunit of ribulose-1,5-bisphosphate carboxylase/oxygenase. *Gene* **20**, 91–102.
Shively, J. B., Ball, F., Brown, D. H. and Saunders, R. E. (1973). Functional organelles in prokaryotes: polyhedral inclusions (carboxysomes) of *Thiobacillus neapolitanus*. *Science* **182**, 584–586.
Shively, J. B., Bock, E., Westphal, K. and Cannon, G. C. (1977). Icosahedral inclusions (carboxysomes) of *Nitrobacter agilis*. *J. Bacteriol.* **132**, 673–675.
Shively, J. M. (1974). Inclusion bodies of prokaryotes. *Annu. Rev. Microbiol.* **28**, 167–187.
Shively, J. M., Saluja, A. and McFadden, B. A. (1978). Ribulose bisphosphate carboxylase from methanol-grown *Paracoccus denitrificans*. *J. Bacteriol.* **134.** 1123–1132.
Siebert, K., Schobert, P. and Bowien, B. (1981). Purification, some catalytic and molecular properties of phosphoribulokinase from *Alcaligenes eutrophus*. *Biochim. Biophys. Acta* **658**, 35–44.

Sirevåg, R. (1974). Further studies on carbon dioxide fixation in *Chlorobium*. *Arch. Microbiol.* **98**, 3–18.
Sirevåg, R. and Castenholz, R. W. (1979). Aspects of carbon metabolism in *Chloroflexus*. *Arch. Microbiol.* **120**, 151–153.
Sirevåg, R. and Ormerod, J. (1970). Carbon dioxide fixation in green sulphur bacteria. *Biochem. J.* **120**, 399–408.
Sirevåg, R., Buchanan, B. B., Berry, J. A. and Troughton, J. H. (1977). Mechanisms of CO_2 fixation in bacterial photosynthesis studied by the carbon isotope fractionation technique. *Arch. Microbiol.* **112**, 35–38.
Smith, A. J. (1973). Synthesis of metabolic intermediates. In "The Biology of Blue-Green Algae" (Eds. N. G. Carr and B. A. Whitton), pp. 1–38. Blackwell Oxford.
Smith, A. J. (1981). Cyanobacterial contributions to the heterotrophic connection. In "Microbial Growth on C_1 Compounds" (Ed. H. Dalton), pp. 122–130. Heyden, London.
Smith, A. J. and Hoare, D. S. (1977). Specialist phototrophs, lithotrophs and methylotrophs: unity among a diversity of prokaryotes? *Bacteriol. Rev.* **41**, 419–448.
Smith, S. M. and Ellis, R. J. (1981). Light-stimulated accumulation of transcripts of nuclear and chloroplast genes for ribulosebisphosphate carboxylase. *J. Mol. Appl. Genet.* **1**, 127–136.
Snead, R. M. and Shively, J. M. (1978). D-ribulose-1,5-bisphosphate carboxylase from *Thiobacillus neapolitanus*. *Curr. Microbiol.* **1**, 309–314.
Stanier, R. Y. and Cohen-Bazire, G. (1977). Phototrophic prokaryotes: the cyanobacteria. *Annu. Rev. Microbiol.* **31**, 225–274.
Stanier, R. Y., Adelberg, E. A. and Ingraham, J. L. (1977). "General Microbiology," 4th ed. Macmillan, London.
Stewart, R., Auchterlonie, C. C. and Codd, G. A. (1977). Studies on the subunit structure of ribulose 1,5-bisphosphate carboxylase from the blue-green alga *Microcystis aeruginosa*. *Planta* **136**, 61–64.
Stewart, W. D. P. (1973). Nitrogen fixation by photosynthetic microorganisms. *Annu. Rev. Microbiol.* **27**, 283–316.
Stewart, W. D. P. (1980). Some aspects of structure and function in N_2-fixing cyanobacteria. *Annu. Rev. Plant Physiol.* **34**, 497–536.
Stewart, W. D. P. and Codd, G. A. (1975). Polyhedral bodies (carboxysomes) of nitrogen-fixing blue-green algae. *Br. Phycol. J.* **10**, 273–278.
Størro, I. and McFadden, B. A. (1981). Glycolate excretion by *Rhodospirillum rubrum*. *Arch. Microbiol.* **129**, 317–320.
Strohl, W. R. and Larkin, J. M. (1978). Enumeration, isolation and characterization of *Beggiatoa* from fresh-water sediments. *Appl. Environ. Microbiol.* **36**, 755–770.
Strohl, W. R., Cannon, G. C., Shively, J. M., Gude, H., Hook, L. A., Lane, C. M. and Larkin, J. M. (1981). Heterotrophic carbon metabolism by *Beggiatoa alba*. *J. Bacteriol.* **148**, 572–583.
Tabita, F. R. (1980). Pyridine nucleotide control and subunit structure of phosphoribulokinase from photosynthetic bacteria. *J. Bacteriol.* **143**, 1275–1280.
Tabita, F. R. (1981). Molecular regulation of carbon dioxide assimilation in autotrophic microorganisms. In "Microbial Growth on C_1 Compounds" (Ed. H. Dalton), pp. 70–82. Heyden, London.
Tabita, F. R. and Colletti, C. (1979). Carbon dioxide assimilation in cyanobacteria: regulation of ribulose 1,5-bisphosphate carboxylase. *J. Bacteriol.* **140**, 452–458.
Tabita, F. R. and McFadden, B. A. (1972). Regulation of ribulose 1,5-diphosphate

carboxylase by 6-phospho-d-gluconate. *Biochem. Biophys. Res. Commun.* **48**, 1153–1160.

Tabita, F. R. and McFadden, B. A. (1974a). One-step isolation of microbial ribulose 1,5-diphosphate carboxylase. *Arch. Mikrobiol.* **99**, 231–240.

Tabita, F. R. and McFadden, B. A. (1974b). D-ribulose 1,5-diphosphate carboxylase from *Rhodospirillum rubrum*. I. Levels, purification and effects of metallic ions. *J. Biol. Chem.* **249**, 3453–3459.

Tabita, F. R. and McFadden, B. A. (1976). Molecular and catalytic properties of ribulose 1,5-bisphosphate carboxylase from the photosynthetic extreme halophile *Ectothiorhodospira halophila*. *J. Bacteriol.* **126**, 1271–1277.

Tabita, F. R., McFadden. B. A. and Pfennig, N. (1974). D-ribulose 1,5-diphosphate carboxylase in *Chlorobium thioshlfatophilum* Tassajara. *Biochim. Biophys. Acta* **341**, 187–194.

Takabe, T. (1977). Ribulose 1,5-bisphosphate carboxylase from the blue-green alga *Anabaena cylindrica*. *Agric. Biol. Chem.* **41**, 2255–2260.

Takabe, T. and Akazawa, T. (1975a). Further studies on the subunit structure of *Chromatium* ribulose-1,5-bisphosphate carboxylase. *Biochemistry* **14**, 46–50.

Takabe, T. and Akazawa, T. (1975b). Molecular evolution of ribulose—1,5-bisphosphate carboxylase. *Plant Cell Physiol.* **16**, 1049–1060.

Takabe, T. and Akazawa, T. (1977). A comparative study of the effect of O_2 on photosynthetic carbon metabolism by *Chlorobium thiosulfatophilum* and *Chromatium vinosum*. *Plant Cell Physiol.* **18**, 753–765.

Takabe, T., Nishimura, M. and Akazawa, T. (1976). Presence of two subunit types in ribulose 1,5-bisphosphate carboxylase from blue-green algae. *Biochem. Biophys. Res. Commun.* **68**, 537–544.

Takabe, T., Osmond, C. B., Summons, R. E. and Akazawa, T. (1979). Effect of oxygen on photosynthesis and biosynthesis of glycolate in photoheterotrophically grown cells of *Rhodospirillum rubrum*. *Plant Cell Physiol.* **20**, 233–241.

Talling, J. F. (1976). The depletion of the carbon dioxide from lake water by phytoplankton. *J. Ecol.* **64**, 79–121.

Taylor, S. (1977). Evidence for the presence of ribulose 1,5-bisphosphate carboxylase and phosphoribulokinase in *Methylococcus capsulatus* (Bath). *FEMS Microbiol. Lett.* **2**, 305–307.

Taylor, S. C. and Dow, C. S. (1980). Ribulose-1,5-bisphosphate carboxylase from *Rhodomicrobium vannielii*. *J. Gen. Microbiol.* **116**, 81–87.

Taylor, S. C., Dalton, H. and Dow, C. S. (1981). Ribulose 1,5-bisphosphate carboxylase/oxygenase and carbon assimilation in *Methylococcus capsulatus* (Bath). *J. Gen. Microbiol.* **122**, 89–94.

Trench, R. K., Pool, R. R., Logan, M. and Engelland, A. (1978). Aspects of the relation between *Cyanophora paradoxa* (Korschikoff) and its endosymbiotic cyanelles *Cyanocyta Korschikoffiana* (Hall and Claus). I. *Proc. R. Soc. London, Ser. B.* **202**, 423–443.

Trüper, H. G. and Pfennig, N. (1978). Taxonomy of the Rhodospirillales. *In* "The Photosynthetic Bacteria" (Eds. R. K. Clayton and W. R. Sistrom), pp. 19–27. Plenum, New York.

Warburg, P. (1920). Über die Geschwindigkeit der photochemischen Kohlensäurezersetung in lebenden Zellen. II. *Biochem. Z.* **100**, 188–217.

Watson, S. W. and Waterbury, J. B. (1971). Characteristics of two marine nitrate oxidizing bacteria. *Arch. Mikrobiol.* **77**, 203–230.

Whatley, J. M. (1977). The fine structure of *Prochloron*. *New Phytol.* **79**, 309–313.

Whitman, W. B., Martin, M. N. and Tabita, F. R. (1979). Activation and regulation of ribulose biphosphate carboxylase-oxygenase in the absence of small subunits. *J. Biol. Chem.* **254**, 10184–10189.

Whittenbury, R. and Dow, C. S. (1977). Morphogenesis and differentiation in *Rhodomicrobium vannielii* and other budding and prosthecate bacteria. *Bacteriol. Rev.* **41**, 754–808.

Whittenbury, R. and Kelly, D. P. (1977). Autotrophy: a conceptual phoenix. *Symp. Soc. Gen. Microbiol.* **27**, 121–149.

Wright. R. T. and Shah, N. M. (1975). The trophic role of glycolic acid in coastal seawater. I. Heterotrophic metabolism in seawater and bacterial cultures. *Mar. Biol. (Berlin)* **33**, 175–183.

Wright, R. T. and Shah, N. M. (1977). The trophic role of glycolic acid in coastal seawater. II. Seasonal changes in concentration and heterotrophic use in Ipswich Bay, Massachussetts, U.S.A. *Mar. Biol. (Berlin)* **43**, 257–263.

Wullenweber, M., Koops, H. P. and Harms, H. (1977). Polyhedral inclusion bodies in cells of *Nitrosomonas* spec. *Arch. Microbiol.* **112**, 69–72.

Zavarzin, G. A. and Nozhevnikova, A. N. (1976). CO oxidizing bacteria. *In* "Microbial Production and Utilization of Gases" (Eds. H. G. Schlegel, G. Gottschalk and N. Pfennig), pp. 207–213. E. Galtze K. G., Göttingen.

Zurawski, G., Perrot, R., Bottomley, W. and Whitfield, P. R. (1981). The structure of the gene for the large subunit of ribulose 1,5-bisphosphate carboxylase from spinach chloroplast DNA. *Nucleic Acids Res.* **9**, 3251–3270.

6

The Ecology and Adaptive Strategies of Benthic Cyanobacteria

MOSHE SHILO AND ALI FATTOM*

*Division of Microbial and Molecular Ecology
Life Science Institute
Hebrew University
Jerusalem, Israel*

Introduction

The aquatic cyanobacteria in ecosystems fall into two distribution groups in regard to their positioning in the euphotic zone. There are *planktonic* types immersed and floating in the water column, and *benthic* forms, which adhere to submerged solid surfaces and attach to interfaces such as bottom sediment (epipelic), stones and rocky shores at the littoral fringe (epilithic), or algae and aquatic plants (epiphytic).

Although extensive work has been done on the physiology and ecology of the planktonic forms, the attached types are difficult to study, and relatively little work has been done to characterise the unique features that cause these cyanobacteria to differ from their planktonic relatives. It is important to unravel the mechanism which underlies the adhesion of the benthic cyanobacteria to their substratum, allowing them to withstand detachment by currents and water turbulence. We also have to understand the different adaptations to the conditions and stresses to which benthic cyanobacteria are exposed in their ecological niche.

The benthic cyanobacteria should possess mechanisms to overcome the limitation of light, caused by the fact that radiation reaching the sediment layer is markedly reduced by absorption in the water column overlayering the sediment. This is especially marked in turbid waters rich in suspended clay particles and planktonic microorganisms. Furthermore, the benthic layer is in a dynamic state, since continuous sedimentation of particles and agglutinated cells from the water columns cover the organisms forming the benthic layers. Although the rain of

*Present address: Department of Biology and Biochemistry, Birzeit University, Birzeit, West Bank, Via Israel.

particles creates a continuous flow of nutrients adsorbed onto these particles, it reduces markedly the light reaching the benthic cyanobacteria and creates a change in the delicate balance of oxygenated parts of the sediment layer and its underlying anaerobic H_2S-rich layers.

In addition, all attached cyanobacteria must provide for dispersal of their progeny through the water column so as to colonise new suitable benthic surfaces. They must also have considerable metabolic versatility to survive and grow in the fluctuating environmental conditions prevailing at the water–soil interface in regard to O_2, H_2S, pH, and E_h. The planktonic cyanobacteria, on the other hand, should have adaptations for survival and growth in the water column. These should include mechanisms of buoyancy and its regulation (Walsby and Booker, 1980), means to cope with the oligotrophic conditions prevailing in the oceans and lakes, and the ability to regulate their pigment composition and content in accordance with the changes in the light gradient of the water column. Furthermore, the cyanobacteria which form surface blooms and are exposed to high light intensities should also have protective mechanisms to withstand the drastic photo-oxidative conditions (Eloff *et al.*, 1976).

The Mechanism of Adhesion of the Benthic Cyanobacteria

In recent years, great emphasis has been put on the importance of adhesion of bacteria to surfaces in understanding ecological phenomena and their role in pathogenesis (Marshall, 1976).

A variety of cell surface structures and compounds have been suggested as being related to or responsible for adhesion. We therefore compared the envelope characteristics of the benthic and planktonic cyanobacteria. Different physicochemical interaction forces come into play when bacteria approach submerged surfaces. These include van der Waals forces, electrostatic interactions, hydrophobic interactions, and hydrogen bonding, as well as adhesion due to polysaccharides (Fletcher and Floodgate, 1973) or proteinaceous material (Danielsson *et al.*, 1977).

In the adhesion of benthic cyanobacteria, hydrophobicity seems to be of great importance. Electrostatic interaction, on the other hand, seems to be of only minor importance, since adhesion is also found to hydrophobic plastics such as polyethylene or polystyrene, which have little or no surface charge.

All benthic cyanobacteria tested, unicellular and filamentous alike, collected from fresh, brackish, marine, and hypersaline waters, showed a high degree of cell surface hydrophobicity. The benthic samples were obtained from widely differing habitats including fish ponds, swamp drainage channels, marshes, and cyanobacterial mats from Solar Lake.

All planktonic types, on the other hand, obtained from culture collections or collected from fresh or brackish water, exhibited hydrophilicity of the cell surfaces (Shilo, 1982; Fattom and Shilo, 1984).

6. ECOLOGY AND ADAPTIVE STRATEGIES OF BENTHIC CYANOBACTERIA

In their surface hydrophobicity the benthic cyanobacteria resemble the neuston microorganisms, which adhere to the water–air interface of oceans and lakes (Norkrans, 1980; Dählback *et al.*, 1981), and the oil-degrading bacteria, which adhere to the liquid–liquid interfaces of oil spills (Gutnick and Rosenberg, 1977; Rosenberg and Gutnick, 1981). We used several methods to demonstrate and quantify the hydrophobicity or hydrophilicity of the cell surface of cyanobacteria. These methods included selective partitioning of the cells in a biphasic system of water and immiscible nonpolar solvents (Rosenberg *et al.*, 1980), retention and adsorption on hydrophobic columns (Hjerten *et al.*, 1974), and measurement of the contact angles of liquid droplets (van Oss *et al.*, 1975) in contact with cell layers spread on Millipore filters.

The sites of hydrophobicity on the cell surface were localised microscopically with the aid of probes (sphingosine isothiocyanate) which bind to the hydrophobic surface sites and fluoresce under ultraviolet (UV) light. The marked difference between the benthic and planktonic cyanobacteria, which expresses the hydrophobic or hydrophilic nature of their cell surfaces, was also shown in their behaviour when mixed with hydrophobic globules of phenylsepharose (Pharmacia) or oil droplets. Only the benthic forms were strongly bound by these globules; the planktonic forms remained free in the aqueous phase. One could visualise the adhesion of the cells or filaments in the microscope and use it to separate the benthic from the planktonic cyanobacteria.

The distribution of hydrophobic sites on the cell surface determines the orientation of the cells at interfaces. Thus, microscopically, the filaments of many of the benthic cyanobacteria, such as *Phormidium* spp. (strain J-1) adhered throughout their entire length to the oil or heptane droplets, indicating that the hydrophobic sites must be distributed along the entire length of the filament. The sphingosine isothiocyanate probe added to mature filaments of *Phormidium* spp. (strain J-1,) led to the appearance of fluorescent sites in UV light. The hydrophobic sites were found to be regularly spaced along the entire filament at equal distances.

Loss of the cell wall upon conversion of *Phormidium* J-1, *Calothrix desertica* (strain 7102), and *Plectonema boryanum (strain 6306)* filaments into spheroplasts resulted in loss of hydrophobicity, confirming that the cell envelope components are solely responsible for cell hydrophobicity (Fattom and Shilo, 1984).

The cell envelope characteristics which are responsible for hydrophobicity or hydrophilicity are genetically controlled. Mutants of cyanobacteria are now available in which these properties differ from those of the wild-type parent strain. Several mutants obtained from the hydrophilic wild-type strain of *Spirulina platensis* (Riccardi *et al.*, 1981) which exhibited clumping have also become hydrophobic. On the other hand, mutants which have lost their hydrophobic cell envelopes and have become hydrophilic have also been observed (e.g., in *Aphanothece halopytica*). Expression of hydrophobicity of the benthic cyanobacteria requires the presence of a minimal concentration of divalent cations such as Mg^{2+} or more than a 100-fold concentration of monovalent cations such as sodium. The cations may play a role in creating bridges between negatively charged surfaces and negatively charged cyanobacterial cells, thus overcoming electrostatic repulsion (Olsson et al., 1976).

Adaptation of Benthic Cyanobacteria to Overcome Light Limitation

In the benthic cyanobacterial mats, light penetration is very limited. Only 1% of the surface light intensity is found at a depth of 2–4 mm in the sediment (Fenchel and Straarup, 1971).

Two adaptive strategies common to many benthic cyanobacteria seem to be of great importance in overcoming the conditions of light limitation inherent in the benthic niche.

One strategy is their ability to react to light by a positive phototactic gliding motility. This property allows them to reach the sediment–water interface layer in spite of continuous coverage by sedimentary particles. Simon (1981) has shown, by comparison of a gliding *A. halophytica* taken from Solar Lake (Sinai) with its nonmotile mutant, that loss of the glycoprotein surface component may play an important role in gliding. Furthermore, unpublished studies by Simon showed that concomitantly with the loss of motility, the mutants had lost the cell envelope hydrophobicity which characterised the wild type and became hydrophilic.

A second strategy found in several benthic filamentous cyanobacteria tested was the formation and excretion of a bioflocculant, which causes particle flocculation and clarification of the overlayering water column. The precipitation of clay particles and clarification of a major drainage channel in the Hula swamp have been described (Zur, 1979; Avnimelech et al., 1982), and the suggestion was made that a bioflocculant produced by the cyanobacterial mat was the responsible agent.

Bioflocculant production and excretion were obtained in axenic cultures of several of the benthic cyanobacteria and were studied under different physiological conditions and at different growth phases. The isolated, purified bioflocculant was found to be of high molecular weight ($>200,000$) and to contain protein, lipid, and polysaccharide moieties (Fattom, 1983).

Dispersal Mechanisms of Attached Cyanobacteria

In many of the cyanobacteria attached to surfaces, multimorphic life-cycles are found which always include dispersal stages that are released into the aqueous phase. Filamentous benthic cyanobacteria produce hormogonia, whereas cyanobacteria belonging to the Pleurocapsa group, characteristically attached to surfaces, produce baeocytes which are released into the water (Waterbury and Stanier, 1978). It seems possible that these developmental cycles may involve changes in the cell envelope and loss of surface hydrophobicity, allowing for the spread of the organisms attached to the soil–water interface.

A number of benthic hormogonia-forming filamentous cyanobacteria were therefore chosen and their surface characteristics studied. In all species tested, including *Plectonema boryanum* (strain 6306), *C. desertica* (strain 7102), *Fremyella diplosiphon,* and *Anabaenopsis circularis* (kindly given by Dr. R. Rippka of the Pasteur Institute, Paris), hormogonia showed a high degree of surface hydrophilicity, whereas the mature filaments were highly hydrophobic. The hormogonial trichomes in these experiments were separated from the mature filaments by differential centrifugation in a Ficoll gradient or by the water–hexadecane biphasic partitioning system. Use of the fluorescent hydrophobic probe in *C. desertica* showed that whereas the mature cells bound the probe and their envelope became fluorescent in UV light, the hormogonial trichomes did not bind the probe.

The transition from the hormogonial stage to mature filaments occurred within 24 to 48 hr after formation of the hormogonia. The loss of surface hydrophilicity and the change into a hydrohobic cell envelope required *de novo* protein synthesis and active photosynthesis. Addition of 3-(3,4-dichlorophenyl)-1,1-dimethyl urea (DCMU) or chloramphenicol prevented the transformation of hormogonia to mature filaments and the change in cell surface characteristics.

Metabolic Versatility of Benthic Cyanobacteria

Environmental conditions in the benthic layers are characterized by rapid fluctuations in space and time. These changes involve the pH at the sediment surface (Gnaiger *et al.*, 1978), as well as the concentration of the oxygen, H_2S, and the E_h of the benthic layers (Jørgensen *et al.*, 1979a). The elegant use of microelectrodes to measure *in situ* pH, H_2S, and O_2 simultaneously (Revsbech *et al.*, 1980a,b, 1981) have given us a detailed picture of these rhythmic diurnal fluctuations in such benthic cyanobacterial mats. Similar studies have also shown rapid fluctuation in the above-mentioned parameters in the pycnocline of the stratified Solar Lake (Jørgensen *et al.*, 1979b).

In these fluctuating ecosystems, cyanobacteria are dominant (Shilo, 1979, 1980, 1982). In fact, cyanobacteria are widely recognized as the most typical mat formers. They are widespread in marshes, rice paddies, marine swamps, and mangroves (Brock, 1973; Fogg *et al.*, 1973). Laminated benthic cyanobacterial mats originated in the Precambrium era, when these organisms dominated large areas of our planet (Schopf, 1974). Ancient stromatolites, 3.5–4.0 billion years old, with clear fossilised cyanobacterial remains, have been found in many parts of the world, showing a striking similarity to the presently found filamentous cyanobacteria. The characteristics of the benthic cyanobacteria thus have a long evolutionary history.

One of the unique adaptation mechanisms found in many of the benthic cyanobacteria, which enables them to cope with fluctuations in their environment, is their ability to shift from oxygenic to anoxygenic photosynthesis (see the reviews of Padan, 1979a,b), *Oscillatoria limnetica* isolated from Solar Lake near Elat (Cohen et al., 1975) Israel, was the first cyanobacterium in which anoxygenic photosynthesis was shown. In this organism, when photosystem II was inhibited by DCMU or not excited by actinic light (700 nm), the photoassimilation of CO_2 was driven by photosystem I only, with Na_2S as the sole electron donor (Cohen et al., 1975a,b). Two sulphide molecules were oxidised to elemental sulphur for each CO_2 molecule photoassimilated, and the sulfur granules were deposited outside the cyanobacterial cells.

It rapidly became evident that *O. limnetica* is not exceptional in its ability to shift from oxygenic to anoxygenic photosynthesis. Garlick et al. (1977) proved that other cyanobacteria, unicellular as well as filamentous types, shared this capability. Among the strains which were found to be positive, many were benthic, representing the dominant species from the cyanobacterial mats of Solar Lake, Baja California; the Bardawill Lagoon near Port Said and Wadi Natrun in Egypt; and as cyanobacteria from the pycnocline of Solar Lake, Israel.

The strains capable of anoxygenic photosynthesis each exhibit a different range of H_2S concentration, permitting anoxygenic photosynthesis to take place. These differences can be attributed to the different apparent sulphide affinities of the sulphide utilising systems, as well as the varying degrees of tolerance to the toxic effects of the sulphide. Thus, the positioning of different cyanobacteria along the sulphide gradients in stratified lakes, as well as their zonation in the sediment mats, may be explained by their different requirements for and tolerance of H_2S.

Only scattered *in situ* measurements have been made to determine the overall role of anoxygenic photosynthesis in the total primary production. In the pycnocline of the Solar Lake, the fluctuations in O_2, sulphur, and H_2S indicate that *O. limnetica* may utilise alternatively oxygenic and anoxygenic photosynthesis at different times in the daily cycle (Jørgensen et al., 1979b). Anoxygenic photosynthesis may also be of importance in cyanobacterial mats, where H_2S reaches at night up to the surface layer and where extensive light-dependent sulphide oxidation was found (Revsbech et al.. 1984, mentioned in Revsbech et al., 1981).

In cyanobacteria as well as in tobacco chloroplast preparation, addition of sulphide leads to inhibition of photosynthetic electron transport between water and photosystem II (Oren et al., 1979).

The shift from oxygenic to anoxygenic photosynthesis, and similarly, in the absence of CO_2, the hydrogenase-mediated H_2 evolution, are sulphide and light dependent and involve an induction process with a requirement for *de novo* protein synthesis (Belkin and Padan, 1978). It seems likely that the induction process is connected to the protection of the photosynthetic pathway from the toxic effect of H_2S. The fact that the induction lag is clearly correlated with the H_2S concentration and that it is prolonged at lowered pH in correlation with the relative amount of the undissociated form of H_2S, which is normally the toxic species, supports this view.

Work has shown that the strong reductant dithionite, which is not capable of electron donation, can mimic or circumvent this induction process (Belkin and Padan, 1982a). It thus appears that the sulphide-vulnerable site is modified or bipassed in the presence of $Na_2S_2O_4$. H_2S was found to donate electrons directly to the electron transfer chain involved in photosynthesis, and the site at which its electrons are transferred is at the plastoquinone level (Belkin and Padan, 1982a,b). This conclusion was reached since the plastoquinone inhibitor 2,5-dibromo-3-methyl-6-isopropy-*p*-benzoquinone (BDM1B) severely inhibited the electron transfer from sulphide. In the presence of N,N,N¹,N¹-tetramethyl-*p*-phenylenediamine (TMPD), known to introduce electrons at the plastocyanin level, no induction lag was observed with sulphide in the presence or absence of dithionite. Thus, the induction process must involve components preceding the TMPD donation site.

The enzyme responsible for sulphide oxidation can also be excluded from involvement in the induction process, since it is probably involved in the reverse reaction, namely, the reduction of sulphur to sulphide, which was found to be constitutive (Oren and Shilo, 1979).

In *O. limnetica,* the electrons donated by H_2S to the photosynthesis electron transport chain can, under different conditions, be used alternatively in three different ways:

1. To photoassimilate CO_2
2. In the absence of CO_2 to evolve hydrogenase-mediated hydrogen (Belkin and Padan, 1978a,b,c)
3. In the absence of a nitrogen source to fix atmospheric dinitrogen after the induction of nitrogenase (Belin *et al.*, 1979).

Anaerobic DCMU-insensitive dinitrogen fixation by nonheterocystous cyanobacteria other than *O. limnetica* has been found by Stewart and Lex (1970) and Rippka and Waterbury (1977), and in microaerophilic conditions by Stal and Krumbein (1981). This process, overlooked until recently, may play an important role in the economy of rice cultivation in rice paddies.

In addition to the ability to shift from oxygenic to anoxygenic photosynthesis, some benthic cyanobacteria have an alternative mechanism to cope with environments rich in sulphide. Y. Cohen and B. B. Jørgensen (personal communication) have shown that *Microcoleus chtonoplastis* from the cyanobacterial mats of

Solar Lake can continue their oxygenic photosynthesis even when H_2S concentrations reached 2 mM. Thus, these organisms are resistant to the toxic effect of sulphide.

The metabolic versatility of benthic cyanobacteria is also expressed in their ability to use multiple alternative pathways for energy generation in the dark. *Oscillatoria limnetica* was found to respire its reserve polyglucose under aerobic conditions, but under anaerobic conditions it could also ferment it to lactic acid or, in the presence of elemental sulphur, reduce the latter to sulphide by anaerobic respiration (Oren and Shilo, 1979).

An additional adaptation found in benthic cyanobacteria to changes in the oxygen concentration in the environment is their ability to increase the level of superoxide dismutase (SOD) when shifted from anaerobic to aerobic conditions. In *O. limnetica,* with a shift to aerobic conditions, the SOD level increased eightfold (Friedberg *et al.,* 1979).

Adaptations of Epilithic and Epiphytic Cyanobacteria Attached to Littoral Algae and Aerial Roots of Mangroves

Cyanobacteria attached to the rocky littoral fringe, to algae in the littoral zone, and to aerial roots of mangrove trees are exposed to extreme conditions of desiccation, salinity fluctuations, and often lethal photo-oxidation.

The colonisation of the different parts of aerial roots of mangroves (*Avicennia marina*) in the Sinai Peninsula shows a very distinct pattern of species distribution (Por *et al.,* 1977; Potts, 1979). At the upper borderline of the intertidal zone on the avicennia roots, we find *Scytonema* spp. colonies which are exposed during low tide to very high temperatures, extremely high salinities, desiccation, and extremely strong photo-oxidative conditions. The adaptive mechanisms of these cyanobacteria have hardly been studied, but we already have indications of the existence of several important survival strategies. The outer surface of the *Scytonema* colonies has a dark brown pigmentation, which may provide protection against photo-oxidation, and dinitrogen fixation in this organism is rhythmically induced upon wetting at high tide (Potts, 1979).

Conclusions

Hydrophobicity of the cell surface of the benthic cyanobacteria seems to be the major mechanism responsible for adhesion to the sediment surface. This extends the notion that hydrophobic interactions are responsible for a wide range of attachment mechanisms in aquatic ecosystems (Fletcher and Loeb, 1979; Marshall, 1976; Dählback *et al.,* 1981).

The benthic cyanobacteria, as a consequence of their adherence to the water–

soil interface, possess adaptation mechanisms which allow them to survive and multiply in the unique conditions prevailing in their niche. These include mechanisms to secure a sufficient supply of light by phototaxis and by the production of extracellular bioflocculants. In addition, they have complex polymorphic life cycles, including a hydrophilic stage to secure their dispersal, as well as metabolic versatility to allow for growth in their fluctuating environment.

The cyanobacteria adhering to surfaces in the littoral zone, which are exposed to the drastic changes connected with tidal cycles, also possess, in addition to their hydrophobic cell surface, mechanisms of adaptation to cope with the drying and wetting cycles, with extreme photo-oxidative conditions, and with the periodic drastic changes in salinity.

The benthic cyanobacteria have their origin in the early Precambrian era, and they may even be relics of their ancient fossil predecessors. We may thus search for ancient traits and metabolic pathways which have been retained in these organisms. Their ability to shift from oxygenic to anoxygenic photosynthesis; their ability to evolve H_2 under conditions of overreduction; their tolerance of sulphides; and their ability to function in conditions with extremely low redox potential may be examples of such ancient characteristics. Analysis of these unique properties of the benthic cyanobacteria and study of their fossil record, encoded in their macromolecular composition and structure, should help us to unravel their past evolutionary history.

Acknowledgments

This work was supported by the Wolfson Foundation and by a grant from Solmat Systems Ltd. We thank Dr. I. Kahane at the Institute of Microbiology of Hebrew University for his help in the use of fluorescent hydrophobic probes.

References

Avnimelech, Y., Troeger, B. W. and Reed, L. W. (1982). Mutual flocculation of algae and clay: evidence and implications. *Science* **216**, 63–65.
Belkin, S. and Padan, E. (1978a). Sulfide dependent hydrogen evolution in the cyanobacterium *Oscillatoria limnetica*. *FEBS Lett.* **94**, 291–294.
Belkin, S. and Padan, E. (1978b). Sulfide dependent hydrogen evolution and CO_2 photoassimilation by the cyanobacterium *Oscillatoria limnetica*. *In* "Hydrogenases: their Catalytic Activity, Structure and Function" (Eds. H. G. Schlegel and K. Schneider), pp. 381–394. E. Goltze, K. C. Göttingen
Belkin, S. and Padan E. (1979). Hydrogen metabolism in the facultative anoxygenic cyanobacteria (blue-green algae) *Oscillatoria limnetica* and *Aphanotece halophytica*. *Arch. Microbiol.* **116**, 109–111.

Belkin, S. and Padan, E. (1982a). Low redox potential elicits sulfide resistance and promotes photosynthetic sulfide utilization by the cyanobacterium *Oscillatoria limnetica*. *Plant Physiol.*

Belkin, S. and Padan, E. (1982b). Low redox potential promotes sulfide and light dependent hydrogen evolution maintained by hydrogenase in *Oscillatoria limnetica*. *J. Gen. Microbiol.*

Belkin, S., Cohen, Y. and Padan, E. (1979). Sulfide dependent activities of the cyanobacterium *Oscillatoria limnetica*- further observations. *Abstr. Int. Symp. Photosynth. Prokaryotes, Third*, p. B17.

Brock, T. D. (1973). Evolutionary and ecological aspects of the cyanophytes. *In* "The Biology of Blue-Green Algae" (Eds. N. G. Carr and B. A. Whittman), pp. 487–500. Blackwell Oxford.

Cohen, Y., Jørgensen, B. B., Padan, E. and Shilo, M. (1975a). Sulfide dependant anoxygenic photosynthesis in the cyanobacterium *Oscillatoria limnetica*. *Nature (London)* **257**, 489–492.

Cohen, Y., Padan, E. and Shilo, M. (1975b). Facultative anoxygenic photosynthesis in the cyanobacterium *Oscillatoria limnetica*, *J. Bacteriol.* **123**, 855–861.

Dählback, B., Hermannsson, M., Kjelleberg, S. and Norkrans, B. (1981). The hydrophobicity of bacteria— an important factor in their initial adhesion at the air-water interface. *Arch. Microbiol.* **128**, 267–270.

Danielsson, A., Norkrans, B. and Bjornsson, A. (1977). On bacterial adhesion- the effect of certain enzymes on adhered cells of a marine Pseudomonas sp. *Bot. Mar.* **20**, 13–17.

Eloff, J. N., Steinitz, Y. and Shilo, M. (1976). Photooxidation of cyanobacteria in natural conditions. *Appl. Environ. Microbiol.* **31**, 118–126.

Fattom, A. (1983). Bioflocculant Production by Benthic Cyanobacteria. Ph.D. Thesis, Hebrew University, Jerusalem.

Fattom, A., and Shilo, M. (1984). Hydrophobicity as an adhesion mechanism of benthic cyanobacteria. *Appl. Environ. Microbiol.* **47**, 135–143.

Fenchel, T. and Staarup, B. J. (1971). Vertical distribution of photosynthetic pigments and the penetration of light in marine sediments. *Oikos* **22**, 172–182.

Fletcher, M. and Loeb, G. I. (1979). Influence of substratum characteristics on the attachment of a marine Pseudomonad to solid surfaces. *Appl. Environ. Microbiol.* **37**, 67–82.

Fogg, G. F., Stewart, W. D. P., Fay, P. and Walsby, A. E. (1973). "The Blue-Green Algae." Academic Press, London.

Friedberg, D., Fine, M. and Oren, A. (1979). Effect of oxygen on the Cyanobacterium *Oscillatoria limnetica*. *Arch. Microbiol.* **123**, 311–313.

Garlick, S., Oren, A. and Padan, E. (1977). Occurrence of facultative anoxygenic photosynthesis among filamentons and unicellular cyanobacteria. *J. Bacteriol.* **128**, 623–629.

Gnaiger, E., Glinth, B. and Wieser, W. (1978). pH fluctuations in an intertidal beach in Bermuda. *Limnol. Oceanogr.* **23**, 851–857.

Gutnick, D. L., and Rosenberg, E. (1977). Oil tanker and pollution: a microbiological approach. *Annu. Rev. Microbiol.* **31**, 379–396.

Hjerten, S., Rosengren, J. and Palman, S. (1974). Hydrophobic interaction chromatography: the synthesis and the use of some alkyl and aryl derivatives of agarose. *J. Chromatogr.* **101**, 281–288.

Jørgensen, B. B., Revsbach, N. P., Blackburn, T. H. and Cohen, Y. (1979a). Diurnal cycle of oxygen and sulfide microgradients and microbial photosynthesis in a cyanobacterial mat sediment. *Appl. Environ. Microbiol.* **38,** 46–58.

Jørgensen, B. B., Kuenen, J. G. and Cohen, Y. (1979b). Microbial transformation of sulfur compounds in a stratified lake (Solar Lake, Sinai). *Limnol. Oceanogr.* **24,** 799–822.

Marshall, K. C. (1976). "Interface Microbiology." Harvard Univ. Press, Cambridge, Massachusetts.

Norkrans, B. (1980). Surface microlayers in aquatic environments. *Adv. Microb. Ecol.* **4,** 51–85.

Olsson, J., Glantz, P. O. and Krasse, B. (1976). Surface potential and adherence of oral steptococci to solid surfaces. *Scand. J. Dent. Res.* **84,** 240–242.

Oren, A. and Shilo, M. (1979). Anaerobic heterotrophic dark metabolism in the cyanobacterium Oscillatoria limnetica: sulfur respiration and lactate fermentation. *Arch. Microbiol.* **122,** 77–84.

Oren, A., Padan, E. and Malkin, S. (1979). Sulfide inhibition of photosystem II in cyanobacteria (blue-green algae) and tobacco chloroplasts. *Biochim. Biophys. Acta* **546,** 270–279.

Padan, E. (1979a). Impact of facultatively anaerobic photoautographic metabolism on ecology of cyanobacteria (Blue-green algae). *Adv. Microb. Ecol.* **3,** 1–48.

Padan, E. (1979b). Facultative anoxygenic photosynthesis in cyanobacteria. *Annu. Rev. Plant Physiol.* **30,** 27–40.

Por, F. D., Dor, I. and Amiv, A. (1977). The mangal of Sinai: limits of an ecosystem. *Helgol. Wiss. Meeresunters.* **30,** 295–314.

Potts, M. (1979). Nitrogen fraction associated with communities of heterocystous and non-hecterocyctous blue-green algae in Mangrove Forest of Sinai. *Oceologia* **39,** 359–379.

Revsbech, N. P., Jørgensen, B. B. and Blackburn, J. H. (1980a). Oxygen in the seabottom measured with a microelectrode. *Science* **207,** 1355–1356.

Revsbech, N. P., Jørensen, J., Blackburn, T. H. and Lamholt, J. P. (1980b). Distribution of oxygen in marine sediments measured with microelectrodes. *Limnol. Oceanogr.* **25,** 403–411.

Revsbech, N. P., Jørgensen, B. B. and Brix, O. (1981). Primary production of microalgae in sediments measured by oxygen microprofile. $H^{14}CO_3$ fixation and oxygen exchange methods. *Limnol. Oceanogr.* **26,** 717–730

Revsbech, N. P. et al. (1984). In preparation.

Riccardi, G., Sanangelantoni, A. M., Carbonera, A. and Ciferri, D. (1981). Characterization of mutants of *Spirulina platensis* resistant to amino acid analogues. *FEMS Microbiol. Lett.* **12,** 333–336.

Rippka, R. and Waterbury, J. B. (1977). The synthesis of nitrogenase by nonheterocystous cyanobacteria. *FEMS Microbiol. Lett.* **2,** 83–86.

Rosenberg, E. and Gutnick, D. L. (1981). The hydrocarbon-oxidizing bacteria *In* "The Prokaryotes: A Handbook on Habitats, Isolation, and Identification of Bacteria" (Eds. M. P. Starr, H. Stolp, H. G. Truper, A. Balows and H. Schlegel), pp. 903–912. Springer-Verlag, Berlin and New York.

Rosenberg, M., Gutnick, B. and Rosenberg, E. (1980). Adherence of bacteria to hydrocarbons: Simple method for measuring co-surface hydrophobicity. *FEMS Microbiol. Lett.* **9,** 29–33

Schopf, J. W. (1974). Paleobiology of the Precambrian: the age of the blue-green algae. *Evol. Biol.* **7,** 1–43.

Shilo, M. (1979). Factors that affect distribution patterns of aquatic microorganisms. *In*

"Aquatic Microbial Ecology (Eds. R. Colwell and J. Foster), pp. 5–11. Maryland Sea Front Publication, University of Maryland, College Park.

Shilo, M. (1980). Strategies of adaptation to extreme conditions in aquatic microorganisms. *Naturwissenschaften* **67**, 384–389.

Shilo, M. (1982). Photosynthetic microbial communities in aquatic ecosystems. *Philos. Trans. R. Soc. London* **297**, 565–567.

Simon, R. (1981). Gliding motility in *Aphanothece halophytica:* Analysis of wall proteins in mot mutants. *J. Bacteriol.* **148**, 315–321.

Stal. L. J. and Krunbein, W. E. (1981). Aerobic nitrogen fixation in pure cultures of a benthic marine Oscillatoria (Cyanobacteria). *FEMS Microbiol. Lett.* **11**, 295–298.

Stewart, W. D. P. and Lex, M. (1970). Nitrogenase activity in the blue-green algae *plectonema boryannum. Arch. Mikrobiol.* **73**, 250–260.

Van Oss, C. J., Gillmann, C. F. and Neumann, H. W. (1975). In "Phagocytic Engulfment and Cell Adhesiveness" (Ed. H. D. Isenberg), pp. 1–160. Dekker, New York.

Walsby, A. E. and Booker, M. J. (1980). Changes in buoyancy of a planktonic blue-green algae in response to light intensity. *Br. Phycol. J.* **15**, 311–319.

Waterbury, J. and Stanier, R. Y. (1978). Patterns of growth and development in pleurocapsalean cyanobacteria. *Microbiol. Rev.* **42**, 2–44.

Zur, R. (1979). Interaction between Algae and Inorganic Suspended Solids. M.Sc. Thesis, Technion Israel, Haifa.

7

Studies on the Regulation and Genetics of Enzymes of *Alcaligenes eutrophus*

H. G. SCHLEGEL

Institute for Microbiology
University of Göttingen
Göttingen, Federal Republic of Germany

Introduction

The title of this lecture does not try to construct relationships between the title of this symposium and the topics to be discussed in this chapter. In fact, ecology and microbial genetics are rarely brought together. To bridge the gap, one should remember that the aim of ecology is to understand the relationships of organisms to their environment. The question pertains to those properties of an organism which make it occupy a distinct niche and live in a distinct ecosystem. The recognition of these properties is relatively easy in the case of those organisms which inhabit ecosystems with strong environmental characteristics such as acid mine waters, hot springs, or alkaline brines. These extreme environments are usually inhabited by only a few species, which are present in high numbers. In contrast, the recognition of the main features which enable an organism to live and survive in an ecosystem of indistinct physicochemical and nutritional characteristics, such as the soil, is not so easy. Nonextreme environments are usually inhabited by a large number of diverse species which are present in only low numbers. In this case, not only the abiotic physical and chemical components of the ecosystem, but also the diverse groups of inhabiting organisms, have to be considered (Schlegel and Jannasch, 1981). The number of interacting species and parameters is enormously high.

Being interested in the aerobic hydrogen-oxidising bacteria, we wish to know their habitats and ecological niches. It is easy to imagine that these facultatively lithoautotrophic bacteria have an advantage in their ability to utilise hydrogen as an energy source and carbon dioxide as a carbon source. The question arises: Where in nature do ecosystems or microenvironments exist which are permanently or transiently supplied with gaseous hydrogen?

Biological production and utilisation of hydrogen are confined essentially to two kinds of environments: anaerobic and aerobic. In the anaerobic ecosystems, much more hydrogen is produced than in aerobic ones. However, the proportion of H_2 released to the atmosphere is negligible compared to the release from aerobic ecosystems. This is due to the role of hydrogen in the anaerobic food chain, which starts with the fermentation of polysaccharides and mainly causes the release of methane and hydrogen sulphide. "Interspecies hydrogen transfer" describes the close coupling of H_2-evolving and H_2-consuming bacteria in space and metabolic routes.

Aerobic environments have not been studied as closely as they deserve to be. There are many anaerobic microenvironments and locations in which fermentative hydrogen production probably occurs transiently and which are in close contact with evidently well-aerated environments, especially in the soil. Compost and conventional silage heaps are such locations. A surprisingly high proportion of the hydrogen reaching the aerobic zone of the soil in gaseous form is released from the root nodules of legumes. On the basis of existing data, we can suppose that aerobic locations, which are in close contact with anaerobic microenvironments or root nodules, are habitats of the aerobic hydrogen-oxidising bacteria. In these locations, the ability to use hydrogen as an energy source and carbon dioxide as a carbon source may be of advantage. However, relevant studies are just beginning (Conrad and Seiler, 1979, 1980a,b).

Ecological studies require detailed knowledge of the inhabitants of the respective ecosystems. Not only the basic physiological properties, nutritional requirements, and growth rates, but also the tolerance of adverse environmental conditions, longevity of cells, regulatory mechanisms, and even the organization of the genome of the cell, are of importance.

In the following paragraphs, I present new results on our research on the aerobic hydrogen-oxidising bacteria (hydrogenotrophs). Some of these results lend themselves to discussion about the relationships between bacteria and their habitats.

The Aerobic Hydrogen-Oxidising Bacteria

The aerobic hydrogen-oxidising bacteria are a taxonomically diverse, physiologically well-defined group which comprises at least 25 species belonging to nine bacterial genera (Aragno and Schlegel, 1981; Bowien and Schlegel, 1981). There are gram-positive and gram-negative members, mesophiles, and thermophiles. Some can grow on carbon monoxide (Meyer and Schlegel, 1983), other C_1 compounds, or thiosulphate as electron donors; others fix nitrogen. Chemolithoautotrophic growth with hydrogen requires two specific metabolic systems: hydrogenase and the components of the Calvin cycle. With regard to the hydrogenases, three types of enzyme complement have to be distinguished: the majority of the hydrogen-oxidising bacteria contain only a membrane-bound hydrogenase which does not reduce NAD: a few strains contain, in addition to the membrane-bound enzyme, a cytoplasmic, NAD-reducing hydrogenase. A few strains dispose of only the latter enzyme. The most intensively studied hydrogen-oxidising bacterium is *Alcaligenes eutrophus,* which is characterized by the possession of two types of hydrogenase. This chapter will deal mainly with various strains of this bacterium, such as type strain TF93, strains H16, N9A, B19, G27, and G29, and some others kept in culture collections. These are the classic strains of *A. eutrophus.*

Two further strains have been recognized to possess two types of hydrogenase. Strain CH34 was originally described as *Pseudomonas palleronii* (Mergeay *et al.,* 1978); it is resistant to Co^{2+}, Ni^{2+}, Zn^{2+}, and Cd^{2+}, and grows as well as *A. eutrophus;* however, it differs slightly with regard to the diversity of utilisable organic substrates. The other strain, *Alcaligenes hydrogenophilus* (Ohi *et al.,* 1979), differs from *A. eutrophus* apparently only in minor properties. Both strains have been used for comparison and will be mentioned occasionally.

Formation of the Hydrogenases

Hydrogenases are not required for heterotrophic growth. However, contrary to expectations, hydrogenases have often been encountered in cells grown heterotrophically on organic substrates. As beautifully shown by Lascelles and Rittenberg (Rittenberg, 1969), the hydrogenase activity of *A. eutrophus* is correlated with the quality of the substrate. When lactate-grown cells were transferred to glutamate or fructose medium, the total hydrogenase activity increased by factors of 6 and 11, respectively, and in stationary phase by up to a factor of 20. On the basis of various studies, it was postulated that derepression of hydrogenases is caused by a ''shortage'' of reducing power (Schlegel and Eberhardt, 1972). The regulation of hydrogenases was reinvestigated (Friedrich *et al.,* 1981a), which led to the design of an efficient system for the derepression of hydrogenases in *A. eutrophus* (Friedrich *et al.,* 1981b). Whereas fructose is a medial substrate enabling the cells to grow with a doubling time of 145 min, glycerol is a poor carbon source resulting in growth at a t_d of 8 hr. The specific activity of the cytoplasmic hydrogenase is 0.2 units per milligram of protein in fructose-grown cells and 0.5 units per milligram of protein in glycerol-grown cells. Growth on a mixture of 0.2% (w/v) fructose and 0.2% (w/v) glycerol results in biphasic growth (Fig. 1). During the transition, the formation of the hydrogenases is coordinately derepressed. The specific activities of the enzymes are 5- to 10-fold higher than those of autotrophically grown cells.

Similar results were obtained when the cells were grown in the chemostat. Succinate is a good growth substrate, and no hydrogenase is formed in cells

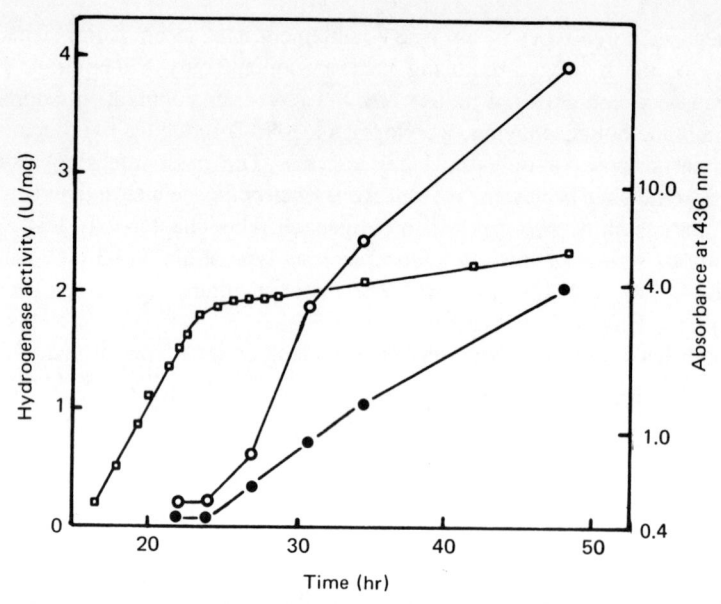

Fig. 1. Increase of hydrogenase activity in cells of *A. eutrophus* during growth on fructose and glycerol. Cells were grown in mineral salt medium containing 0.2% (w/v) fructose and 0.2% (w/v) glycerol. Growth was recorded by measuring the absorbance at 436 nm (□). Samples were taken at intervals, and the soluble and particulate cell fractions were prepared to determine the activities of the NAD-reducing hydrogenase (○) and the membrane-bound hydrogenase (●). Reprinted from Friedrich et al., 1981b.

growing in the presence of excesses of this carbon source. However, cells growing in succinate-limited continuous culture contain high activities of both hydrogenases. The data suggest that hydrogenases are formed when reducing equivalents limit growth (Friedrich, 1982). The gratuitous formation of the hydrogenases raises the question of the mechanism of derepression at the molecular level. On the other hand, the studies presented support the assumption that in natural habitats, where organic substrates limit growth, the cells always dispose of a basic level of hydrogenases.

Composition of Hydrogenases

Both hydrogenases of *A. eutrophus* H16 have been purified and characterized. One enzyme is soluble, contains flaviumononucleotide as a chromophore, reduces NAD, and has a complex structure (Schlegel and Schneider, 1978; Schneider and Schlegel, 1981). The second enzyme is membrane bound, smaller, flavin free, and unable to react with nicotinamide nucleotides (Schink and Schlegel, 1979, 1980). New aspects arose when the soluble hydrogenase was compared with the hydrogenase of *Nocardia opaca* strain 1b (Schneider et al., 1984).

Nocardia opaca, a gram-positive bacterium, contains a single cytoplasmic hydrogenase which is linked to NAD reduction and requires Ni^{2+} and Mg^{2+} ions for activity *in vitro* (Aggag and Schlegel, 1974). The *N. opaca* hydrogenase exhibited a peculiarity not shown by the *A. eutrophus* enzyme (Fig. 2). If the purified hydrogenase was separated by polyacrylamide gel electrophoresis (PAGE) in the presence of 0.5 mM $NiCl_2$, only a single homogeneous band was formed. In the absence of nickel, two bands were formed: an upper dark yellow

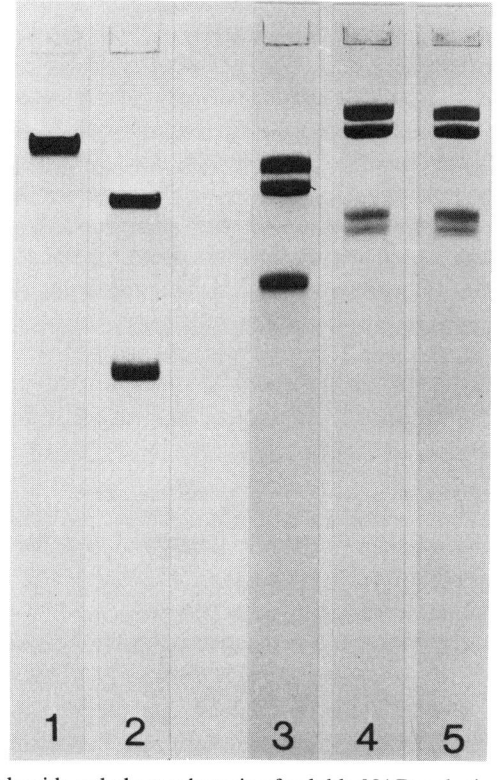

Fig. 2. Polyacrylamide gel electrophoresis of soluble NAD-reducing hydrogenases purified from *N. opaca* 1b and *A. eutrophus* H16. Track 1, *Nocardia* enzyme (25 μg) in the presence of 0.5 mM $NiCl_2$ and 5 mM $MgSO_4$ in a 7.5% (w/v) acrylamide gel; track 2, *Nocardia* enzyme in the absence of magnesium and nickel salts (conditions as for track 1); track 3, *Nocardia* enzyme (60 μg) in the presence of 0.1% (w/v) SDS and 8 M urea in a 5% (w/v) acrylamide gel; track 4, *Nocardia* enzyme (60 μg) in the presence of 0.2% (w/v) SDS and 8M urea in a 7.5% (w/v) acrylamide gel; track 5, *Alcaligenes* enzyme (50 μg) (conditions as for track 4). From K. Schneider (unpublished data).

component and a lower faint yellow component. Both bands were further separated by sodium dodecyl sulphate polyacrylamide gel electrophoresis (SDS-PAGE) into subunits with molecular weights of 64,000 and 31,000, as well as 56,000 and 27,000, respectively. This observation suggests that the *N. opaca* enzyme is composed of two dimers which dissociate in the absence of nickel. Dissociation of the enzyme into dimers can be suppressed not only by nickel ions but also by high concentrations of salts such as 0.5 *M* ammonium sulphate or 0.5 *M* potassium phosphate.

Studies have aimed at a comparison of the NAD-reducing hydrogenases of *A. eutrophus* and *N. opaca* 1b (Fig. 2). If denaturation and PAGE of the *Nocardia* enzyme were carried out under conditions elaborated for the soluble hydrogenase of *A. eutrophus*, namely, in the presence of 5% (w/v) polyacrylamide, 0.1% (w/v) sodium dodecyl sulphate (SDS), and 8 *M* urea, the *Nocardia* enzyme exhibited three protein bands, just like the *Alcaligenes* enzyme. When the conditions used for denaturation and electrophoresis were varied, a cleavage of the small subunit band was detected. This cleavage occurred with both hydrogenases when the enzymes were subjected to SDS-PAGE in the presence of 0.2% (w/v) SDS and increased concentrations of polyacrylamide (7.5–10.0%). According to this electrophoresis pattern, the hydrogenases consist of tetramers composed of four nonidentical subunits with molecular weights of 64,000, 56,000, 31,000, and 27,000 for the *Nocardia* enzyme and 63,000, 56,000, 30,000, and 26,000 for the *A. eutrophus* enzyme in a 1:1:1:1 molar ratio. This complex structure, detected in two completely different bacteria, is apparently characteristic of the type of NAD-reducing hydrogenase. On the basis of this analysis, the total molecular weight of both hydrogenases is 175,000–180,000. This coincides with a molecular weight of 178,000 determined by sucrose density gradient centrifugation of the native hydrogenase of *N. opaca*.

Nickel as a Component of the Hydrogenases

As indicated in 1962 by Bartha, *A. eutrophus* strains H1 and H16 require trace elements for lithoautotrophic growth. Subsequent studies resulted in the discovery of nickel as an essential trace element for growth of both strains on CO_2 and both H_2, but not for heterotrophic growth (Bartha and Ordal, 1965). This was the first report on the effect of nickel ions on a specific biological function in microorganisms. It required 15 years before these studies were extended to other genera of hydrogen-oxidising bacteria, with the application of a convenient plate diffusion assay and EDTA as a chelating agent (Tabillion *et al.*, 1980).

When the conditions leading to maximum hydrogenase formation during heterotrophic growth were studied, the reason for the nickel requirement was revealed (Friedrich *et al.*, 1981b). In the absence of nickel, the hydrogenase activity of *A. eutrophus* reached only 20% compared to a culture containing 12 µ*M* nickel chloride in fructose-glycerol medium. Nickel could not be replaced by cobalt, copper, or manganese. An effect of nickel on CO_2 fixation and the enzymes of the Calvin cycle was excluded by determining the respective enzyme activities and nickel-independent growth on formate. Thus, nickel was recognised to be required for the synthesis of catalytically active hydrogenase. Continuation of these studies revealed nickel to be a constituent of the soluble and membrane-bound hydrogenases of strain H16 (Friedrich *et al.*, 1982). This conclusion was based on three lines of evidence: (1) the uptake of ^{63}Ni and the formation of catalytically active hydrogenase in fructose–glycerol medium occurred coordinately; (2) chromatography of the crude extract of ^{63}Ni-grown cells on DEAE-cellulose demonstrated the coincidence of the radioactivity peaks and the activity peaks of the soluble as well as the membrane-bound hydrogenase; and (3) x-ray fluorescence analysis of the homogeneously purified, soluble hydrogenase indicated the presence of nickel—specifically, two nickel atoms per mole of enzyme.

Nickel has been reported to be a constituent of the hydrogenases of other bacteria, including *Methanobacterium thermoautotrophicum* (Graf and Thauer, 1981), *Desulfovibrio gigas* (Le Gall *et al.*, 1982), and *Clostridium thermoaceticum* (L. Ljungdahl, personal communication). The function of nickel in the enzyme protein remains to be elucidated. First results indicate that in the *A. eutrophus* soluble hydrogenase, nickel has no effect on the association of the subunits.

The CO_2 Fixation System

In cells of *A. eutrophus*, the only enzymes required for carbon dioxide fixation, ribulose bisphosphate carboxylase-oxygenase (RuBisCO) and phosphoribulokinase (PRKase), are present at high activity levels in autotrophically and formate-grown cells. There are a few substrates, such as acetate, pyruvate, and succinate, which do not promote the synthesis of these enzymes. During growth on gluconate and fructose, intermediate activities are measured (Friedrich *et al.*, 1981a). These results indicate that, like the formation of the hydrogenases, the formation of the key enzymes of the Calvin cycle depends on the quality of the organic substrate. Previous data were discussed by Bowien and Schlegel (1981). More recent experiments have provided data obtained by measurements of the enzyme activity as well as the amount of enzyme protein (Table 1); the latter data are based on rocket immunoelectrophoresis (Leadbeater *et al.*, 1982). In autotrophically grown cells, RuBisCO and PRKase constitute 7.8% and 0.68%, respectively, of the soluble cell protein. As indicated by the protein and molar ratios of the enzymes, RuBisCO is synthesized in large excess over PRKase, and

Table 1. *Activities and protein levels of RuBisCO and PRKase in Alcaligenes eutrophus*

Substrate	RuBisCO specific activity (units/mg cell protein)	PRKase (units/mg cell protein)	Activity ratio (RuBisCO/PRKase)	RuBPC content (% soluble cell protein)	PRK content (% soluble cell protein)	Molar ratio (RuBisCO/PRKase)[a]
CO_2/H_2	0.123	0.243	0.51	7.8	0.68	5.5
Gluconate	0.050	0.041	1.22	4.5	0.24	9.0
Fructose	0.035	0.011	3.18	1.54	0.09	8.2

[a] Based on the molecular weights of 534,000 for RuBisCO and 256,000 for PRKase.

the RuBisCO/PRKase ratio in the heterotrophically grown cells is higher than in those grown autotrophically. This means that the synthesis of the two enzymes is not strictly coordinated. These findings are especially intriguing since each of the enzymes can perform its function only in the presence of the other enzyme. The solution of this apparent paradox awaits genetical analysis.

When *A. eutrophus* H16 was grown in the chemostat at various growth rates and under varied substrate limitations, significant results were obtained (Friedrich, 1982). RuBisCO activities are not correlated with those of the hydrogenases. Succinate limitation results in intermediate RuBisCO activities. CO_2 limitation during autotrophic growth gives the highest RuBisCO activities, whereas energy limitation leads to very low enzyme activities. As indicated by the results of this study, RuBisCO is subject to a complex regulation: Derepression apparently occurs if reducing power and energy are excessive and growth is limited by the availability of carbon compounds.

RuBisCO catalyzes both the carboxylation of ribulose-1,5-bisphosphate to yield 3-phosphoglycerate and its oxygenolytic cleavage, resulting in the formation of 2-phosphoglycollate (see Chapter 5, this volume). Glycollate excretion by *A. eutrophus* H16 occurred when autotrophically grown cells were incubated under hydrogen and oxygen in the absence of carbon dioxide, especially at an elevated oxygen partial pressure in the atmosphere (Codd et al., 1976). This investigation on glycollate excretion was extended by an elegant study using a glycollate oxidoreductase-deficient mutant of the *A. eutrophus* type strain, and the competitive nature of carbon dioxide fixation and glycollate excretion has been elegantly demonstrated (King and Andersen, 1980, and Chapter 5, this volume).

Carbon dioxide fixation via the Calvin cycle requires the activity of several enzymes which are not involved solely in this cycle, but serve basic metabolic functions in the cell. The following question arises: Under autotrophic growth conditions are isoenzymes such as glyceraldehyde phosphate dehydrogenase and fructose bisphosphatase formed? These enzymes may be subject to activity control. However, relevant problems so far have not been studied in detail for any autotrophic bacterium. Only in autotrophically grown *N. opaca* have isoenzymes of fructose bisphosphatase (forms A and B) been found. In heterotrophically grown cells, only one enzyme (form B) was present (Amachi and Bowien, 1979).

Defective Mutants in Lithoautotrophic Metabolism

Studies aimed at the differentiation of the lithoautotrophic capabilities on the genome of *A. eutrophus* H16 have been initiated by the isolation of mutants (Aut$^-$) unable to grow with $H_2 + CO_2$. A few such mutants had been isolated after mutagenic treatment (Pfitzner, 1974; Schink and Schlegel, 1978), and it was shown that whereas both enzymes can function in energy supply to the cell, a lack of the membrane-bound hydrogenase scarcely affected autotrophic growth, a lack of the cytoplasmic hydrogenase resulted in a drastic decrease in the autotrophic growth rate.

The autotrophic character of *A. eutrophus* H16 is very stable. When cells of strains H16, G27, N9A, and the type strain (TF93) were repeatedly subcultured on complex medium and screened for mutants defective in autotrophic metabolism, Aut$^-$ mutants were found relatively seldom. Some of the spontaneous mutants were unable to oxidise hydrogen (Hox$^-$), and were defective in both the soluble, NAD-reducing hydrogenase and the membrane-bound (particulate) hydrogenase (Friedrich *et al.*, 1981c). They were able to grow on formate, indicating the functioning of the ribulose bisphosphate cycle (Friedrich *et al.*, 1979). This type of mutant was most frequently obtained. More recently, the transposon Tn5 was used for mutagenesis of strain H16. Among the Tn5-induced mutants were derivatives impaired in lithoautotrophic metabolism (Srivastava *et al.*, 1982).

The fact that in some of the Hox$^-$ mutants the Hox character had been lost irreversibly pointed to the possible involvement of plasmid DNA in hydrogen metabolism. In fact, agarose gel electrophoresis of cell lysates has revealed the presence of a large plasmid designated pHG (molecular weight 270×10^6) in all of the wild-type strains of *A. eutrophus* studied so far. Furthermore, the large plasmid was present in all revertible Hox$^-$ or leaky mutants. Only the nonrevertible Hox$^-$ mutants of strain TF93 lacked the plasmid (Friedrich *et al.*, 1981c). Physical evidence for the large plasmid was also presented by Andersen *et al.* (1981). These results have indicated that the ability to oxidise hydrogen in *A. eutrophus* is encoded by a plasmid.

Among the Aut$^-$ mutants was a large group defective in carbon dioxide fixation (Cfx$^-$), recognisable by the inability to use formate as a substrate for growth (Friedrich *et al.*, 1979). The Cfx$^-$ mutants appeared either after conventional treatment by mutagenic agents (Andersen, 1979; Friedrich *et al.*, 1979) or after transposon mutagenesis (Srivastava et al., 1982). Among the defective mutants, various types have been distinguished, such as those lacking the enzyme protein or containing large quantities of a catalytically inactive enzyme protein (Andersen, 1979). Another type was isolated on the basis of its inability to grow on succinate. One mutant (BB4) of *A. eutrophus* H16 lacked phosphoglycerate mutase activity and was unable to grow autotrophically (Reutz et al., 1982).

Conjugational Transfer of the Native Plasmid

In order to develop a conjugation system for hydrogen bacteria and to achieve the transfer of genes involved in lithotrophic metabolism, the broad host range plasmid RP4 was transferred to various strains of *A. eutrophus*. The transfer of the Hox character from RP4-containing donor cells to appropriate recipient cells was successful. However, the proper controls showed that the frequency of Hox$^+$ transconjugants was the same with RP4-free and RP4-harbouring donor strains. These experiments revealed Hox$^+$ to have a self-transmissible character. When the plasmid-free mutant TF97, derived from the type strain, was used as the recipient, the transconjugant frequency was about 10^{-2} per donor. As shown by plasmid analysis, the Hox$^+$ transconjugants contained the large plasmid. When plasmid-harbouring How$^-$ mutants of strains H16, N9A, or TF93 were used as recipients, the frequency of transconjugants was lower (about 10^{-4} per donor), indicating incompatibility. The Hox$^+$ transconjugants had acquired the soluble and particulate hydrogenase activities, as shown by enzyme and serological analysis (Friedrich et al., 1981c).

The current model presented by B. Friedrich at the International Symposium on the Genetics of Industrial Microorganisms (1982) suggests that the structural genes of both hydrogenases are located on the plasmid. In addition, there is apparently a regulatory gene, the product of which controls the transcription of structural genes. However, experimental evidence not discussed here suggests that an additional chromosomal component is required for the expression of plasmid-determined genes (Hogrefe et al., 1984; Friedrich et al., 1984).

The Distribution of Plasmids in Hydrogen-Oxidising Bacteria

The high frequency of the spontaneous, irreversible loss of autotrophy in *N. opaca* strain 1b, and the transfer of the autotrophic character from autotrophic donor strains to *Rhodococcus erythropolis* and to Aut$^-$ mutants of *N. opaca*, provided the first indication that the ability to grow as a lithoautotroph may be encoded on a plasmid (Reh and Schlegel, 1975, 1981). The detection of a conjugative plasmid-encoding hydrogen oxidation ability in *A. eutrophus* (Friedrich *et al.*, 1981c) created the desire to find out whether plasmid DNA is present in other genera and species of the hydrogen-oxidising bacteria described. Almost all of the strains have been subjected to plasmid analysis by agarose gel electrophoresis (Gerstenberg *et al.*, 1982).

Of particular interest are those hydrogen bacteria which are characterised by the possession of two types of hydrogenase. This group comprises the classic strains of *A. eutrophus* isolated more than 20 years ago near Göttingen, such as strain H16 (Wilde, 1962) and strains N9A, B19, G27, and G29, as well as the strains isolated in the United States (Davies *et al.*, 1970), such as the type strain (TF93) and others. These wild-type strains are almost identical with regard to the diversity of utilisable substrates, degradation pathways, and other biochemical properties. *Alcaligenes ruhlandii* (Packer and Vishniac, 1955; Aragno and Schlegel, 1977), strain CH34, characterized by heavy metal resistance (Mergeay *et al.*, 1978), and *A. hydrogenophilus* (Ohi *et al.*, 1979) belong to this group.

Contrary to expectations, not all of the wild-type strains that share the ability to form two types of hydrogenase have a plasmid pattern in common (Table 2). Evidence for the large plasmid pHG (molecular weight about 270×10^6) was found in the classic strains and *A. hydrogenophilus*. The latter strain has a smaller plasmid in addition (B. Friedrich, personal communication). *Alcaligenes ruhlandii* and strain CH34 lack the large plasmid and possess medium-sized plasmids. On the basis of the present studies, the plasmid patterns show that the classic strains are identical with regard to the possession of a large plasmid. However, the presence of two hydrogenases is not closely correlated with the presence of this plasmid. The sole gram-positive bacterium which contains a NAD-reducing hydrogenase, *N. opaca* 1b, even possesses several small and medium-sized plasmids, none of which is identical to the large plasmid of *A. eutrophus* H16 (M. Reh, unpublished data).

Table 2. *Hydrogen-oxidising bacteria able to form the NAD-reducing cytoplasmic hydrogenase*

Species and strains	Plasmid designation	Molecular weight (millions)
Alcaligenes eutrophus H16	pHG1	270
A. eutrophus CH34	pHG13-a	110
	pHG13-b	145
A. ruhlandii	pGH15	63
A. hydrogenophilus	—	270
	—	~230
Nocardia opaca 1b		110, 96, 33, 21.3, 10.6

Table 3. *Plasmids in hydrogen-oxidising bacteria[a]*

Species and strain	Number of plasmids	Plasmid designation	Molecular weight (millions)
Alcaligenes eutrophus H16	1	pHG1	270
A. eutrophus CH34	2	pHG13-a	110
		pHG13-b	145
A. ruhlandii	1	pHG15	63
Paracoccus denitrificans "Stanier"	2	pHG16-a	50
		pHG16-b	>300
P. denitrificans "Morris"	1	pHG18	>300
P. denitrificans "Vogt"	1	pHG19	>300
Pseudomonas facilis	1	pHG20	230

[a] No plasmids were detected in *Alcaligenes paradoxus, Aquaspirillum autotrophicum, Arthrobacter* strain 11/X, *Pseudomonas palleronii, P. pseudoflava* strain GA3, or *P. saccharophila.*

All of the other hydrogen-oxidising bacteria, which form only a membrane-bound hydrogenase, exhibit various patterns of plasmids or contain none (Table 3). Large plasmids with molecular weights above 300×10^6 are present in the three strains of *Paracoccus denitrificans* studied. Among these, the old strain isolated by Beijerinck and Minkman (1910) contains a second medium-sized plasmid. *Pseudomonas facilis,* in which the presence of a plasmid was predicted (Pootjes, 1977), was shown to contain a single plasmid with a molecular weight of 230×10^6. There are several strains which are apparently plasmid free.

Among the nitrogen-fixing, hydrogen-oxidising bacteria (Table 4), *Alcaligenes latus* (Palleroni and Palleroni, 1978) stands out by possessing at least six differently sized plasmids, one with a molecular mass above 300×10^6. *Derxia gummosa* is characterized by two large plasmids. The two slowly growing bacteria *Microcyclus aquaticus* and *Renobacter vacuolatum* resemble each other by containing two medium-sized plasmids. Surprisingly, in *Xanthobacter autotrophicus,* for which a conjugational gene transfer system has been described (Wilke, 1980), no plasmid DNA band was detected.

Among the carbon monoxide-oxidising bacteria, which as a group share the ability to grow on $H_2 + CO_2$ with the hydrogen-oxidising bacteria, only three out of five species examined proved to contain plasmids (Table 5). One species, *Achromobacter carboxydus,* harbours a large plasmid. Two species, *Pseudomonas carboxydovorans* and *P. carboxydoflava,* contain one or two medium-sized plasmids, respectively, and two species possess none.

Considering the physiological group of the hydrogen-oxidising bacteria as a whole, the diversity of plasmid patterns is remarkable. There are many strains

Table 4. *Plasmids in nitrogen-fixing hydrogen bacteria[a]*

Species and strain	Number of plasmids	Plasmid designation	Molecular weight (millions)
Alcaligenes latus	8	pNG1-a	50
		pNG1-b	120
		pNG1-c	130
		pNG1-d	150
		pNG1-e	218
		pNG1-f	280
		pNG1-g	300
		pNG1-h	>300
Derxia gummosa	2	pNG2-a	300
		pNG2-b	>300
Microcyclus aquaticus	2	pNG3-a	120
		pNG3-b	160
Renobacter vacuolatum	2	pNG4-a	95
		pNG4-b	160

[a] No plasmids were detected in *Xanthobacter autotrophicus* strains 7cSF and GZ29 or *Xanthobacter flavus* 301.

which are apparently free from plasmids. The plasmids found vary in size within a wide range. Correlations between plasmid sizes, numbers, and patterns and potential peripheral metabolic activities such as hydrogen oxidation, nitrogen fixation, autotrophic carbon dioxide fixation, and carbon monoxide and thiosulfate utilisation, aromatic carbon compound degradation or the type of hydrogenase could not be detected. For the majority of strains, only physical evidence for the plasmids was presented. The cryptic nature of these plasmids does not allow speculation on their genetic information and transmissibility. Discussion of the role of extrachromosomal DNA in lithoautotrophic bacteria is certainly not yet justified.

Table 5. *Plasmids in carbon monoxide-oxidising bacteria[a]*

Species and strain	Number of plasmids	Plasmid designation	Molecular weight (millions)
Achromobacter carboxydus	3	pCG1-a	100
		pCG1-b	270
		pCG1-c	305
Pseudomonas carboxydoflava	2	pCG2-a	25
		pCG2-b	77
P. carboxydovorans OM5	1	pCG3	85

[a] No plasmids were detected in *Comamonas compransoris* and *Pseudomonas carboxydohydrogena*.

Transfer of Plasmid Encoding Herbicide Degradation

As mentioned above, *A. eutrophus* is a recipient for IncP1 group plasmids. (Friedrich *et al.*, 1981c; Srivastava *et al.*, 1982). The detection of a P1 group herbicide-degrading plasmid (pJP4) in a soil bacterium, strain JMP134, which was assigned to *A. eutrophus* (Don and Pemberton, 1981), resulted in a study on the transfer and expression of this plasmid in hydrogen-oxidising bacteria (Friedrich *et al.*, 1983). The original host which was isolated from enrichment cultures with 2,4-dichlorophenoxyacetic acid (2,4-D) as substrate neither grew lithoautotrophically nor formed hydrogenases and RuBisCO. However, morphologically, and with regard to its pattern of utilisable substrates, it was very similar to *A. eutrophus*. The best recipients among the hydrogen bacteria were the strains of *A. eutrophus* (Table 6). These all formed transconjugants when agar mating was performed on nutrient broth, fructose medium, the selective 2,4-D medium. The donor was *Escherichia coli* strain JMP397, harbouring the plasmid pJP4. Unlike *E. coli,* in which only mercury resistance and phage susceptibility were expressed, in the lithoautotrophic strains of *A. eutrophus* all four characteristics encoded by pJP4 in the original host JMP134 were expressed. These are the ability to grow on 2,4-D and 3-chlorobenzoate, resistance to mercuric ions, and susceptibility to the male-specific phage PR11, with a few exceptions.

All transconjugants were able to act as donors in intraspecific crosses. Strain CH34 was chosen as a recipient in these crosses because it was outstanding as a recipient by a high transfer frequency when used as recipient (2×10^{-2} transconjugants per donor with JMP397 as donor), and it offered a convenient marker for counterselection because of its ability to grow in the presence of 1 mM nickel ions. The donor's abilities correlated well with the stability of the plasmid pJP4.

Table 6. *Transfer of plasmid pJP4 from Escherichia coli JMP397 to various autotrophic bacteria*

Recipient	Transfer frequency per donor	Recipient	Transfer frequency per donor
A. eutrophus		*A. eutrophus*	
CH34	2×10^{-2}	A 701	6×10^{-3}
H16	9×10^{-4}	A 702	1×10^{-6}
N9A	2×10^{-3}	A 704	2×10^{-6}
G27	1×10^{-3}	A 705	1×10^{-5}
G29	1×10^{-3}	A 707	4×10^{-3}
B19	1×10^{-4}	TF93	4×10^{-3}

Physical evidence for the presence of plasmid pJP4 was obtained by subjecting the cell lysates of all donors, recipients, and transconjugants to agarose gel electrophoresis. Bands of the plasmid DNA (50 × 10^6) were recovered in all donors and transconjugants. It was shown that pJP4 coexisted with the resident plasmids. In some hosts, pJP4 had a reduced size.

As reported (Pemberton et al., 1979), liquid enrichment for 2,4-D-utilising bacteria was very successful and resulted in fairly well-growing isolates from 60% of the soil samples used. This indicates a wide distribution of bacteria harbouring the 2,4-D plasmid in the soil. As the plasmid can be easily transferred to *Rhizobium* sp., *Agrobacterium tumefaciens*, *Pseudomonas putida*, *Pseudomonas fluorescens*, *Acinetobacter calcoaceticus*, and *Rhodopseudomonas sphaeroides*, curiosity arises as to whether this plasmid is present in these bacteria in nature, too. So far, we have tested only 12 of the lithoautotrophic wild-type strains of *A. eutrophus*, and found neither growth on the two chlorinated aromatics and mercury resistance nor physical evidence for a plasmid similar to pJP4 in addition to the native plasmid pHG. However, as mentioned, the plasmid pJP4 is not maintained in a stable state in most strains tested. This means that even if we had isolated hydrogen bacteria containing a herbicide plasmid 20 years ago, the plasmid would have been lost during transfer under nonselective conditions in the culture collection. Fresh isolates should therefore be examined for plasmid DNA, as well as plasmid-encoded peripheral pathways and other properties, to estimate the incidence of plasmids within soil bacteria in their habitats. It is possible that bacteria owe their survival in the soil to one or the other plasmid, which, due to low stability, becomes lost on isolation in laboratory culture and thus escapes detection.

Derepression and Regulation of Lactate Dehydrogenase

Another physiological/genetic trait of *A. eutrophus* not previously considered in other bacteria may be of ecological significance. This trait concerns survival under anoxic conditions. *Alcaligenes eutrophus* is a strictly aerobic bacterium with exclusively respiratory metabolism. It is not a fermentative organism and does not grow anaerobically. Aerobically grown cells do not contain fermentation enzymes such as NAD-linked lactate dehydrogenase, alcohol dehydrogenase, or butanediol dehydrogenase. The respective genes were assumed to be lacking. However, when the cells were kept for prolonged periods of time (5–20 hr) under low or very low aeration conditions, allowing cellular respiration to occur only at 1/10th or 1/20th of its maximum rate, lactate, butanediol, and ethanol were excreted (Vollbrecht and Schlegel, 1978). The cells harvested contained the corresponding enzymes (Schlegel and Vollbrecht, 1980; Schlegel

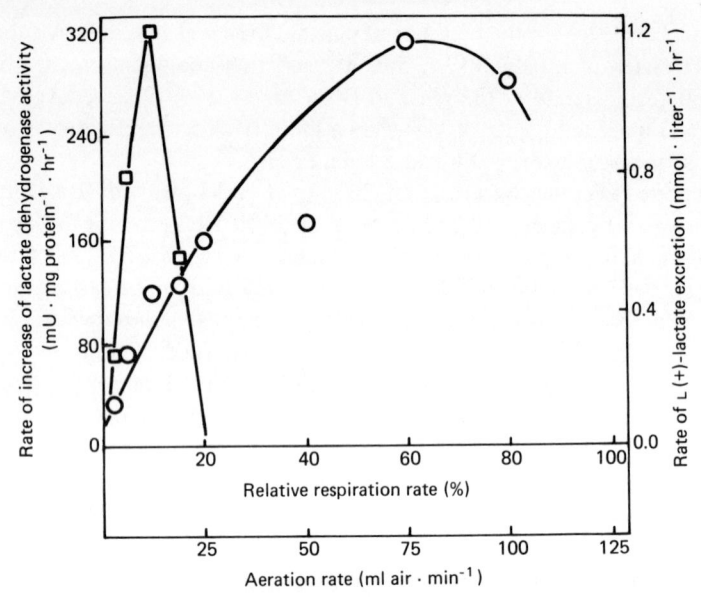

Fig. 3. Dependence of the rates of L(+)-lactate excretion and the increase of lactate dehydrogenase activity on the relative respiration rate of *A. eutrophus* cells incubated under varied conditions of oxygen supply. *Alcaligenes eutrophus* N9A-PHB⁻02-HB⁻1 was grown in eight 4-litre fermentors, each containing 3.8 litres mineral medium with 1.0% (w/v) sodium gluconate as substrate. All fermentors were aerated with 500 ml air × min⁻¹, agitated at 630 rpm, and kept at 30°C. After the cells growing exponentially with an unrestricted oxygen supply had reached approximately 2 g dry weight per litre, 1.5% (w/v) sodium gluconate was added, and the aeration rate was decreased to allow only a restricted respiration rate to occur. Each fermentor was run at a different aeration rate, which resulted in varied relative respiration rates within a range of 3 to 80% of the maximum respiration rates (abscissa). Samples were taken at various time intervals, and in the cells lactate dehydrogenase activity was determined and lactate concentration was measured in the culture supernatant. From these data, the initial rates of increase of both enzyme activity (○) and lactate concentration (□) were calculated and plotted (ordinates).

and Steinbüchel, 1981). This indicates the presence of "dormant genes" (Riley and Anilionis, 1978) which are derepressed only under conditions differing from ordinary cultural conditions.

Derepression of lactate dehydrogenase and lactate excretion occur at low respiration rates. However, whereas lactate dehydrogenase is formed at a wide range of respiration rates corresponding to 3–80% of the maximum rates, lactate is excreted only at very low rates, such as 3–10% of the maximum respiration rates (Fig. 3). These relationships indicate an effective cellular control mechanism that impedes the lactate dehydrogenase from functioning and produces lactate. The lactate dehydrogenase purified from *A. eutrophus* N9A-PHB⁻02 is indeed extremely sensitive to oxaloacetate. Reduction of pyruvate is inhibited almost completely by 1 μM oxaloacetate, and 90% inhibition occurs at 0.1 μM. This type of inhibition is considered to be of physiological importance, allowing the function of lactate dehydrogenase only at low oxaloacetate concentrations, and consequently at high NADH/NAD ratios and very low oxygen supply rates, respectively (Steinbüchel and Schlegel, 1983a,b,c).

It should be mentioned that derepression of lactate dehydrogenase occurred in several strains of *A. eutrophus,* and also in Hox$^-$ mutants which had lost the large native plasmid such as strains TF97 and N9AF06. The respective genes are therefore localized on the bacterial chromosome (Steinbüchel *et al.,* 1983).

Current investigations point to a widespread ability among strict aerobes to form fermentation enzymes when exposed to conditions of restricted respiration. This ability may be useless for the cell and an evolutionary relic. There are many examples of the retention of dormant genetic material in the genome which one might expect to be eliminated as an unnecessary burden (Riley and Anilionis, 1978). On the other hand, fermentation enzymes and modest fermentative energy production may be of ecological significance, and may confer a selective advantage on strictly aerobic bacteria temporarily exposed to anoxic conditions. To examine this hypothesis, we are presently concerned with measuring the duration of cells viability under various conditions.

Conclusions and Encouragement

Microbial biochemistry and genetics have been advancing at an extraordinarily rapid pace during the last 30 years. Many incentives for the study of regulation have arisen from ecological observations. The microbiological literature dating back to about 1900 contains many indications of phenomena such as enzyme induction and repression, enzyme inhibition, and metabolic consequences, of course, but without clear ideas about the mechanism involved and precise usage of terms. Now the basic mechanisms of these regulatory processes are fairly well known, and their ecological significance is becoming increasingly clear.

However, a new dimension was added to microbiology by the discovery of bacterial conjugation, the bacterial sex factor, and other extrachromosomal elements. The bacterial genome is divided into two parts, the chromosome and plasmids. The plasmids carry information which is nonessential for survival of the cells under the culture conditions provided in the laboratory. Plasmids are more mobile than the chromosome; most if not all plasmids of gram-negative bacteria can be conjugally transferred. Conjugative plasmids promote their own conjugal transfer, and nonconjugative plasmids are frequently mobilised from cells which contain a conjugative plasmid.

During the past decade, many plasmids have been described that encode the ability to catabolise "exotic" substrates such as aromatic carbon compounds, xenobiotics, terpenes, and halogenated compounds. Many of these compounds are natural substances; others, such as herbicides, fungicides, and insecticides, are synthetic compounds. They finally enter the soil where they are degraded. The presence of relevant plasmids confers a selective advantage on the cells by enabling them to propagate and survive in otherwise low-nutrient environments. Present knowledge of the biochemical and genetic capabilities of bacteria enables us to ask precise questions about the relevance of plasmid-encoded capabilities of bacteria for their maintenance and propagation in the soil. How widely is a distinct plasmid distributed among the potential host soil bacteria? Does intercell plasmid transfer occur in the soil? Is plasmid-encoded information acquired transiently by bacteria in which the plasmids are not maintained in a stable state?

To express my speculations more clearly, I wish to consider our model system, *A. eutrophus*. As shown, the Hox plasmid pHG is conjugatively transferred with very high frequency. This raises the question of whether this transfer has any significance in nature. Is *A. eutrophus* accompanied by a bacterium which depends on the plasmid but cannot maintain it when grown in the laboratory? Are there heterotrophic bacteria which benefit from the plasmid-encoded information for expressing the hydrogenase genes or other information? Hox$^+$ phenotypes without the ability to fix CO_2 would be imaginable and would be the reciprocal pendant to *Pseudomonas oxalaticus,* which corresponds to a hydrogen bacterium defective with regard to the ability to form hydrogenase (Hox$^-$, Cfx$^+$). Furthermore, it is still questionable whether the ability to oxidise hydrogen is really the ecological niche for *A. eutrophus*. It is possible that due to its ability to act as a recipient in conjugational plasmid transfer, the bacterium takes advantage of biochemical degradative abilities which are not expressed in the laboratory, due to the loss of the plasmid under nonselective conditions.

One can ask the questions about many of the bacteria in the soil, especially highly specialized ones such as nitrifiers, sulphur oxidisers, iron oxidisers, cellulose-utilising, lignin-degrading, and other bacteria. Studies on these lines of speculation should be rewarding and encouraged. The experimental methods are available and lend themselves to ecological studies.

Acknowledgments

This chapter contains results contributed by many collaborators over many years. I thank all of them for their lasting devotion to the hydrogen-oxidising bacteria. I owe thanks to Mrs. Bärbel Friedrich for reading the manuscript. The recent investigations, on which large parts of the report are based, were supported by grants from the Deutsche Forschungsgemeinschaft and by Förderungsmittel des Landes Niedersachsen.

References

Aggag, M. and Schlegel, H. G. (1974). Studies on a Gram-positive hydrogen bacterium, *Nocardia opaca* 1b. III. Purification, stability, and some properties of the soluble hydrogen dehydrogenase. *Arch. Microbiol.* **100**, 25–39.

Amachi, T. and Bowien, B. (1979). Characterization of two fructose bisphosphatase isoenzymes from the hydrogen bacterium *Nocardia opaca* 1b. *J. Gen. Microbiol.* **113**, 347–356.

Andersen, K. (1979). Mutations altering the catalytic activity of a plant-type ribulosebisphosphate carboxylase/oxygenase in *Alcaligenes eutrophus*. *Biochim. Biophys. Acta* **585**, 1–11.

Andersen, K., Tait, R. C. and King, W. R. (1981). Plasmids required for utilization of molecular hydrogen by *Alcaligenes eutrophus*. *Arch. Microbiol.* **129**, 384–390.

Aragno, M. and Schlegel, H. G. (1977). *Alcaligenes ruhlandii* (Packer and Vishniac) comb. nov. a peritrichous hydrogen bacterium previously assigned to *Pseudomonas*. *Int. J. Syst. Bacteriol.* **27**, 279–281.

Aragno, M. and Schlegel, H. G. (1981). The hydrogen-oxidizing bacteria. *In* "The Prokaryotes: A Handbook on Habitats, Isolation and Identification of Bacteria" (Eds. M. P. Starr, H. Stolp, H. G. Trüper, A. Balows and H. G. Schlegel), Vol. 2, pp. 865–893. Springer-Verlag, Berlin and New York.

Bartha, R. (1962). Physiologische Untersuchungen über den chemolithotrophen Stoffwechsel neu isolierter *Hydrogenomonas*-Stämme. *Arch. Microbiol.* **41**, 313–350.

Bartha, R. and Ordal, E. J. (1965). Nickel-dependent chemolithotrophic growth of two *Hydrogenomonas* strains. *J. Bacteriol.* **89**, 1015–1019.

Beijerinck, M. and Minkman, D. C. J. (1910). Bildung und Verbrauch von Stickoxydul durch Bakterien. *Zentralbl. Bakteriol., Parasitenk., Infektionskr. Hyg. Abt. 2, Naturwiss.: Allg., Landwirtsch. Tech. Mikrobiol.* **25**, 30–63.

Bowien, B. and Schlegel, H. G. (1981). Physiology and biochemistry of aerobic hydrogen-oxidizing bacteria. *Annu. Rev. Microbiol.* **35**, 405–452.

Codd, G. A., Bowien, B. and Schlegel, H. G. (1976). Glycollate production and excretion by *Alcaligenes eutrophus*. *Arch. Microbiol.* **110**, 167–171.

Conrad, R. and Seiler, W. (1979). The role of hydrogen bacteria during the decomposition of hydrogen by soil. *FEMS Microbiol. Lett.* **6**, 143–145.

Conrad, R. and Seiler, W. (1980a). Contribution of hydrogen production by biological nitrogen fixation to the global hydrogen budget. *JGR, J. Geophys. Res.* **85**, 5493–5498.

Conrad, R. and Seiler, W. (1980b). Decomposition of atmospheric hydrogen by soil microorganisms and soil enzymes. *Soil. Biol. Biochem.* **13**. 43–49.

Davies, D. H., Stanier, R. Y., Doudoroff, M. and Mandel, M. (1970). Taxonomic studies on some Gram-negative polarly flagellated "hydrogen bacteria" and related species. *Arch. Mikrobiol.* **70**, 1–13.

Don, R. H. and Pemberton, J. M. (1981). Properties of six pesticide degradation plasmids isolated from *Alcaligenes paradoxus* and *Alcaligenes eutrophus*. *J. Bacteriol.* **145**, 682–686.

Friedrich, C. G. (1982). Derepression of hydrogenase during limitation of electron donors and derepression of ribulosebisphosphate carboxylase during carbon limitation of *Alcaligenes eutrophus*. *J. Bacteriol.* **149**, 203–210.

Friedrich, C. G., Bowien, B. and Friedrich, B. (1979). Formate and oxalate metabolism in *Alcaligenes eutrophus*. *J. Gen. Microbiol.* **115**, 185–192.

Friedrich, C. G., Friedrich, B. and Bowien, B. (1981a). Formation of enzymes of autotrophic metabolism during heterotrophic growth of *Alcaligenes eutrophus. J. Gen. Microbiol.* **122**, 69–78.
Friedrich, B., Heine, E., Finck, A. and Friedrich, C. G. (1981b). Nickel requirement for active hydrogenase formation in *Alcaligenes eutrophus. J. Bacteriol.* **145**, 1144–1149.
Friedrich, B., Hogrefe, C. and Schlegel, H. G. (1981c). Naturally occurring genetic transfer of hydrogen-oxidizing ability between strains of *Alcaligenes eutrophus. J. Bacteriol.* **147**, 198–205.
Friedrich, C. G., Schneider, K. and Friedrich, B. (1982). Nickel—a component of catalytically active hydrogenase of *Alcaligenes eutrophus. J. Bacteriol.* **152**, 42–48.
Friedrich, B., Meyer, M. and Schlegel, H. G. (1983). Transfer and expression of the herbicide-degrading plasmid pJP4 in aerobic autotrophic bacteria. *Arch. Microbiol.* **134**, 92–97.
Friedrich, C. G., Friedrich, B. and Bowien, B. (1981a). Formation of enzymes of hydrogenase in *Alcaligenes* spp. is altered by interspecific plasmid exchange. *J. Bacteriol.* **158**, 331–333.
Gerstenberg, C., Friedrich, B. and Schlegel, H. G. (1982). Physical evidence for plasmids in autotrophic, especially hydrogen-oxidizing bacteria. *Arch. Microbiol.* **133**, 90–96.
Graf, E. G. and Thauer, R. K. (1981). Hydrogenase from *Methanobacterium thermoautotrophicum*, a nickel-containing enzyme. *FEBS Lett.* **136**, 165–169.
Hogrefe, Ch., Römermann, D. and Friedrich, B. (1984). *Alcaligenes eutrophus* hydrogenase genes. *J. Bacteriol.* **158**, 43–48.
King, W. R. and Andersen, K. (1980). Efficiency of CO_2 fixation in a glycollate oxidoreductase mutant of *Alcaligenes eutrophus* which exports fixed carbon as glycollate *Arch. Microbiol.* **128**, 84–90.
Leadbeater, L., Siebert, K., Schobert, P. and Bowien, B. (1982). Relationship between activities and protein levels of ribulosebisphosphate carboxylase and phosphoribulokinase in *Alcaligenes eutrophus. FEMS Microbiol. Lett.* **14**, 263–266.
LeGall, J., Ljungdahl, P. O., Moura, I., Peck, H. D., Jr., Xavier, A. V., Moura, J. J. G., Teixera, M., Huynh, B. H. and DerVartanian, D. V. (1982). The presence of redox-sensitive nickel in the periplasmic hydrogenase from *Desulfovibrio gigas*. *Biochem. Biophys. Res. Commun.* **106**, 610–616.
Mergeay, M., Houba, C. and Gerits, J. (1978). Extrachromosomal inheritance controlling resistance to cadmium, cobalt, and zinc ions: evidence from curing in a *Pseudomonas. Arch. Int. Physiol. Biochim.* **86**, 440–441.
Meyer, O. and Schlegel, H. G. (1983). Biology of aerobic carbon monoxide-oxidizing bacteria. *Annu. Rev. Microbiol.* **37**, 277–310.
Ohi, K., Takada, N., Komemushi, S., Okazaki, M. and Miura, Y. (1979). A new species of hydrogen-utilizing bacterium. *J. Gen. Appl. Microbiol.* **25**, 53–58.
Packer, L. and Vishniac, W. (1955). Chemosynthetic fixation of carbon dioxide and characteristics of hydrogenases in resting cell suspensions of *Hydrogenomonas ruhlandii* nov. spec. *J. Bacteriol.* **70**, 216–233.
Palleroni, N. J. and Palleroni, A. V. (1978). *Alcaligenes latus*, a new species of hydrogen bacteria. *Int. J. Syst. Bacteriol.* **28**, 416–424.
Pemberton, J. M., Corney, B. and Don. R. H. (1979). Evolution and spread of pesticide degrading ability among soil microorganisms. In "Plasmids of Medical, Environmental, and Commercial Importance" (Eds. K. N. Timmis and A. Pühler), pp. 287–299. Elsevier/North-Holland Biomedical Press, Amsterdam.

Pfitzner, J. (1974). Ein Beitrag zum H_2-O_2-Oxidoreductase-system von *Hydrogenomonas eutropha* Stamm H16: Hydrogenasedefekte Mutanten. *Zentralbl. Bakteriol., Parasitenk., Infektionskr. Hyg., Abt. 1: Orig., Reihe A* **228**, 121–127.
Pootjes, C. F. (1977). Evidence for plasmid coding of the ability to utilize hydrogen gas by *Pseudomonas facilis*. *Biochem. Biophys. Res. Commun.* **76**, 1002–1006.
Reh, M. and Schlegel, H. G. (1975). Chemolithoautotrophie als eine übertragbare, autonome Eigenschaft von *Nocardia opaca* 1b. *Nachr. Akad. Wiss. Goettingen, Math.-Phys. Kl., 2A: Math.-Phys./Chem. Abt.* **12**, 207–216.
Reh, M. and Schlegel, H. G. (1981). Hydrogen autotrophy as a transferable genetic character of *Nocardia opaca* 1b. *J. Gen. Microbiol.* **126**, 327–336.
Reutz, I., Schobert, P. and Bowien, B. (1982). Effect of phosphoglycerate mutase deficiency on heterotrophic and autotrophic carbon metabolism of *Alcaligenes eutrophus*. *J. Bacteriol.* **151**, 8–14.
Riley, M. and Anilionis, A. (1978). Evolution of the bacterial genome. *Annu. Rev. Microbiol.* **32**, 519–560.
Rittenberg, S. C. (1969). The roles of exogenous organic matter in the physiology of chemolithotrophic bacteria. *Adv. Microb. Physiol.* **3**, 159–196.
Schink, B. and Schlegel, H. G. (1978). Mutants of *Alcaligenes eutrophus* defective in autotrophic metabolism. *Arch. Microbiol.* **117**, 123–129.
Schink, B. and Schlegel, H. G. (1979). The membrane-bound hydrogenase of *Alcaligenes eutrophus*. I. Solubilization, purification, and biochemical properties. *Biochim. Biophys. Acta* **567**, 315–324.
Schink, B. and Schlegel, H. G. (1980). The membrane-bound hydrogenase of *Alcaligenes eutrophus*. II. Localization and immunological comparison with other hydrogenase systems. *Antonie van Leeuwenhoek* **46**, 1–14.
Schlegel, H. G. and Eberhardt, U. (1972). Regulatory phenomena in the metabolism of Knallgasbacteria. *Adv. Microb. Physiol.* **7**, 205–242.
Schlegel, H. G. and Schneider, K. (1978). Introductory Report: Distribution and physiological role of hydrogenases in microorganisms. In "Hydrogenases: Their catalytic Activity, Structure and Function" (Eds. H. G. Schlegel and K. Schneider), pp. 15–44. Verlag Erich Goltze KG, Göttingen.
Schlegel, H. G. and Jannasch, H. W. (1981). Prokaryotes and their habitats. In "The Prokaryotes: A Handbook on Habitats, Isolation, and Identification of Bacteria" (Eds. M. P. Starr, H. Stolp, H. G. Trüper, A. Balows and H. G. Schlegel), pp. 43–82. Springer-Verlag, Berlin and New York.
Schlegel, H. G. and Steinbüchel, A. (1981). Die relative Respirationsrate (RRR), ein neuer Belüftungsparameter. In "Fermentation" (Ed. R. M. Lafferty), pp. 10–26. Springer-Verlag, Berlin and New York.
Schlegel, H. G. and Vollbrecht, D. (1980). Formation of the dehydrogenases for lactate, ethanol, and butanediol in the strictly aerobic bacterium *Alcaligenes eutrophus*. *J. Gen. Microbiol.* **117**, 475–481.
Schneider, K. and Schlegel, H. G. (1981). Production of superoxide radicals by soluble hydrogenase from *Alcaligenes eutrophus* H16. *Biochem. J.* **193**, 99–107.
Schneider, K., Schlegel, H. G. and Jochim, K. (1984). Effect of nickel on activity and subunit composition of purified hydrogenase from *Nocardia opaca* 1b. *Eur. J. Biochem.* **138**, 533–541.
Srivastava, S., Urban, M. and Friedrich, B. (1982). Mutagenesis of *Alcaligenes eutrophus* by insertion of the drug-resistance transposon Tn5. *Arch. Microbiol.* **131**, 203–207.

Steinbüchel, A. and Schlegel, H. G. (1983a). Nicotinamide adenine dinucleotide-linked L(+)-lactate dehydrogenase from the strict aerobe *Alcaligenes eutrophus*. I. Purification and properties. *Eur. J. Biochem.* **130,** 321–328.

Steinbüchel, A. and Schlegel, H. G. (1983b). Nicotinamide adenine dinucleotide-linked L(+)-lactate dehydrogenase from the strict aerobe *Alcaligenes eutrophus*. II. Kinetic properties and inhibition by oxaloacetate. *Eur. J. Biochem.* **130,** 329–334.

Steinbüchel, A. and Schlegel, H. G. (1983c). The rapid purification of lactate dehydrogenase from *Alcaligenes eutrophus* in a two-step procedure. *Eur. J. Appl. Microbiol. Biotechnol.* **17,** 163–167.

Steinbüchel, A., Kuhn, M., Niedrig, M. and Schlegel, H. G. 1983). Fermentation enzymes in strictly aerobic bacteria: Comparative studies on strains of the genus *Alcaligenes* and on *Nocardia opaca* and *Xanthobacter autotrophicus*. *J. Gen. Microbiol.* **129,** 2825–2835.

Tabillion, R., Weber, F. and Kaltwasser, H. (1980). Nickel requirement for chemolithotrophic growth in hydrogen-oxidizing bacteria. *Arch. Microbiol.* **124,** 131–136.

Vollbrecht, D. and Schlegel, H. G. (1978). Excretion of metabolites by hydrogen bacteria. II. Influences of aeration, pH, temperature, and age of cells. *Eur. J. Appl. Microbiol. Biotechnol.* **6,** 157–166.

Wilde, E. (1962). Untersuchungen über Wachstum und Speicherstoffsynthese von *Hydrogenomonas*. *Arch. Mikrobiol.* **43,** 109–137.

Wilke, D. (1980). Conjugational gene transfer in *Xanthobacter autotrophicus* GZ29. *J. Gen. Microbiol.* **117,** 431–436.

Supplementary Readings*

Cammack, R., Lalla-Maharajh, W. V. and Schneider, K. (1982). Epr studies of some oxygen-stable hydrogenases. *In* "Electron Transport and Oxygen Utilization" (Ed. Ch. Ho), pp. 411–413. Elsevier/North-Holland Biomedical Press, Amsterdam/New York.

Friedrich, C. G. and Friedrich, B. (1983). Regulation of hydrogenase formation is temperature sensitive and plasmid coded in *Alcaligenes eutrophus*. *J. Bacteriol.* **153,** 176–181.

Im, D.-S. and Friedrich, C. G. (1983). Fluoride, hydrogen, and formate activate ribulosebisphosphate carboxylase formation in *Alcaligenes eutrophus*. *J. Bacteriol.* **154,** 803–808.

Lissolo, T., Cocquempot, M.-F., Thomas, D., LeGall, J., Schneider, K. and Schlegel, H. G. (1983). Hydrogen production using chloroplast membranes without oxygen scavengers: An assay with hydrogenases from aerobic hydrogen-oxidizing bacteria and flavodoxins from *Desulfovibrio* sp. *Eur. J. Appl. Microbiol. Biotechnol.* **17,** 158–162.

Payen, B., Segui, M., Monsan, P., Schneider, K., Friedrich, C. G. and Schlegel. H. G. (1983). Use of cytoplasmic hydrogenase from *Alcaligenes eutrophus* for NADH regeneration. *Biotechnol. Lett.* **5,** 463–468.

Pinkwart, M., Schneider, K. and Schlegel, H. G. (1983). The hydrogenase of a thermophilic hydrogen-oxidizing bacterium. *FEMS Microbiol. Lett.* **17,** 137–141.

*To enable the reader to keep abreast of relevant studies, the most recent papers that were not mentioned in the text are quoted here.

Pinkwart, M., Schneider, K., and Schlegel, H. G. (1983). Purification and properties of the membrane-bound hydrogenase from N_2-fixing *Alcaligenes latus*. *Biochim. Biophys. Acta* **745**, 267–278.

Podzuweit, H. G., Schneider, K., and Schlegel, H. G. (1983). Autotrophic growth and hydrogenase activity of *Pseudomonas saccharophila* strains. *FEMS Microbiol. Lett.* **19**, 169–173.

Schlesier, M. and Friedrich, B. (1982). Histidine utilization by *Alcaligenes eutrophus*: Regulation of histidase formation under heterotrophic conditions of growth. *Arch. Microbiol.* **132**, 254–259.

Schlesier, M. and Friedrich, B. (1982). Effect of molecular hydrogen on histidine utilization by *Alcaligenes eutrophus*. *Arch. Microbiol.* **132**, 260–265.

Schneider, K., Patil, D. S. and Cammack, R. (1983). ESR properties of membrane-bound hydrogenases from aerobic hydrogen bacteria. *Biochim. Biophys. Acta* **748**, 353–361.

Schneider, K., Pinkwart, M. and Jochim, K. (1983). Purification of hydrogenases by affinity chromatography on Procion Red-agarose. *Biochem. J.* **213**, 391–398.

8

Ecology of the Colourless Sulphur Bacteria

DON P. KELLY

Department of Environmental Sciences
University of Warwick
Coventry CV4 7AL, England
and

J. GIJS KUENEN

Laboratory of Microbiology
Delft University of Technology
Julianalaan 67A, Delft 8, The Netherlands

Introduction

In our treatment of the ecology of the colourless sulphur bacteria, we have been rather selective in our choice of illustrative material. We have attempted to find a median point between the ideals of seeking to familiarize the reader wholly unfamiliar with these organisms and of dealing in great depth with specific environments and organisms or with laboratory techniques such as mixed chemostat culture that have been used to test interactions (such as competition) between sulphur bacteria. The term "colourless sulphur bacteria" is the classic term used to describe one of the three main groups of "sulphur bacteria." The sulphur bacteria comprise the phototrophic bacteria (including some cyanobacteria), the sulphur- and sulphate-reducing bacteria, and the colourless sulphur bacteria. The interaction of these groups is essential to the sustained activity of the global sulphur cycle. The colourless sulphur bacteria are able (or in some cases, believed to be able) to generate metabolically useful energy from the oxidation of reduced inorganic sulphur compounds. In other words, the sulphur compounds act as an electron donor and a respiratory energy source for growth. Different species of colourless sulphur bacteria can use oxygen or oxidized nitrogen compounds (nitrate, nitrite, nitrous oxide) as electron acceptors for sulphur compound oxidation. The term "colourless" is sometimes rather a misnomer, as concentrated samples of bacteria such as the thiobacilli may be highly pigmented with cytochromes. None of the colourless sulphur bacteria contain photosynthetic pigments. Some of the phototrophic bacteria are able to grow not only phototrophically, with sulphide as an electron donor for carbon dioxide fixation, but also in the dark using the oxidation of reduced inorganic sulphur compounds as the sole source of energy. Clearly, such organisms are not included as members of the colourless sulphur bacteria group. The difference between the sulphur- or sulphate-reducing bacteria and the colourless sulphur bacteria is, of course, that the former organisms use organic compounds as electron donors and the latter are used only in respiratory electron acceptors for energy conservative (Postgate, 1979; Peck and Le Gall, 1982; Pfennig and Widdel, 1982).

In general, we have taken the view that the best approach to understanding the ecology of these bacteria is through their physiology. Ecological interactions are generally competitive or cooperative interactions between physiological types. The type of biochemistry being expressed is not necessarily unique to a morphologically or taxonomically discrete group of organisms. In many cases, the study of microbial ecology is less the study of morphological identification and enumeration of specific types than the evaluation of chemical manifestations of the activity of those bacteria, possibly functioning in a cooperative or synergistic manner in a given environment.

It has become apparent in recent years that the microbiology of sulphur oxidation is rather complex: The thiobacilli and filamentous types such as *Beggiatoa* are part of a system containing many physiological types, several of very restricted ecological adaptability. We are fortunate that some aspects of the ecology of sulphur bacteria have recently been reviewed (Jørgensen, 1982), as have the microbiology and biochemistry of sulphur oxidation (Kelly, 1982; Kuenen and Beudeker, 1982), as well as the role of sulphur bacteria in sulphur geochemistry and the global sulphur cycle (Kuenen, 1975; Pfennig, 1975; Kelly, 1980; Trudinger, 1982). There have also been numerous comprehensive studies on special environments and the importance of metabolic flexibility as an ecological factor (Jørgensen *et al.*, 1979; Gottschal and Kuenen, 1981; Gottschal *et al.*, 1979, 1981a; Smith and Kelly, 1979; Beudeker *et al.*, 1982).

Physiological Flexibility Among the Colourless Sulphur Bacteria and the "Dark Oxidation" of Sulphur in the Natural Environment

The differences in physiological types among the nonphotosynthetic sulphur compound-oxidising bacteria are summarised in Table 1, which illustrates that the physiological types among these organisms can be grouped under four headings. The confusion sometimes experienced by newcomers to the terminology of these bacteria is avoidable, especially if the proliferation of descriptive subter-

Table 1. Physiological types among the nonphototrophic sulphur compound-oxidising bacteria[a]

Physiological type[b]	Synonyms or alternative name commonly used	Energy source (electron donor)		Carbon source	
		Inorganic sulphur compound oxidation	Organic compound oxidation	CO_2	Organic compound
A. Obligate chemolithoautotroph	Obligate chemolithotroph Obligate autotroph Obligate chemoautotroph Obligate lithotroph	+	–	+	–
B. Facultative chemolithoautotroph	Facultative chemolithotroph Facultative autotroph Facultative chemoautotroph Facultative lithotroph Mixotroph	+	+	+	+
C. "Symbiont"	—	+	?	+	?
D. Chemolithoheterotroph	Heterotroph able to obtain energy from oxidation of an inorganic sulphur compound	+	+	–	+
E. Chemoorganoheterotroph	Heterotroph able to oxidise sulphur compounds but unable to obtain energy	–	+	–	+

[a] By definition, the colourless sulphur bacteria belong to the first four groups.
[b] Chemo, metabolic energy obtained from chemical oxidation rather than from photosynthesis; litho, inorganic rather than organic (organo-) sources of energy; auto, carbon dioxide used as the carbon source (in contrast to heterotrophic use of organic carbon sources).

minology is avoided. Thus, most of the bacterial world is composed of "chemoorganotrophic heterotrophs" (normally simply called "heterotrophs"). Some chemolithotrophic autotrophs are also capable of growth as chemoorganotrophic heterotrophs (type B) and are frequently referred to as "facultative autotrophs," a term which defines precisely the easy choice of these two types of physiology in response to environmental conditions. As these organisms can grow on mixtures of organic and inorganic compounds, they may also be called "mixotrophs." Some chemolithotrophic autotrophs (e.g., *Thiobacillus neapolitanus* and *T. denitrificans;* see Table 2) are "obligate" (type A), that is, they are capable of growth only in this physiological mode, although they may have a limited heterotrophic capacity to assimilate some exogenous organic matter at the expense of energy from sulphur compound oxidation and to metabolise intracellular polyglucose for energy generation (Kelly, 1971, 1981; Rittenberg, 1972; Kuenen and Beudeker, 1982). The occurrence and biochemistry of these two types have been considered in detail (Kuenen and Tuovinen, 1981; Kelly, 1982; Kuenen and Beudeker, 1982). Those capable of true mixotrophy (type B) seem to have the ability to sustain regulated metabolism using organic and inorganic sources of carbon and energy, and to have a considerable survival and competitive capacity in the environment (Gottschal *et al.,* 1979; Smith *et al.,* 1980; Perez and Matin, 1980; Wood and Kelly, 1981; Kuenen and Beudeker, 1982).

Table 2. Examples of sulphide-oxidising bacteria including the colourless sulphur bacteria, the phototrophic bacteria, and the inorganic sulphur compound-oxidising bacteria

Physiological type	Example
Obligate chemolithoautotroph	*Thiobacillus neapolitanus*
	T. denitrificans, including marine strains
	Thiomicrospira pelophila
Facultative chemolithoautotroph	*T. novellus*
	Thiobacillus A2
	Sulfolobus acidocaldarius
	Beggiatoa sp.
	Chromatium sp.
	Thiocapsa sp.
Symbiont able (at least) to generate energy from S^{2-} oxidation and to fix carbon dioxide	Symbionts of *Riftia pachyptila, Solemya,* and symbionts with other eukaryotes or prokaryotes?
Chemolithotrophic heterotroph	*Thiobacillus perometabolis*
	Beggiatoa sp.
Chemoorganotroph	*Pseudomonas aeruginosa* and many other soil and marine microorganisms

Among the facultative species, only *Thiobacillus* A2 and *T. novellus* have been shown capable of growth as chemoorganotrophic autotrophs. Thus, both grow on formate as an energy source but fix carbon dioxide by the Calvin cycle (Chandra and Shethna, 1977; Kelly *et al.*, 1979), and *Thiobacillus* A2 has been proved to oxidise methanol as an energy substrate while obtaining all of its carbon from carbon dioxide (Kelly and Wood, 1982). This organism also shows a physiological mixture of chemoorganotrophic heterotrophy and autotrophy when cultured on mixtures of formate and glucose (Wood and Kelly, 1981).

In Table 1, the third type of organism (C) comprises the symbiotic organisms of unknown physiological status, which may prove to be similar to that of type A or B. The type D organisms, which we shall discuss in some detail, are essentially chemoorganotrophic heterotrophs that are able to use sulphur compounds as (additional) energy sources for growth. Finally, a whole range of chemoorganotrophic heterotrophs is known to be able to oxidise reduced inorganic sulphur compounds, but they cannot obtain energy from such oxidations.

We have emphasised the physiological possibilities available to many of the colourless sulphur bacteria (including the less studied filamentous forms and a diversity of eubacterial forms that probably remain to be isolated or reisolated) because they are likely to experience mixed substrate environments in natural habitats, so that the regulatory processes and environmental pressures operating to determine their physiological behaviour are probably more complex than the relatively simple responses elicited by laboratory culture under single substrate-defined conditions.

Before proceeding to a detailed study of the types of bacteria, we have summarised in Table 2 the types of bacteria likely to affect the "dark oxidation" of inorganic sulphur in diverse environments. For completeness, the sulphur-oxidising photolithotrophs have been included because they are now known to carry out significant oxidation of sulphide in the dark under microaerophilic conditions. In quantitative terms, the greatest activity of all of these organisms is probably exhibited in environments where sulphide and oxygen (or nitrate) coexist. They are thus likely to be preponderant in oxic–anoxic interface habitats. Such environments are diverse in character, ranging from the broad sulphide-oxygen coexistence layers in stratified lake and ocean systems (Richards, 1965; Sorokin, 1972; Tuttle and Jannasch, 1973; Jørgensen, 1982; Indrebø *et al.*, 1979a,b; Jørgensen *et al.*, 1979; Kelly, 1980) to the narrower interface bands found in some sediments or in bacterial mats (Jørgensen, 1982). The general occurrence of the different metabolic types in soil and freshwater environments and marine (estuarine) ecosystems indicates that niches for the different types must be available in all of these habitats. Their ubiquity even in the open ocean further indicates their potential ability to scavenge both locally produced sulphide and (dilute) supplies of dissolved compounds such as thiosulphate or locally available sources of particulate elemental sulphur of natural or agricultural origin.

Heterotrophic Sulphur-Oxidising Bacteria

The first reports of heterotrophic bacteria that oxidise inorganic sulphur compounds dealt with their ability to develop in media containing thiosulphate. The thiosulphate was often quantitatively converted into tetrathionate, and no sulphate was formed (Trudinger, 1967). In other cases, heavy precipitates of elemental sulphur were generated during growth. At the same time, organic compounds were required for growth. It was assumed that these organisms could obtain metabolically useful energy from this oxidation process, but evidence for this was inconclusive (Starkey, 1935, 1966). Over the years, numerous papers have reported the presence of high numbers of such organisms, particularly in soils (Swaby and Vitolins, 1969; Kelly, 1972) and later in marine environments (Tuttle and Jannasch, 1972, 1973; Tuttle *et al.*, 1974). The heterotrophic sulphur compound-oxidising bacteria described were usually unable to oxidise any compound other than thiosulphate. Aerobic media with sulphide as a sulphur source will inevitably contain thiosulphate (Kuenen, 1975), so that even if growth (or oxidation) on sulphide was reported, this may in fact have been growth on thiosulphate. The often raised questions were (1) whether these organisms were able to obtain energy from this oxidation step and (2) whether the oxidation of thiosulphate to tetrathionate has any ecological significance. To date, these questions have only partially been answered, and a great need exists for further details and quantitative analysis of the metabolism of these organisms. In the 1960s, Trudinger (1967) presented data to show that some thiosulphate-oxidising heterotrophs oxidised thiosulphate to tetrathionate, but the bacteria were apparently unable to generate energy from the oxidation. *Pseudomonas aeruginosa* was reported to oxidise thiosulphate and other sulphur compounds to sulphate via tetrathionate (Schook and Berk, 1979), but in further experiments with this strain in our laboratory (J. G. Kuenen, unpublished data), we were unable to demonstrate that this oxidation process produced metabolically useful energy.

The first well-described example of a chemolithotrophic heterotroph was *Thiobacillus perometabolis* (London and Rittenberg, 1967), an organism which grew best in mixtures of organic substrates and thiosulphate. The thiosulphate was converted into sulphuric acid, but autotrophic growth was not possible. However, Katayama-Fujimura *et al.* (1982) reported that this organism (ATCC 23370) could be grown autotrophically. Apparently, further work is necessary to prove that carbon is indeed obtained from CO_2 and not from traces of organic compounds.

In the 1970s, Tuttle (1980) and Tuttle and Jannasch (1977) studied large numbers of heterotrophic bacteria, isolated from the sulphide–oxygen interface of the Black Sea, that were able to oxidise thiosulphate to tetrathionate. In a series of papers, these authors presented evidence for the ability of some of their strains to obtain energy from this oxidation step. In many cases, experiments designed to demonstrate increases in growth yields on addition of thiosulphate to heterotrophic media showed marginal and pH-dependent increases. Later results indicated unrealistically high yields from the single step of oxidation of thiosulphate to sulphate (for a discussion, see Kuenen and Beudeker, 1982). On the other hand, in Tuttle's laboratory (Tuttle, 1980) and also in our own (J. G. Kuenen, unpublished), it has been clearly demonstrated that on addition of thiosulphate to starving cells of these heterotrophs, the ATP pools of the cells rapidly increase. In further experiments on one Black Sea isolate, it was shown that $^{14}CO_2$ fixation was clearly thiosulphate dependent (Tuttle and Jannasch, 1977). CO_2 incorporation was at a level of 30% of a control experiment with an obligately chemolithoautotrophic *Thiobacillus* strain. Later work with the same strain (Tuttle, 1980) clearly demonstrated that the organism was a heterotroph that could use thiosulphate as an additional energy source for organic carbon assimilation. This finding seems to contradict the previous one unless this organism carried out very active heterotrophic CO_2 fixation, for example, by means of PEP carboxylase.

To avoid further confusion in this field of research, it seems extremely important that before claims are made, quantitative studies should be done on (1) yields obtained from the oxidation of sulphur compounds, preferably in continuous culture, and (2) levels of CO_2-fixing enzymes, not only those specific for autotrophic CO_2 fixation [notably ribulose 1,5-bisphosphate (RuBP) carboxylase and phosphoribulokinase] but also anaplerotic enzymes such as PEP and pyruvate carboxylase. Those strains showing substantial $^{14}CO_2$ fixation under some experimental conditions should obviously be used for such tests.

Gottschal and Kuenen (1980) developed a new method to isolate various metabolic types of sulphur-oxidizing bacteria from freshwater sources by enrichments in a continuous culture. The technique allows selection at will for obligate, facultative, or heterotrophic sulphur-oxidizing bacteria. Depending on the ratio of organic and inorganic sulphur compounds fed into the continuous culture (chemostat), the different physiological types will be selected according to Fig. 1. In the actual experiments, all metabolic types have been enriched for, with the possible exception of the heterotrophs that are able to oxidise sulphur compounds but not to generate energy. This latter type may have been present in the enrichments but was not identified. One organism was apparently a chemolithotrophic heterotroph. The organism could not be grown autotrophically but, in contrast to the organisms mentioned before, could oxidise thiosulphate completely to sulphate. This strain proved to be extremely sensitive to sulphite, which was easily

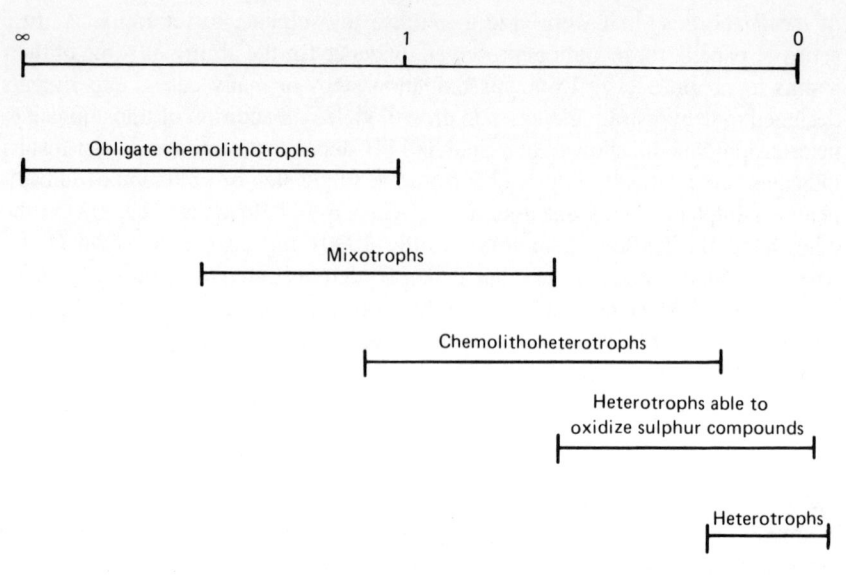

Fig. 1. Representation of the turnover ratio of inorganic compounds and the occurrence of different physiological types among sulphur-oxidising bacteria.

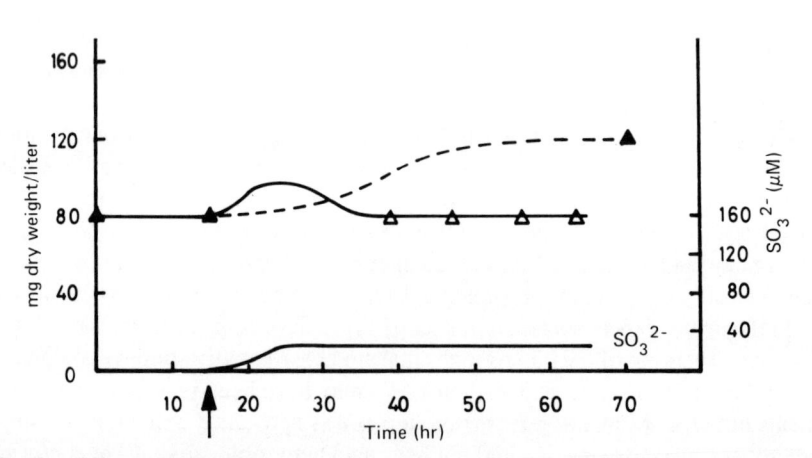

Fig. 2. Effect of thiosulphate on heterotrophic sulphur-oxidising bacteria in acetate-limited chemostat culture. ▲, $D = 0.02$; △, $D = 0.05$. Arrow indicates addition of 20 mM thiosulphate.

produced from incomplete oxidation of thiosulphate. Figure 2 shows the dry weight (equivalent to bacterial protein) as a function of time in a chemostat under acetate limitation. When thiosulphate was added to the inflowing medium while the dilution rate remained constant, an initial increase in cell density was observed. This was interpreted as evidence for the ability of the organism to obtain energy from the oxidation of the thiosulphate. However, the cell density returned

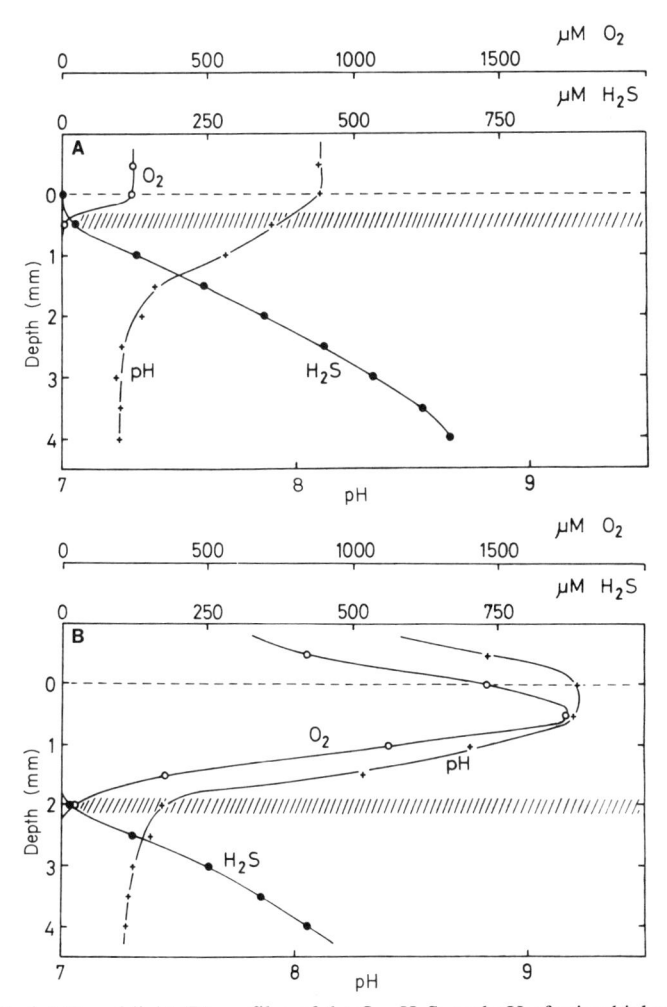

Fig. 3. Dark (A) and light (B) profiles of the O_2, H_2S, and pH of microbial mats from Solar Lake, Sinai. Dashed line indicates mat surface. Data were obtained with microelectrodes. In B, oxygenic photosynthesis creates a supersaturated oxygen concentration in the microbial mats. Note the extremely narrow zone of coexisting O_2 and H_2S both in the light and in the dark. With permission from Jørgensen (1983).

to its initial value after another 12 hr. After this period, thiosulphate remained untouched, whereas all acetate was still utilised. In a second experiment, the same sequence of substrate supply was followed, but on addition of the thiosulphate to the inflowing medium, the dilution rate was reduced in order to provide thiosulphate to the culture at a slower rate than in the first experiment. Figure 2 shows that in that case, the increase in yield persisted. The increase was of the same order of magnitude as that found for mixotrophic thiobacilli (Perez and Matin, 1980; Gottschal and Kuenen, 1980). It seems that if the thiosulphate concentration in the culture increased too rapidly, complete oxidation did not occur and sulphite accumulated in the culture, which subsequently irreversibly inhibited the thiosulphate oxidation. Small amounts of sulphite would continue to be produced, as a result of which the culture remained unable to oxidize thiosulphate. This interpretation could also explain why this organism, when grown in batch culture in the presence of excess thiosulphate, never showed substantial thiosulphate oxidation, since under such conditions sulphite would certainly have been produced.

In our laboratory we have observed that most, if not all, commercial thiosulphate preparations contain sulphite, which may inhibit the activity of heterotrophic sulphur-oxidising bacteria. This may explain why, in many cases, results of growth experiments in batch culture have been irreproducible. Finally, it should be realized that some heterotrophic sulphur-oxidising bacteria tend particularly to produce sulphite, whereas the specialist obligate chemolithotrophs can rapidly oxidise the sulphite to sulphate. This may also be of ecological importance. In mixtures of specialised obligate chemolithotrophic bacteria with heterotrophs, sulphite may not accumulate due to the activity of the specialists.

Curiously, in our hands, the continuous culture enrichment technique was not successful for the enrichment of facultative chemolithotrophs from the marine environment. Instead, large numbers of thiosulphate-oxidising heterotrophs mixed with specialists were obtained from enrichments with mixtures of organic substrates and inorganic sulphur compounds. However, Smith and Finazzo (1981) successfully isolated a marine strain of the facultatively chemolitho(auto)-trophic *Thiobacillus intermedius* in a chemostat enrichment culture, showing that differences in inoculum material and perhaps also other experimental conditions may greatly influence the outcome of the experiments.

Fieldwork with Colourless Sulphur Bacteria

The colourless sulphur bacteria can be found almost anywhere in nature where inorganic sulphur compounds and oxygen or other suitable electron acceptors, such as nitrate, are available. Such is the case at the sulphide–oxygen interface of marine and freshwater sediments. Figure 3 shows an example of a cyanobacterial mat with such an interface. In sediments the oxygen–sulphide interface is not necessarily limited to a narrow horizontal zone, but can also reach deeper into sediment due to channels made by burrowing animals. Furthermore, isolated anaerobic pockets with active sulphate reduction can be found in otherwise oxic sediments (Jørgensen, 1977). Other well-known examples of habitats for the sulphur-oxidising bacteria are sulphur deposits, sulphur springs in areas with geothermal activity (Brock et al., 1972), and waste treatment plants receiving sulphur compounds such as sulphide or thiocyanate (Woodard et al., 1976).

Another important habitat of sulphur-oxidising bacteria is the oxygen–sulphide interface of stratified water bodies. Notable examples are the Black Sea, the Cariaco Trench, deep freshwater lakes throughout the world, and unusual environments such as hypersaline Solar Lake (Sinai). Figure 4 gives a typical example of such a layer of coexisting S^{2-} and O_2 where sulphur-oxidising bacteria may thrive. Since the literature on the role of colourless sulphur bacteria was reviewed by Kuenen (1975), relatively few advances have been made in field techniques to detect specifically the activity of colourless sulphur bacteria in

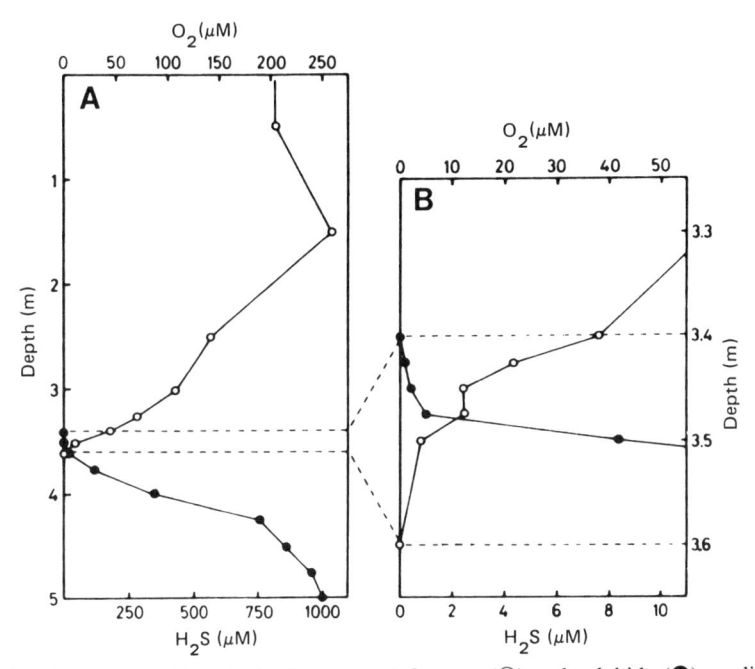

Fig. 4. Gradients in Solar Lake (Sinai). (A) Oxygen (○) and sulphide (●) gradients. (B) Oxygen–sulphide interface from (A) shown with expanded scales to demonstrate the zone of coexisting O_2 and H_2S (16 February). From Jørgensen et al. (1979).

the natural environment. Also, techniques to enumerate sulphur bacteria have not been improved significantly. The methods available in the literature are

1. Thiosulphate or elemental sulphur oxidation.
2. Decline in pH due to the production of sulphuric acid from sulphur, sulphide, or thiosulphate.
3. Rise in pH due to the production of tetrathionate from thiosulphate.
4. Production and/or turnover rates of intermediates from sulphide or thiosulphate labelled with ^{35}S.
5. Fixation of $^{14}CO_2$ in the dark.
6. Direct measurement of RuBP carboxylase in samples from permanently dark environments.
7. Most probable number (MPN) counts combined with (1), (2), or (3).
8. Direct count and measurements of colourless sulphur bacteria which can be directly seen and measured under the microscope.
9. Microelectrode measurements in sediments and aquatic environments.
10. Fluorescent antibody techniques.

The advantages and disadvantages of many of these methods have been discussed in detail by Kuenen (1975), with the exception of the last two methods. The microelectrode technique is a new and interesting field of research now being developed by Jørgensen's group.

The conclusion was, and still is, that in most cases no one of these methods will give a reliable estimate of the role of colourless sulphur bacteria in nature. A combination of these methods will certainly increase their meaningfulness, but even in that case, a quantitative estimate of microbial activity will be possible only in those (natural) environments where sulphide-oxidising bacteria dominate the total population. Such is the case, for example, in acid hot sulphur springs (Mosser et al., 1973, 1974) and some highly acidic soils (Fliermans and Brock, 1972; Kuenen, 1975), and in environments dominated by the unusually large colourless sulphur bacteria such as *Beggiatoa* or *Thiovulum,* which can be recognized under the microscope (Jørgensen, 1982). A few examples will serve to make this point.

Beggiatoa

For a long time, *Beggiatoa* has been considered one of the most typical colourless sulphur bacteria in that it can generate energy from the oxidation of inorganic sulphur compounds and may therefore be able to grow autotrophically. Unchallenged is its ability to oxidise sulphur compounds, but the role of sulphur oxidation may be only indirectly linked to energy generation. Only one report (Kowallik and Pringsheim. 1966) has described autotrophic growth. Nelson and Castenholz (1980a,b) discovered that their strain of *Beggiatoa* was a heterotroph which used its "intracellular" elemental sulphur as an electron acceptor for the oxidation of organic compounds under anaerobic conditions. This sulphur is located between the cellular membrane and the cell wall of *Beggiatoa* (Strohl and Larkin, 1978). During the oxidation of organic compounds, this sulphur was converted to sulphide. Subsequently, *Beggiatoa* used its high mobility to move in the sediment from the anaerobic environment to microaerophilic conditions, where the available sulphide was reoxidised to elemental sulphur. The *Beggiatoa* returned with stored sulphur to the anaerobic zone, where organic substrates such as acetate were available for the next round of oxidation. In this way, *Beggiatoa* shuttled between aerobic and anaerobic conditions to renew its internal electron acceptor. These observations, however, do not disprove the idea that *Beggiatoa* may also be able to oxidise elemental sulphur to sulphate, thereby obtaining chemolithotrophic energy.

In the same period, Strohl *et al.* (1981; Strohl and Larkin, 1978) reported on the heterotrophic abilities of *Beggiatoa* strains, which leaves little doubt that these strains are indeed heterotrophs. The same workers (Güde *et al.*, 1981) also reported "mixotrophic growth" of *Beggiatoa* on a mixture of sulphide and acetate. Under these conditions, sulphide may be oxidised, at least in part, to sulphate. Sulphide stimulated growth, but it was pointed out (Kuenen and Beudeker, 1982) that the increase in yield observed on addition of sulphide to acetate cultures may be too high to be explained by sulphur oxidation. CO_2 fixation in these organisms was demonstrated but was extremely small. Quantitative balances of input and output of organic compounds (oxidised, assimilated, etc.), CO_2, and/or sulphur metabolism will be needed to establish a good basis for further work on *Beggiatoa*. It should be emphasized here that it is most important for the initial physiological work to concentrate on a few strains, since it may very well be that, like the thiobacilli, the *Beggiatoa* comprise the complete spectrum of obligate chemolithotrophs, facultative chemolithotrophs, and chemolithotrophic heterotrophs.

As *Beggiatoa* are conspicuous organisms which can be recognized directly under the microscope, they can be studied, counted, and observed in samples taken from nature. Jørgensen (1977) reported large numbers of *Beggiatoa* in a coastal marine sediment. The organisms were hardly present in sandy sediments, but were abundant in muds aggregated around faecal pellets. It was postulated that *Beggiatoa* would be able to move more freely in the muddy than in the sandy sediment, and might therefore not be able to live in the sandy environment. Furthermore, the presence of sulphide-producing pockets (faecal pellets) in the softer, otherwise aerobic sediments would provide an ideal niche for these organisms. On the basis of the observed rate of oxidation of intracellular elemental sulphur in *Beggiatoa* (rate of disappearance of sulphur under aerobic conditions) and the biomass measured, an estimate was made of the potential rate of oxidation of sulphur in the sediments. The calculated value of 5–15 mmol S^{2-} m^{-2}

day $^{-1}$ was of the same order as the local sulphide production (sulphate reduction), which was approximately 10 mmol S^{2-} m^{-2} day^{-1}. This indicated that *Beggiatoa* could play an important role in the oxidation of sulphur compounds in the marine sediment. Microelectrodes were used to study the behaviour of populations of *Beggiatoa* and *Thiovulum* species in the microgradients in and above sulphide-containing sediment. Both populations located themselves very accurately at the interface between O$_2$ and S^{2-}, and maintained a steep gradient of sulphide diffusing upwards and oxygen diffusion downwards (Jørgensen, 1982). In the case of *Thiovulum*, the individual cells very actively moved in and out the gradient, until eventually the organisms arranged themselves in their veils of slime where the exact gradient could be maintained. As pointed out, these experiments show that the environments in which these gradient organisms live are extremely dynamic and obviously easy to disturb.

Competition Between Biological and Spontaneous ("Chemical") Oxidation of Sulphide with Oxygen

The H$_2$S formed in anoxic water bodies or sediments diffuses upwards towards the interface with oxygen, where it can react spontaneously with oxygen. Oxygen and sulphide coexist in salt environments at concentrations between 10^{-4} and 10^{-6}M. The half-life of the spontaneous oxidation of sulphide ranges from a few minutes to an hour. Oxidation rates are dependent on a variety of environmental parameters such as temperature, salt concentration, presence of trace metals, and catalytic amounts of other sulphur compounds (Chen and Morris, 1972; Jørgensen et al., 1979; Cline and Richards, 1969). The interesting question is whether bacteria would be able to compete successfully with the spontaneous reaction at low concentrations of O$_2$ and sulphide. Products of spontaneous oxidation are sulphur (10%), thiosulphate (30%), sulphite (30%) and sulphate (30%) (Chen and Morris, 1972). All of the available fieldwork seems to indicate that this is indeed so. For example, microelectrode measurements in *Beggiatoa* and *Thiovulum* veils show that these bacteria can oxidise sulphide at such a rate that neither sulphide nor oxygen is detectable, whereas without bacteria oxygen and sulphide coexist (Jørgensen, 1982).

Laboratory work with pure cultures of *T. neapolitanus* and *Thiobacillus* A2 also provides strong evidence that these organisms can efficiently and directly oxidise sulphide. This was shown by growing organisms under sulphide limitation in the chemostat. Table 3 shows the production of *T. neapolitanus* biomass during growth on sulphide and presents similar data for thiosulphate-limited growth. In the chemostat at pH 7.0, sulphide was below the level of detection, which is about 5 × 10^{-7} M. The thermodynamics of thiosulphate and sulphide oxidation to sulphate predict that the energy yield of both oxidation reactions should be equal (for further discussion, see Kelly 1982). Furthermore, the rates

Table 3. *Growth of* Thiobacillus neapolitanus *in an aerobic energy substrate-limited chemostat at a dilution rate of 0.05 hr^{-1} in a mineral medium with thiosulphate (40 mM) or sulphide (40 mM) as the growth-limiting nutrient*

	Growth-limiting substrate	
	$S_2O_3^{2-}$	S^{2-}
Biomass (mg dry wt l^{-1})	200	196
Residual concentration of $S_2O_3^{2-}$ or S^{2-} (M)	<10^{-4}	<5 × 10^{-7}

[a] Data from Beudeker *et al.* (1982).

of oxidation of both sulphide and thiosulphate are identical under a variety of growth conditions. One would therefore predict that yields per mole of thiosulphate should be equal to the yield per mole of sulphide oxidised. Table 3 shows that this was indeed the case. If any substantial spontaneous oxidation had taken place, the yield on sulphide should have been lower.

Analogous experiments with *Thiobacillus* A2 gave identical results. Interestingly, this organism is unable to oxidise elemental sulphur. Thus, if substantial spontaneous oxidation had occurred, this would have led to accumulation in the medium of elemental sulphur, which was not the case (Beudeker *et al.*, 1982). These results indicate that thiobacilli can indeed compete very successfully with spontaneous oxidation. This is, however, not always true. A well-documented case was described by Sorokin (1972), who studied the oxidation of H_2S in the sulphur–oxygen interface in the Black Sea. The rate of H_2S disappearance was estimated by adding [^{35}S]sulphide to samples of water and incubating in the absence or presence of chloroform, which was added to inhibit biological oxidation. The rate of sulphide disappearance was the same in the presence of chloroform as in untreated samples. From this observation, it was concluded that the initial reaction of H_2S with oxygen was "chemical" and not "biological." Thiosulphate, a major product of spontaneous oxidation, was, however, oxidised to sulphate at a higher rate in the untreated samples than it was by the chloroform-killed controls. From this it was concluded that bacteria are involved primarily in the oxidation in the chemocline of thiosulphate rather than of sulphide. Although this interpretation may certainly be correct, it could be argued that even if chemical and biological rates of sulphide oxidation are very similar, the microorganisms present *in situ* might still be able to interfere with the chemical reaction by removing polysulphides, which have been shown to act autocatalytically on the chemical oxidation of sulphide (Kuenen, 1975). In the absence of these catalysts, the spontaneous oxidation would proceed at a lower rate and thus allow the organisms to oxidise a greater proportion of the sulphide directly, thereby possibly gaining an enhanced supply of metabolically useful energy.

In this context, it should be mentioned that Tuttle and Jannasch (1972) and J. G. Kuenen (unpublished results), in contrast to Sorokin, were unable to isolate any obligate chemolithotrophic thiobacilli from the Black Sea. Instead, large numbers of sulphur- or thiosulphate-oxidising heterotrophs were isolated. Their properties have already been discussed in this chapter, but one interesting property was not mentioned. It was shown that a number of the thiosulphate-oxidising strains could also use this sulphur compound as electron acceptor for heterotrophic growth under anaerobic conditions (Tuttle and Jannasch, 1977). Such a property would render the organisms extremely suited for life at interfaces where oxygen is always available. It remains unclear, however, why true specialist thiobacilli are apparently not present at the interface. This becomes even more puzzling if it is realized that in the oxic zone thiosulphate is often available.

In the layer of coexisting sulphide and oxygen in the hypersaline Solar Lake (see also Fig. 4), the spontaneous oxidation rate may also be significant. Laboratory simulation experiments showed that the half-life of sulphide in filtered controls was 7.4 min, whereas half-lives of 5.1 and 2.7 min were exhibited by untreated and concentrated samples, respectively, showing that the biological oxidation was somewhat faster. As discussed in the previous paragraph, this may imply that in the presence of microorganisms spontaneous oxidation may be negligible. Indeed, an important role for biological oxidation has been established beyond any doubt. In the interface of Solar Lake, a major role is played by a bloom of sulphide-oxidising cyanobacteria which oxidise the sulphide to elemental sulphur in the light under anaerobic conditions. When the sulphide in the layer becomes depleted, the cyanobacterial layer switches to normal oxygenic photosynthesis.

In an extensive field experiment, Jørgensen et al. (1979) measured light and dark CO_2 fixation over a 36-hr period at various depths around the interface of Solar Lake. Variations of dissolved O_2 and S^{2-} concentrations in the diurnal cycle were also measured. The processes of sulphide-dependent or oxygenic photosynthesis (CO_2 fixation) can be discriminated experimentally by the addition of 3-(3,4-dichlorophenyl)-1,1-dimethylurea (DCMU) (which inhibits photosystem II) to the samples, thereby inhibiting oxygenic- but not sulphide-dependent CO_2 fixation. Figure 5 shows, in the first row of panels, the changes in parameters with depth at 2300–0100 hr (nighttime). Two peaks of dark CO_2 fixation are evident. The lower peak may be due to sulphide-dependent chemolithotrophic CO_2 fixation. The upper peak may also be chemolithotrophic CO_2 fixation dependent on thiosulphate, which was shown to be present at low concentrations at this depth.

After sunrise (middle panels, 0600–0900 hr, Fig. 5), the lower peak of dark CO_2 fixation disappeared, concomitantly with a considerable decrease in the

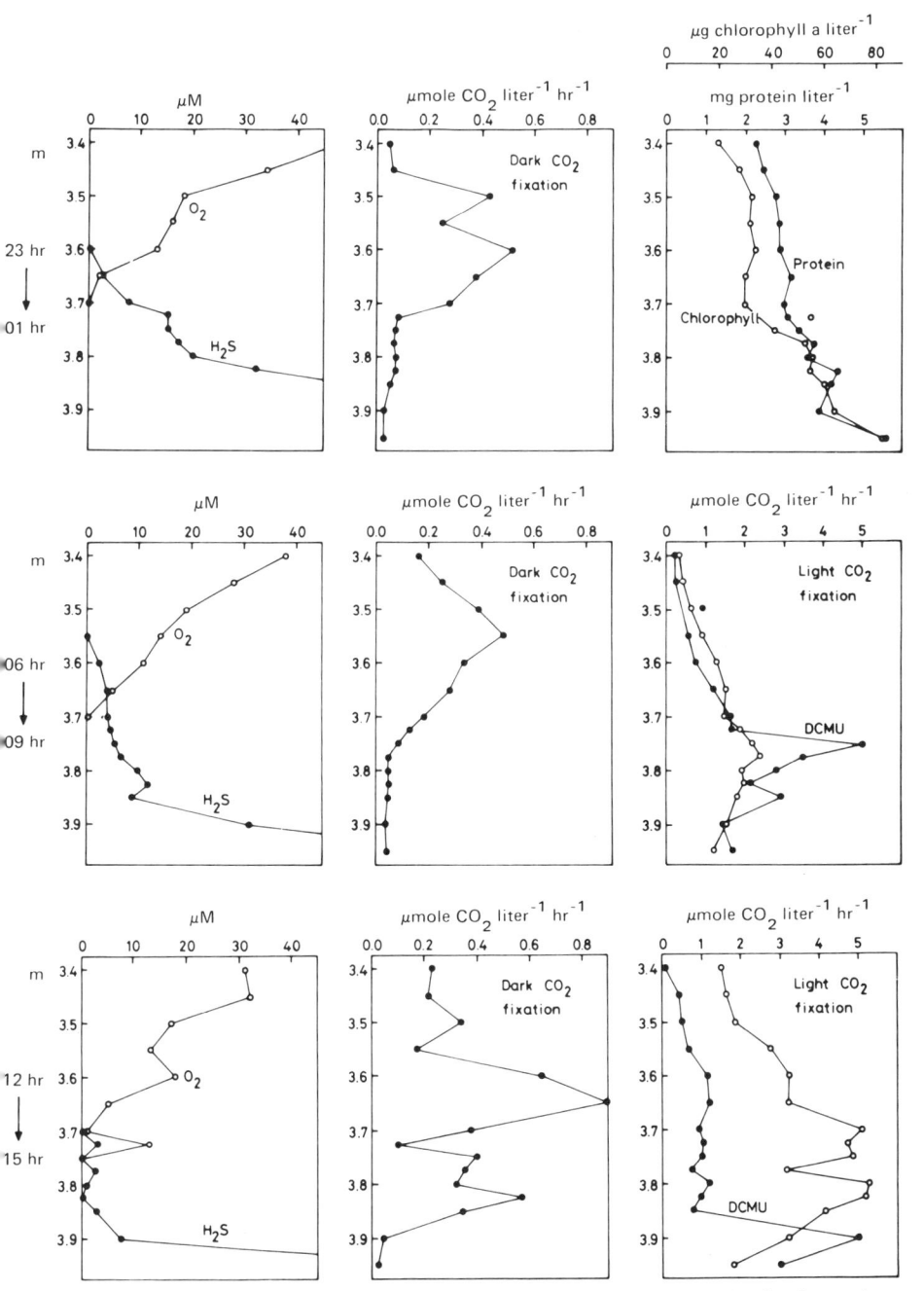

Fig. 5. Distribution in the chemocline of oxygen and sulphide; dark CO_2 fixation and light CO_2 fixation with or without DCMU. *In situ* incubations for 30 min at different times. Protein and chlorophyll a gradient also shown (28–29 March). From Jørgensen *et al.* (1979).

sulphide concentration. The disappearance of sulphide was clearly due to sulphide-dependent photosynthesis (in fact, DCMU stimulated photosynthesis, a phenomenon which remains unexplained).

The decrease in the lower dark CO_2 fixation at this time of the day indicates that the CO_2 fixation was indeed sulphide dependent. Separate experiments confirmed this contention since addition of sulphide to this layer increased dark CO_2 fixation. After midday (1200–1500 hr, Fig. 5), a very interesting phenomenon was observed. At that time, light-dependent CO_2 fixation increased to its maximum and clearly become strongly DCMU sensitive. This means that considerable oxygen production must have taken place at a depth of 3.5–3.9 m. However, at 3.75–3.9 m, no oxygen was detectable; at this depth, dark CO_2 fixation had again risen significantly. This can be taken as evidence for a very rapid, nonphotosynthetic biological oxidation of the sulphide, leading to CO_2 fixation. Particularly interesting is the small peak of oxygen at 3.725 m, which coincided with a low in the dark CO_2 fixation. This might mean that at this depth oxygen/sulphide-dependent dark CO_2 fixation does not occur, perhaps because of the absence of suitable organisms at this very dynamic sulphide–oxygen interface. Additional experiments showed that at 3.55 m the dark CO_2 fixation rate was generally electron donor (S^{2-}, $S_2O_3^{2-}$) limited, whereas at 3.75 m this rate was usually electron acceptor (CO_2) limited. Taken together, these data strongly indicate that an active chemolithoautotrophic population of sulphide-oxidising bacteria is present at the interface in different layers. Further circumstantial evidence for their presence comes from the fact that a chemostat enrichment culture prepared by inoculating interface sample water into a saline-mineral salts medium supplied with growth-limiting thiosulphate or sulphide (input concentration, 50 mM) resulted in a chemolithotrophic population with a yield of about 5 g dry weight (per mole of thiosulphate or sulphide oxidised). This is a typical yield for *Thiobacillus* and *Thiomicrospira* species (Kelly, 1982).

A practical aspect of the work on the activity of chemolithotrophic bacteria growing at the oxygen–sulphide interface is the sampling. The actual layer of coexistence in aqueous systems is often less than a few centimeters, making extremely accurate sampling essential. One method is the use of a pumping system with a specially designed inlet which ensures withdrawal of liquid from a horizontal circular segment of the column at 1- to 2-cm intervals.

The usefulness of this system was already apparent from the result of the Solar Lake work (for a description of the sampling device, see Jørgensen *et al.*, 1979). Later work showed that it can be used successfully not only in the very stable salt gradient of Solar Lake but also in a brackish environment where the density gradient is much less steep. For example, Saelenvaan Lake (near Bergen, Norway) on a still day with little or no wind showed a clear pattern of sulphide and oxygen concentration and of dark CO_2 fixation (Fig. 6). Coexistence of O_2 and S^{2-} was observed only over a few centimeters. At that level, a clear peak of dark

8. ECOLOGY OF THE COLOURLESS SULPHUR BACTERIA

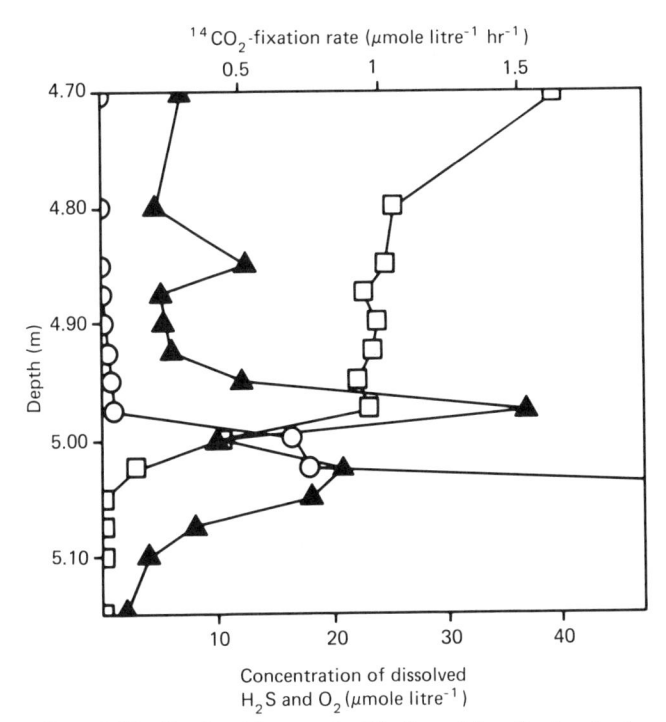

Fig. 6. Profiles of CO_2 fixation (▲, μmole CO_2 litre^{-1} hr^{-1}), dissolved oxygen (□), and dissolved H_2S (○) concentrations (μmole litre^{-1}) in Sælenvaan Lake, Norway, sampled at 0500 hr, 15 August 1978.

CO_2 fixation was also visible (Børsheim, 1979). The important point to be noted here is that sampling instruments with lower resolution, such as those commonly used in fieldwork, will not only fail to give accurate data but will also give a "smear" of dark CO_2 fixation over about 10–20 cm. It is obvious that had this been the case, the peak of dark CO_2 fixation at 4.975 m in the case of Fig. 6 would have disappeared into the background measurements. As a consequence, an erroneous conclusion would have been drawn concerning the possible presence of colourless sulphur bacteria.

An entirely new approach has been chosen with the use of microelectrodes for O_2, pH, and sulphide (Jørgensen et al., 1979; Revsbech and Jørgensen, 1981; Jørgensen, 1982). The use of electrodes has proved very useful for the study of the static and dynamic properties of algal mats. The O_2 microelectrode has a sensitive tip of only 20–30 μm, which permits accurate measurements of oxygen profiles over less than 1 mm. Such profiles are essentially similar to those found at interfaces of water bodies, but on a very much more compressed scale. The algal mats can be removed from the natural environment without disturbing the system, and can subsequently be studied under controlled laboratory or field conditions. By turning the light off and on, the dynamic behaviour of the oxygen profile can be studied, and initial rates of O_2 appearance and disappearance, in light and dark, respectively, can be measured. From such data, *in situ* rates of photosynthetic oxygen production can be calculated. Changes in sulphide content and pH can be measured in the same samples. pH is often very sensitive to changes in the bicarbonate concentration, which in turn is directly related to CO_2 fixation. Once a CO_2 microelectrode is available, this should provide a beautiful opportunity to study not only light but also sulphide/oxygen-dependent dark CO_2 fixation.

These examples have shown that the possibilities of achieving more than a general indication of the role of sulphide-oxidising bacteria are still very limited. This is also due to the fact that appropriate enumeration methods for the spectrum of sulphide-oxidising bacteria are still lacking. Consequently, in most cases, the presence of high numbers of sulphide-oxidising bacteria has almost never been correlated with the activities of these organisms. An interesting exception to this rule was reported by Mosser *et al.* (1974).

In the hot acid springs of Yellowstone Park, a dominant population of *Sulfolobus* species has been found. As sulphide and elemental sulphur are the predominant substrates in these ponds, direct measurements of turnover rates were possible. In these environments, it is likely that the greater part of the sulphide is spontaneously oxidized to sulphur, and that sulphur is then oxidized further to sulphuric acid by *Sulfolobus*.

A most interesting but further complicating phenomenon has been reported by Kondratieva *et al.* (1976) and Kämpf and Pfennig (1980). They found that phototrophic bacteria belonging to the *Chromatiaceae* can grow aerobically in the dark as chemolithotrophic autotrophs. Phototrophs that can also grow very well as chemolithoautotrophs may present a very serious challenge to the colourless sulphur bacteria. It may safely be assumed that the affinity $(\mu)/K_s$ of the phototrophs for sulphide is less than that of the colourless sulphur bacteria, but if a bloom of *Chromatiaceae* develops at the sulphide–oxygen interface, such organisms can outcompete the few thiobacilli in the dark by mere numbers. Therefore, at those interfaces which receive light, chemolithotrophic dark CO_2 fixation may not be due to thiobacilli and other colourless sulphur bacteria, but rather to blooms of the photosynthetic sulphur bacteria.

Colourless Sulphur Bacteria in the Cycles of Elements and in Food Chains

The role of colourless sulphur bacteria in the sulphur cycle hardly needs to be stressed. Their role in the carbon flow is also obvious since they act as primary producers in many ecosystems. For example, protozoa can often be seen in

Beggiatoa mats, their cells filled with ingested filamentous, sulphur-containing bacteria. In these environments, where 50% of the mineralization proceeds by sulphate reduction and subsequent sulphide oxidation by *Beggiatoa*, their contribution to the total energy and carbon flow may indeed be substantial. Principally, however, light remains the primary energy source for the food chains.

An entirely different situation exists in the ecosystems around the hydrothermal vents near the Galapagos Islands. From the vents, sulphide-rich water is forced into the ocean, mixes with oxic seawater (giving initial sulphide concentrations up to 150 μM), and thereby provides an ideal substrate for the growth of colourless sulphur bacteria. Around the vents, which are located at a depth of 2000–2500 m in complete darkness, large numbers of worms, clams, anemones, and even crabs can be found. Since light or imported organic compounds can be ruled out as sources of energy for this ecosystem, it was postulated that this system might live on sulphide as the *primary* energy source. Indeed, around thermal vents, high numbers of bacteria which could have been colourless sulphur bacteria were observed. From water samples taken from the vents, a large number of *Thiobacillus* and *Thiomicrospira* species were isolated. Typical of the *Thiomicrospira* was that it displayed the same high sulphide tolerance observed in the original isolate (Kuenen and Veldkamp, 1972; Ruby *et al.*, 1981; Ruby and Jannasch, 1982). Large numbers of *Hyphomicrobium*-like organisms were also observed (Jannasch and Wirsen, 1981). The genus *Hyphomicrobium* is known to harbour many methylotrophs (Harder and Atkinson, 1978), some of which can denitrify. As the gases in the vent water also contain methane (H. W. Jannasch, personal communication), these organisms may be involved in the metabolism of methane and other methylated compounds formed in the vents.

It is thus likely that in this specialized hydrothermal ecosystem, sulphide-oxidising bacteria could be the major primary producers, being consumed by animals such as clams and possibly providing dissolved organic matter for uptake by the indigenous marine worms.

Even more fascinating is the discovery that bacteria were living in symbiosis with eukaryotic organisms, such as the pogonophore worms, around the vent. One of these worms, the 2-m-long *Riftia pachyptila*, which lacks both mouth and gut, possesses an organ called the "trophosome," which fills most of the inside of the worm. Organisms were found in thin sections of the trophosome which were clearly prokaryotic (Cavanaugh *et al.*, 1981). At the same time, it was discovered that the tissues contained high concentrations of two key enzymes of the Calvin cycle, which occur only in prokaryotes and plants, and not in animals (Table 4). It is now postulated that the prokaryotes in the tissue could be sulphide-oxidising chemolithoautotrophic symbionts which are provided with substrates (S, S^{2-}, O_2, CO_2) by the worm through an elaborate network of blood vessels surrounding the trophosome. The blood has been shown to possess haemoglobin with a very high affinity for oxygen and to be insensitive to sulphide poisoning. Apart from the Calvin cycle enzymes, enzymes for sulphur

Table 4. Enzymes of the Calvin cycle and sulphur metabolism in marine animals living at the sulphide–oxygen interface[a]

Species and habitat	Enzyme activity (μmol min^{-1} per g wet mass of tissue)				
	RuBP carboxylase	Phosphoribulokinase	ATP sulphurylase	APS reductase	Rhodanese
Rift vents					
Riftia pachyptila (worm)	0.22	19.0	74.0	23.3	7.6
Riftia sp. (worm)	1.13	—	133.0	30.1	5.2
Sewage outfall					
Solemya panamensis (bivalve)	2.4	4.4	77.0	4.1	0.7

[a] Adapted from Felbeck et al. (1981).

metabolism were also detected in the tissue, namely, rhodanese, APS reductase, and ATP sulphurylase. These may play an important role in the energy generation of the bacteria (Kelly, 1982; Table 4, this chapter). Subsequently, *Riftia* species and bivalve molluscs such as *Solemya,* living in sulphide-containing mud from other nonhydrothermal habitats, have been shown to contain putatively chemolithotrophic CO_2-fixing bacteria in their tissues (Felbeck *et al.,* 1981). Work on the comparative $^{13}C/^{12}C$ ratios of cellular and available carbon sources further indicates that bacterial CO_2 fixation rather than the assimilation of dissolved organic matter could be a significant factor in providing fixed carbon for the animals (Southward *et al.,* 1981). Studies on small species of Pogonophora from the Bay of Biscay and the Norwegian fjords (Southward and Southward, 1980; Southward *et al.,* 1979) showed that these were able to take up significant quantities of dissolved organic matter from their environment. In some cases at least, it was proposed that the bulk of the carbon nutrition could be provided in this way (Southward and Southward, 1981). Numerous examples of the small species have now been shown to contain intracellular bacteria (Southward, 1982). This observation, together with the fact that ^{13}C depletion values of -35 to $-45‰$ have been found for all the pogonophores so far examined (A. J. Southward and E. C. Southward, personal communication, 1983) and the occurrence of ribulose bisphosphate carboxylase, predominantly in the bacteria-filled tissues in two small species (P. Dando, A. J. Southward and E. C. Southward, personal communication, 1983), all indicate that all the pogonophores have internal bacterial symbionts that provide fixed organic compounds to the animals. The relative importance of sulphur compound oxidation and carbon dioxide fixation is uncertain, as oxidisable ammonia and methane could conceivably also be substrates for the symbionts. Clearly, this study is in its infancy. It is of course possible that diverse symbiotic relationships exist, depending on whether an environment is sulphide or methane rich. Indeed, morphological investigations have indicated that the bacterial symbionts are of several types, including examples with extensive or no internal membranes and of rod or coccoidal shape (Southward, 1982; Southward *et al.,* 1981; E. C. Southward, personal communication; H. W. Jannasch, personal communication). The nature of the symbiosis also needs much clarification. It is not known whether carbon is supplied to the animals by *secretion* of fixed carbon from the bacteria, or whether the bacteria are actually digested at a fixed rate, maintaining a constant population in the tissues. There is some visual evidence for actual digestion of the bacterial cells (Southward, 1982). Using known growth yields for sulphide-oxidising bacteria (e.g., Kelly, 1982) and estimating fixed carbon excretion rates (e.g., from the observed rate of excretion of glycollate by *T. neapolitanus;* Cohen *et al.,* 1979), one could determine what growth rates might be supported for worms growing in habitats in which a constant supply of dissolved sulphide was available. If it is assumed that the symbiotic bacterium in a pogonophore is a sulphide-oxidising chemolithotroph capable of a yield *in vivo* of 2.5 g fixed carbon per mole of H_2S oxidised, and that the pogonophore digests the whole bacterium with an ecological efficiency conversion factor (trophic level 1 to trophic level 2) of 20%, then the production of new worm tissue containing 0.5 g carbon would require the oxidation of 32 g sulphide ion. Given that the wet weight of an individual *Siboglinum fiordicum* ranges from about 0.43 to 1.21 mg (Southward et al., 1979) and that of *S. ekmani* between 0.11 and 0.47 mg (Southward and Southward, 1980), and assuming a carbon content of 10% of the wet weight, one animal would require carbon needing the oxidation of 0.7–7.7 mg S^{2-} to produce the observed size range if all of the carbon were provided by the symbiont. The mean size of *S. fiordicum* at 0.8 mg and its maximum number at around 7000 m^{-2} (Southward et al., 1979) would imply consumption of about 36 g H_2S m^{-2} for their production. This calculation is simplistic, as the habitats for *Siboglinum* contain considerable biomass of other marine animals (Brattegard, 1967; Southward and Southward, 1980) and may have considerable dissolved organic carbon input from external sources (Southward et al., 1979). It does, however, show that the sulphide requirement, given also the slow growth rates of the pogonophores, is not impossibly unrealistic and indicates that sulphide-dependent chemolithoautotrophy could contribute significant amounts (>10%) of fixed carbon for pogonophore development in typical sediment environments in which sulphide production occurs.

It would thus seem that both free-living and symbiotic sulphide-oxidising bacteria may play a significant role as primary producers in food chains in marine environments, especially as the pogonophora are of global distribution (Southward, 1979, 1980). The existence of such symbiosis in freshwater animals or terrestrial environments with available sulphide remains to be established.

Some Concluding Remarks on Competition, Specialization, and Restricted Ecological Niches

We have not dwelt in detail on the topics mentioned in this heading, as they have either been considered extensively elsewhere or are as yet little investigated. The competitive advantage of potentially mixotrophic thiobacilli has been demonstrated by continuous culture methods both under steady state conditions and with fluctuating nutrient supplies (Smith and Kelly, 1979; Gottschal, 1980; Gottschal and Kuenen, 1981; Gottschal et al., 1981b), which illustrates that habitats containing both organic nutrients and oxidisable sulphur are likely to be dominated by mixotrophs. This success is attributable, at least in part, to the enhancement of the growth rate by mixotrophic metabolism (Smith and Kelly, 1979; Wood and Kelly, 1981) and will consequently be influenced by other factors that alter the growth rate, such as pH or the concentration of available nutrients. Thus, a mixotroph may be outcompeted by a specialist chemolithotroph (e.g., *Thiobacillus* A2 versus *T. neapolitanus;* Smith and Kelly, 1979) under autotrophic conditions at a pH most suitable for the specialist, but will dominate at that pH if an organic nutrient is also available.

The web of interactive effects of physical and chemical variables in any habitat explains the apparent occurrence of different organisms of similar physiology in apparent coexistence, when the ecological exclusion principle would predict that one only would survive the competition. In practice, the environments in which colourless sulphur bacteria are abundant contain great physicochemical diversity. These are characteristically gradients of oxygen, sulphide, dissolved sulphur compounds, dissolved carbon dioxide, and organic nutrients such as lactate, acetate, and possibly other organic acids and even sugars. Small differences in pH could influence the nature of the dominant organism in a microenvironment, just as could the absolute concentration of available oxygen or sulphide. Affinity for oxygen and tolerance of sulphide can both be crucial determinants in a stratified environment. Recent work has illustrated how particular types of physiology are found in organisms clearly adapted to fill each "niche" available in sulphide-generating habitats. Thus, the relatively sulphide-sensitive *Thiobacillus thioparus* and *T. neapolitanus,* which are also vigorous aerobes not noticeably stimulated by microaerophilic conditions, are likely to predominate in aerobic situations where subtoxic levels of sulphide occur. They may thus be found on the *surface* of sulphide-generating muds, where oxygen is high and sulphide low. There they may have to compete with mixotrophs such as *Thiobacillus* A2. Deeper in the mud are the more sulphide-tolerant and microaerophilic types such as *Thiomicrospira pelophila* (Kuenen and Veldkamp, 1972). Under anaerobic conditions when nitrate is available, the facultatively aerobic *Thiobacillus denitrificans* would be expected to succeed rather than *T. thioparus* or *Tms. pelophila,* but this expectation is now also complicated by the isolation of an *obligately* anaerobic nitrate reducer, *Thiomicrospira denitrificans* (Timmer-ten Hoor, 1975). The outcome of competition between this organism and *T. denitrificans* (both of which are obligate chemolithotrophs) would presumably be determined by relative sulphide tolerance, affinity for nitrate uptake, or reduction and possibly susceptibility to oxygen toxicity, as *Tms. denitrificans* is remarkably oxygen sensitive (Timmer-ten Hoor, 1975). Further competition for these organisms is indicated by the isolation of *Thiosphaera pantotropha* (Robertson and Kuenen, 1982), which is a *mixotrophic* facultative denitrifier, and is thus likely to have a competitive advantage in environments containing organic nutrients as well as sulphide and nitrate. Again, the dominating organism in any situation is determined by a multiplicity of factors, a particular combination of variants leading to ideal conditions for each organism.

The Future

It is clear that many habitats still need to be investigated (including the tissues of possible host animals—and plants?), and many habitats deserve much more intensive study. Little is known of the physiology of *thermophilic* autotrophic sulphur bacteria (other than *Sulfolobus*), even though the existence of *"Thiobacterium"* and *"Thiospirillum"* has been known for many years (Czurda, 1935, 1937). The approaches are many, including improvement of technique to study the dynamics *in situ* of sulphur transformations. The greater understanding of the microbiology and biochemical ecology of colourless sulphur bacteria will, however, come from selective enrichment culture and isolation (e.g., by continuous culture technique) of new organisms, followed by detailed study of their physiology in order to establish how they adapt to their habitat and the features that enable them to dominate particular environmental niches.

A combination of studies on the chemistry and kinetics of their habitats, with a full understanding of their individual properties will allow the interaction of the colourless sulphur bacteria to be understood. Much can be learnt from the isolated organism in pure culture, but we must heed the warning by Baas Becking and Wood (1955): "It seems, therefore, a hopeless task to reproduce the happenings at a mud surface by means of pure culture . . . in ecology their use is limited. . . . In order to reconstruct such a play as 'Macbeth' it seems unsatisfactory to study Banquo's part alone."

References

Baas Becking, L. G. M. and Wood, E. J. F. (1955). Biological processes in the estuarine environment. I. Ecology of the sulphur cycle. *Proc. K. Ned. Akad. Wet., Ser. B: Phys. Sci.* **58,** 161–172.

Beudeker, R. F., Gottschal, J. C. and Kuenen, J. G. (1982). Reactivity versus flexibility in thiobacilli. *Antonie van Leeuwenhoek* **48,** 39–51.

Børsheim, K. Y. (1979). Karbonsyklus og Svovelsyklus i Sœlenvannet. Doctoral thesis, University of Bergen, Norway.

Brattegard, T. (1967). Pogonophora and associated fauna in deep basin of Sognefjorden. *Sarsia* **29,** 299–306.

Brock, T. D., Brock, K. M., Belly, R. T. and Weiss, R. L. (1972). Sulfolobus: a new genus of sulfur-oxidizing bacteria living at low pH and high temperature. *Arch. Mikrobiol.* **84,** 54–68.

Cavanaugh, C. M., Gardiner, S., Jones, M., Jannasch, H. W. and Waterbury, J. B. (1981). Procaryotic cells in the hydrothermal vent tube worm *Riftia pachyptila* Jones: possible chemoautotrophic symbionts. *Science* **213,** 340–342.

Chandra, T. S. and Shethna, Y. I. (1977). Oxalate, formate, formamide, and methanol metabolism in *Thiobacillus novellus. J. Bacteriol.* **131,** 389–398.

Chen, K. Y. and Morris, J. C. (1972). Oxidation of sulfide by O_2-catalysis and inhibition. *J. Sanit. Eng. Div., Am. Soc. Civ. Eng.* **98,** 215–227.

Cline, J. D. and Richards, F. A. (1969). Oxygenation of hydrogen sulfide in seawater at constant salinity, temperature and pH. *Environ. Sci. Technol.* **3**, 838–843.
Cohen, Y., de Jonge, I. and Kuenen, J. G. (1979). Excretion of glycollate by *Thiobacillus neapolitanus* in continuous culture. *Arch. Microbiol.* **122**, 189–194.
Czurda, V. (1935). Über eine neue autotrophe und thermophile Schwefelbakteriengesellschaft. *Zentralbl. Bakteriol. Parasitenkd., Infektionskr. Hyg., Abt. 2, Naturwiss.: Allg., Landwirtsch. Tech. Mikrobiol.* **92**, 407–414.
Czurda, V. (1937). Weiterer Beitrag zur Kenntnis der neuen autotrophen und thermophilen Schwefelbakteriengesellschaft. *Zentralbl. Bakteriol. Parasitenkd., Infektionskr. Hyg., Abt. 2, Naturwiss.: Allg., Landwirtsch, Tech. Mikrobiol.* **96**, 138–145.
Felbeck, H., Childress, J. J. and Somero, G. N. (1981). Calvin-Benson cycle and sulphide oxidation enzymes in animals from sulphide-rich habitats. *Nature (London)* **293**, 291–293.
Fliermans, C. D. and Brock, T. D. (1972). Ecology of sulfur-oxidizing bacteria in hot acid soils. *J. Bacteriol.* **111**, 343–350.
Gottschal, J. C. (1980). Mixotrophic Growth of *Thiobacillus* A2 and its Ecological Significance. Doctoral Thesis, University of Groningen.
Gottschal, J. C. and Kuenen, J. G. (1980). Selective enrichment of facultatively chemolithotrophic thiobacilli and related organisms in the chemostat. *FEMS Microbiol. Lett.* **7**, 241–247.
Gottschal, J. C. and Kuenen, J. G. (1981). Physiological and ecological significance of facultative chemolithotrophy and mixotrophy in chemolithotrophic bacteria. In "Microbial Growth on C_1-compounds" (Ed. H. Dalton), pp. 92–104. Heyden, London.
Gottschal, J. C., de Vries, S. and Kuenen, J. G. (1979). Competition between the facultatively chemolithotrophic *Thiobacillus* A2, an obligately chemolithotrophic *Thiobacillus*, and a heterotrophic *Spirillum*, for inorganic and organic substrates. *Arch. Microbiol.* **121**, 241–249.
Gottschal, J. C., Pol, A. and Kuenen, J. G. (1981a). Metabolic flexibility of *Thiobacillus* A2 during substrate transitions in the chemostat. *Arch. Microbiol.* **129**, 23–28.
Gottschal, J. C., Nanninga, H. and Kuenen, J. G. (1981b). Growth of *Thiobacillus* A2 under alternating growth conditions in the chemostat. *J. Gen. Microbiol.* **126**, 85–96.
Güde, H., Strohl, W. R. and Larkin, J. M. (1981). Mixotrophic and heterotrophic growth of *Beggiatoa alba* in continuous culture. *Arch. Microbiol.* **129**, 357–361.
Harder, W. and Atkinson, M. M. (1978). Biology, physiology and biochemistry of hyphomicrobia. *Adv. Microb. Physiol.* **17**, 303–359.
Indrebø, G., Pengerud, B. and Dundas, I. (1979a). Microbial activity in a permanently stratified estuary. I. Primary production and sulfate reduction. *Mar. Biol. (Berlin)* **51**, 295–304.
Indrebø, G., Pengerud, B. and Dundas, I. (1979b). Microbial activity in a permanently stratified estuary. II. Microbial activities at the oxic-anoxic interface. *Mar. Biol. (Berlin)*, **51**, 305–309.
Jannasch, H. W. and Wirsen, C. O. (1981). Morphological survey of microbial mats near deep sea thermal vents. *Appl. Environ. Microbiol.* **41**, 528–538.
Jørgensen, B. B. (1977). Distribution of colourless sulfur bacteria (*Beggiatoa* spp.) in a coastal marine sediment. *Mar. Biol. (Berlin)* **41**, 19–28.
Jørgensen, B. B. (1982). Ecology of the bacteria of the sulphur cycle with special reference to anoxic-oxic interface environments. *Philos. Trans. R. Soc. London, Ser. B* **298**, 543–561.
Jørgensen, B. B. (1983). The microbial sulphur cycle. In "Microbial Geochemistry" (Ed. W. E. Krumbein), pp. 91–124. Blackwell, Oxford (in press).

Jørgensen, B. B., Kuenen, J. G. and Cohen, Y. (1979). Microbiological transformations of sulfur compounds in a stratified lake (Solar Lake, Sinai). *Limnol. Oceanogr.* **24**, 799–822.
Kämpf, C. and Pfennig, N. (1980). Capacity of Chromatiaceae for chemotrophic growth. Specific respiration rates of *Thiocystis violacea* and *Chromatium vinosum*. *Arch. Microbiol.* **127**, 125–137.
Katayama-Fujimura, Y., Tsuzaki, N. and Kuraishi, H. (1982). Ubiquinone, fatty acid and DNA base composition determinations as a guide to the taxonomy of the genus *Thiobacillus*. *J. Gen. Microbiol.* **128**, 1599–1611.
Kelly, D. P. (1971). Autotrophy: concepts of lithotrophic bacteria and their organic metabolism. *Annu. Rev. Microbiol.* **25**, 177–210.
Kelly, D. P. (1972). Transformations of sulphur and its compounds in soils. *Int. Symp. Sulphur Agric.* [*Proc.*], *1970* pp. 217–232.
Kelly, D. P. (1980). The sulphur cycle: definitions, mechanisms and dynamics. *Ciba Found. Symp.* **72** (new ser.), 3–18.
Kelly, D. P. (1981). Introduction to the chemolithotrophic bacteria. *In* "The Prokaryotes: A Handbook on Habitats, Isolation, and Identification of Bacteria" (Eds. M. P. Starr, H. Stolp, H. G. Trüper, A. Balows and H. G. Schlegel), Vol. 1, pp. 997–1004. Springer-Verlag, Berlin and New York.
Kelly, D. P. (1982). Biochemistry of the chemolithotrophic oxidation of inorganic sulphur. *Philos. Trans. R. Soc. London, Ser. B* **298**, 499–528.
Kelly, D. P. and Wood, A. P. (1982). Autotrophic growth of *Thiobacillus* A2 on methanol. *FEMS Microbiol. Lett.* **15**, 229–233.
Kelly, D. P., Wood, A. P., Gottschal, J. C. and Kuenen, J. G. (1979). Autotrophic metabolism of formate by *Thiobacillus* A2. *J. Gen. Microbiol.* **114**, 1–13.
Kondratieva, E. N., Zhukov, V. G., Ivanovsky, R. N., Petushkova, Yu. P. and Monsonov, E. Z. (1966). The capacity of phototrophic sulphur bacterium *Thiocapsa roseopersicina* for chemosynthesis. *Arch. Microbiol.* **108**, 287–292.
Kowallik, U. and Pringsheim, E. G. (1966). Oxidation of hydrogen sulfide by *Beggiatoa*. *Am. J. Bot.* **53**, 801–806.
Kuenen, J. G. (1975). Colourless sulfur bacteria and their role in the sulfur cycle. *Plant Soil* **43**, 49–76.
Kuenen, J. G. and Beudeker, R. F. (1982). Microbiology of thiobacilli and other sulphur-oxidizing autotrophs, mixotrophs and heterotrophs. *Philos. Trans. R. Soc. London, Ser. B* **298**, 473–497.
Kuenen, J. G. and Tuovinen, O. H. (1981). The genus *Thiobacillus* and *Thiomicrospira*. *In* "The Prokaryotes: A Handbook on Habitats, Isolation, and Identification of Bacteria" (Eds. M. P. Starr, H. Stolp, H. Trüper, A. Balows, H. G. Schlegel), Vol. 1, pp. 1023–1036. Springer-Verlag, Berlin and New York.
Kuenen, J. G., and Veldkamp, H. (1972). *Thiomicrospira pelophila*, gen.n., sp.n., a new obligately chemolithotrophic colourless sulphur bacterium. *Antonie van Leeuwenhoek* **38**, 241–256.
London, J. and Rittenberg, S. C. (1967). *Thiobacillus perometabolis* nov. sp., a nonautotrophic *Thiobacillus*. *Arch. Mikrobiol.* **59**, 218–225.
Mosser, J. L., Mosser, A. G. and Brock, T. D. (1973). Bacterial origin of sulfuric acid in geothermal habitats. *Science* **179**, 1323–1324.
Mosser, J. L., Bohlool, B. B. and Brock, T. D. (1974). Growth rates of *Sulfolobus acidocaldarius* in Nature *J. Bacteriol.* **118**, 1075–1081.
Nelson, D. C. and Castenholz, R. W. (1980a). Use of reduced sulfur compounds by *Beggiatoa* sp. *J. Bacteriol.* **147**, 140–154.

Nelson, D. C. and Castenholz, R. W. (1980b). Organic nutrition of *Beggiatoa* sp. *J. Bacteriol.* **147**, 236–247.
Peck, H. D. and Le Gall, J. (1982). Biochemistry of dissimilatory sulphate reduction. *Philos. Trans. R. Soc. London, Ser. B* **298**, 443–466.
Perez, R. C. and Matin, A. (1980). Growth of *Thiobacillus novellus* on mixed substrates (mixotrophic growth). *J. Bacteriol.* **142**, 633–638.
Pfennig, N. (1975). The phototrophic bacteria and their role in the sulphur cycle. *Plant Soil* **43**, 1–16.
Pfennig, N. and Widdel, F. (1982). The bacteria of the sulphur cycle. *Philos. Trans. R. Soc. London, Ser. B* **298**, 433–441.
Postgate, J. R. (1979). "The Sulphate-Reducing Bacteria." Cambridge Univ. Press, London and New York.
Revsbech, N. P. and Jørgensen, B. B. (1981). Primary productivity of microalgae in sediments measured by oxygen profile, $H^{14}CO_3$-fixation and oxygen exchange methods. *Limnol. Oceanog.* **26**, 717–730.
Richards, F. A. (1965). Anoxic basins and fjords. *Chem. Oceanog.* **1**, 611–645.
Rittenberg, S. C. (1972). The obligate autotroph: the demise of a concept. *Antonie van Leeuwenhoek* **38**, 457–478.
Robertson, L. A. and Kuenen, J. G. (1982). *Thiosphaera pantotropha*—a denitrifying facultative autotroph. *Soc. Gen. Microbiol., Meet. Programme, 1982* Abstract, 13.
Ruby, E. G. and Jannasch, H. W. (1982). Physiological characteristics of *Thiomicrospira* ℓ-12 isolated from deep-sea hydrothermal vents. *J. Bacteriol.* **149**, 161–165.
Ruby, E. G., Wirsen, C. O. and Jannasch, H. W. (1981). Chemolithotrophic sulfur-oxidizing bacteria from the Galapagos rift hydrothermal vents. *Appl. Environ. Microbiol.* **42**, 317–324.
Schook, L. B. and Berk, R. S. (1979). Partial purification and characterization of thiosulfate oxidase from *Pseudomonas aeruginosa*. *J. Bacteriol.* **140**, 306–308.
Smith, A. L. and Kelly, D. P. (1979). Competition in the chemostat between an obligately and a facultatively chemolithotrophic *Thiobacillus*. *J. Gen. Microbiol.* **115**, 377–384.
Smith, A. L., Kelly, D. P. and Wood, A. P. (1980). Metabolism of *Thiobacillus* A2 grown under autotrophic, mixotrophic and heterotrophic conditions in chemostat culture. *J. Gen. Microbiol.* **121**, 127–138.
Smith, D. W. and Finazzo, S. F. (1981). Salinity requirement of a marine *Thiobacillus intermedius*. *Arch. Microbiol.* **129**, 199–204.
Sorokin, Y. I. (1972). The bacterial population and the process of hydrogen sulphide oxidation in the Black Sea. *J. Cons. Cons. Int. Explor. Mer* **34**, 423–455.
Southward, A. J. and Southward, E. C. (1980). The significance of dissolved organic compounds in the nutrition of *Siboglinum ekmani* and other small species of Pogonophora. *J. Mar. Biol. Assoc. U.K.* **60**, 1005–1034.
Southward, A. J. and Southward, E. C. (1981). Dissolved organic matter and the nutrition of Pogonophora: a reassessment based on recent studies of their morphology and biology. *Kiel Meeresforsch. Sonderh.* **5**, 445–453.
Southward, A. J., Southward, E. C., Brattegard, T. and Bakke, T. (1979). Further experiments on the value of dissolved organic matter as food for *Siboglinum fiordicum* (Pogonophora). *J. Mar. Biol. Assoc. U.K.* **59**, 133–148.
Southward, A. J., Southward, E. C., Dando, P. R., Rau, G. H., Felbeck, H. and Flügel, H. (1981). Bacterial symbionts and low $^{13}C/^{12}C$ ratios in tissues of Pogonophora indicate unusual nutrition and metabolism. *Nature (London)* **293**, 616–620.

Southward, E. C. (1979). Horizontal and vertical distribution of Pogonophora in the Atlantic Ocean. *Sarsia* **64**, 51–55.
Southward, E. C. (1980). Two new species of Pogonophora from Hawaii. *Pac. Sci.* **34**, 371–378.
Southward, E. C. (1982). Bacterial symbionts in Pogonophora. *J. Mar. Biol. Assoc. U.K.* **62**, 889–906.
Starkey, R. L. (1935). Isolation of some bacteria which oxidize thiosulfate. *Soil Sci.* **39**, 197–219.
Starkey, R. L. (1966). Oxidation and reduction of sulfur compounds in soils. *Soil Sci.* **101**, 297–306.
Strohl, W. R. and Larkin, J. M. (1978). Enumeration, isolation and characterization of *Beggiatoa* from freshwater sediments. *Appl. Environ. Microbiol.* **36**, 755–770.
Strohl, W. R., Cannon, G. C., Shively, J. M., Güde, H., Hook, L. A., Lare, C. M. and Larkin, J. M. (1981). Heterotrophic carbon metabolism in *Beggiatoa alba*. *J. Bacteriol.* **148**, 572–583.
Swaby, R. J. and Vitolins, M. I. (1969). Sulfur oxidation in Australian soils. *Trans. Int. Conf. Soil Sci. 9th, 1968*, Vol. 4, pp. 673–681.
Timmer-ten Hoor, A. (1975). A new type of thiosulphate oxidizing, nitrate reducing microorganism: *Thiomicrospira denitrificans* sp. nov. *Neth. J. Sea Res.* **9**, 344–350.
Trudinger, P. A. (1967). Metabolism of thiosulfate and tetrathionate by heterotrophic bacteria from soil. *J. Bacteriol.* **93**, 550–559.
Trudinger, P. A. (1982). Geological significance of sulphur oxidoreduction by bacteria. *Philos. Trans. R. Soc. London, Ser. B* **298**, 563–581.
Tuttle, J. H. (1980). Organic carbon utilization by resting cells of thiosulfate-oxidizing marine heterotrophs. *Appl. Environ. Microbiol.* **40**, 516–521.
Tuttle, J. H. and Jannasch, H. W. (1972). Occurrence and types of *Thiobacillus*-like bacteria in the sea. *Limnol. Oceanogr.* **17**, 532–543.
Tuttle, J. H. and Jannasch, H. W. (1973). Sulfide and thiosulfate-oxidizing bacteria in anoxic marine basins. *Mar. Biol. (Berlin)* **20**, 64–71.
Tuttle, J. H. and Jannasch, H. W. (1977). Thiosulfate stimulation of microbial dark assimilation of carbon dioxide in shallow marine waters. *Microb. Ecol.* **4**, 9–25.
Tuttle, J. H., Holmes, P. E. and Jannasch, H. W. (1974). Growth rate stimulation of marine pseudomonads by thiosulfate. *Arch. Microbiol.* **99**, 1–15.
Wood, A. P. and Kelly, D. P. (1981). Mixotrophic growth of *Thiobacillus* A2 in chemostat culture on formate and glucose. *J. Gen. Microbiol.* **125**, 55–62.
Woodard, A. J., Stafford, D. A. and Callely, A. G. (1976). Biochemical studies on accelerated treatment of thiocyanate by activated sludge using growth factors such as pyruvate. *J. Appl. Bacteriol.* **37**, 277–287.

9

Genetics, Metabolic Versatility, and Differentiation in Photosynthetic Prokaryotes

VENETIA A. SAUNDERS

Department of Biology
Liverpool Polytechnic
Liverpool, England
UK

Introduction

The photosynthetic prokaryotes constitute a diverse biological group that includes, on the one hand, oxygen-evolving photosynthetic prokaryotes (the cyanobacteria and the Prochlorophyta) and, on the other hand, the green and purple bacteria which perform anoxygenic photosynthesis (herein referred to as the "photosynthetic bacteria"). Much interest in these photosynthetic organisms stems from their evolutionary and ecological significance. Furthermore, they provide excellent experimental systems for exploring the mechanics of photosynthesis and nitrogen fixation, and the processes governing morphogenesis and differentiation in prokaryotes.

In recent years, facilities for the genetic manipulation of photosynthetic prokaryotes have been developed and are beginning to add a new dimension to the study of this biological group. Increased availability of such genetic systems, in conjunction with the technologies of DNA cloning and DNA sequencing, is expected to permit extensive genetic analysis of photosynthetic prokaryotes. In turn, this should enhance our understanding of the molecular biology of photosynthesis, nitrogen fixation, and related phenomena. Moreover, such genetic methodology provides the opportunity to identify genetic mechanisms governing the evolution and ecology of photosynthetic organisms. Much attention has been focused on the ecological factors that affect the distribution of photosynthetic prokaryotes (Brock, 1973; Keating, 1977, 1978; Van Gemerden and Beeftink, 1983). Integration of such ecological data with complementary genetic data would allow the development of the ecological genetics of this photosynthetic group. It is the primary intention of this chapter to consider the current state of the genetics of photosynthetic prokaryotes in relation to their development and adaptation.

Morphology and Differentiation

Photosynthetic prokaryotes exhibit considerable morphological and structural diversity, and include species with polymorphic cell cycles. Accordingly, these organisms offer experimentally attractive systems for studying the biology of development in prokaryotes.

Cellular Organization in Photosynthetic Bacteria

Photosynthetic bacteria show much cytological variation. Four families can be recognised: the green and brown sulphur bacteria (Chlorobiaceae), the gliding filamentous green bacteria (Chloroflexaceae), the purple sulphur bacteria (Chromatiaceae), and the purple nonsulphur bacteria (Rhodospirillaceae) (Pfennig, 1977; Trüper and Pfennig, 1978). The photosynthetic apparatus of green bacteria is located in part on the cytoplasmic membrane and in part on intracellular vesicles known as "chlorobium vesicles" (or "chlorosomes"). These vesicles, found solely in green bacteria, contain the bulk accessory light-harvesting pigments. Reaction-centre and light-harvesting bacteriochlorophyll a are found on the cytoplasmic membrane (for reviews, see, for example, Olson, 1980; Pierson and Castenholz, 1978).

In purple bacteria, photosynthetic components are typically associated with various types of intracytoplasmic membrane systems (Drews and Oelze, 1981; Remsen, 1978; Oelze, 1983). The purple nonsulphur bacteria *Rhodopseudomonas sphaeroides* and *R. capsulata* elaborate a vesicular intracytoplasmic membrane system under phototrophic conditions, whereas *R. viridis* and *R. palustris* possess a series of lamellae underlying the cytoplasmic membrane (Pfennig, 1977). Unlike other members of the Rhodospirillaceae, *Rhodospirillum tenue* and *Rhodopseudomonas gelatinosa* are not able to differentiate significant intracytoplasmic membranes under photosynthetic conditions. The photosynthetic apparatus is instead incorporated into the cytoplasmic membrane (Wakim et al., 1978; Weckesser et al., 1969). Pronounced differences are found in the architecture of cytoplasmic membranes from cells of *Rh. tenue* grown chemotrophically in the dark with aeration and phototrophically under anaerobic conditions (Golecki and Oelze, 1980). By contrast, the cytoplasmic membrane of *R. sphaeroides* has essentially the same supramolecular architecture under both sets of growth conditions, with intracytoplasmic membranes forming at specific sites on the cytoplasmic membrane (Remsen, 1978). In certain purple nonsulphur bacteria, it seems that the intracytoplasmic membranes can accommodate respiratory electron transfer components in addition to photosynthetic components (Zannoni et al., 1978).

Certain photosynthetic bacteria exhibit polymorphic cell cycles. *Rhodomicrobium vannielii*, an exospore-forming, budding bacterium, displays complex cell cycles with morphologically distinct cell types (for a detailed review, see Whittenbury and Dow, 1977).

Differentiated Cells of Cyanobacteria

Cyanobacteria include both filamentous and unicellular forms. Many filamentous species are capable of differentiation to produce specialized cell types, and thus provide useful model systems for analysing mechanisms involved in cellular differentiation. Vegetative cells of cyanobacteria harbour the photosynthetic apparatus, which typically comprises a series of flattened membranous sacs (thylakoids) that accommodate light-harvesting and photoreactive pigments and the photosynthetic electron transport carriers (Krogmann, 1973; Stanier and Cohen-Bazire, 1977). Some respiratory electron transport components are also probably located on the thylakoid membrane (Bisalputra et al., 1969; Peschek et al., 1981). The major light-harvesting accessory pigments, however, are located outside the thylakoids in organelles known as "phycobilisomes," which, however, are closely attached to the thylakoids. Inclusion bodies of vegetative cells include glycogen granules, cyanophycin granules, carboxysomes (polyhedral bodies), and gas vesicles (for a review, see Stanier and Cohen-Bazire, 1977). Carboxysomes contain ribulose 1,5-bisphosphate carboxylase, the cardinal enzyme of CO_2 fixation via the Calvin cycle (see Codd, Chapter 5, this volume). There is evidence for the presence of DNA in carboxysomes (Westphal et al., 1979).

Specific environmental stimuli initiate the formation of two differentiated cell types, heterocysts and akinetes, from vegetative cells. Akinetes are viewed as resting cells with a capacity for germination and are formed in ageing cultures (Nichols and Carr, 1978; Simon, 1977; Sutherland et al., 1979). Heterocysts house and serve to protect the nitrogenase from oxygen inactivation (Wolk, 1973). They are the major site of nitrogen fixation in heterocystous cyanobacteria. Heterocyst structure, function, and development have been extensively reviewed (Adams and Carr, 1981; Carr, 1983; Haselkorn, 1978; Stanier and Cohen-Bazire, 1977; Stewart, 1977; Wilcox et al., 1975a; Wolk, 1975). Development of heterocysts is triggered by the deprivation of a combined nitrogen source (Fogg, 1949). A regular pattern of heterocysts, dictated by genetic and environmental factors, develops throughout cyanobacterial filaments (Wolk, 1975). The precise molecular events governing heterocyst differentiation and spacing are not completely clear. Functional activity of the heterocyst itself, in terms of N_2 fixation, does not apparently regulate the spatial pattern since the normal "proheterocyst" pattern emerges even in the absence of N_2 fixation (Bradley and Carr, 1976; Wilcox et al., 1973a). It is more likely that the heterocyst pattern is a consequence of heterocyst development, whereby establishing heterocysts prevent adjacent cells from becoming heterocysts (Wilcox et al., 1973a, 1975a). Exogenous cyclic AMP (cAMP) has been shown to alter the pattern of heterocyst spacing in *Anabaena variabilis* (Smith and Ownby, 1981). Possibly the cAMP interferes with the establishment of inhibitory zones around differentiating heterocysts. Unequal cell division is implicated in the maintenance of the heterocyst pattern in filamentous cyanobacteria (Wilcox et al., 1975a). Similarly, in *Chlorogloeopsis fritschii*, asymmetrical division precedes heterocyst development (Foulds and Carr, 1981). A delicately balanced metabolic interdependence exists between heterocysts and adjacent vegetative cells. Intercytoplasmic channels, termed "microplasmodesmata," which establish cytoplasmic continuity between neighbouring cells in filaments (Giddings and Staehelin, 1978, 1981), appear to be crucial to the proper development and functioning of heterocysts. Synthesis of various proteins of vegetative cells is curtailed, and new proteins form during heterocyst differentiation (Haselkorn, 1978). Heterocysts apparently possess a modified photosynthetic apparatus that is incapable of normal photosynthetic function (Tel-Or and Stewart, 1977). Mature heterocysts are regarded as terminal cells, being unable to dedifferentiate or divide (Carr and Bradley, 1973; Wilcox et al., 1973b).

Photosynthetic and Respiratory Metabolism

The photosynthetic prokaryotes display considerable metabolic versatility. Various types of energy conversion have been described for this biological group. Cyanobacteria are typically photoautotrophs capable of oxygenic photosynthesis. CO_2 fixation occurs predominantly by the Calvin cycle (Pelroy and Bassham, 1972). Some cyanobacteria can grow heterotrophically; both chemoheterotrophy and photoheterotrophy have been reported (e.g., Khoja and Whitton, 1971; Rippka, 1972). However, photoautotrophy is the most efficient growth mode. All cyanobacteria possess an endogenous dark respiratory metabolism, which apparently involves the utilisation of glycogen via the oxidative pentose phosphate cycle (Lehmann and Wober, 1976). This presumably provides a means of survival under conditions of light limitation or intermittent darkness. Such endogenous metabolism is, however, subjected to severe inhibition in the light (Brown and Webster, 1953). (For more extensive coverage of the metabolism of cyanobacteria, refer to the reviews of Stanier and Cohen-Bazire, 1977, Smith and Hoare, 1977, and Carr and Whitton, 1982.)

The photosynthetic mechanisms of photosynthetic bacteria are exclusively anoxygenic. Typical representatives of the Chlorobiaceae are obligate photoautotrophs, whereas the Chloroflexaceae includes photoheterotrophs that are also capable of chemoheterotrophic growth under aerobic conditions. Purple bacteria can be divided into two physiological groups: the purple sulphur bacteria (Chromatiaceae), which are typically photoautotrophs fixing CO_2 with concomitant oxidation of sulphur compounds, and the purple nonsulphur bacteria (Rhodospirillaceae), which are typically facultative photoheterotrophs that are also capable of growing as chemoheterotrophs under aerobic conditions. Representatives of this family also exhibit photoautotrophy (Ormerod and Sirevag, 1983; Pfennig, 1967, 1977; Trüper and Pfennig, 1978; Van Niel, 1944, 1954). Certain members of the Rhodospirillaceae and Chromatiaceae can grow aerobically as chemoautotrophs (Kampf and Pfennig, 1980; Kondratieva et al., 1976; Madigan and Gest, 1979). Furthermore, some purple nonsulphur bacteria can grow anaerobically in darkness by fermentative metabolism (Madigan et al., 1980; Schultz and Weaver, 1981; Uffen and Wolfe, 1970; Yen and Marrs, 1977).

A fundamental feature that distinguishes the cyanobacteria from the photosynthetic bacteria is the possession by cyanobacteria of oxygenic photosynthesis. Moreover, the photosynthetic machinery of cyanobacteria is more akin to that of eukaryotic plant cells than to that of the green and purple bacteria. There are two photosystems (PS I and PS II), and chlorophyll *a* and phycobiliproteins serve as major light-harvesting pigments. By contrast, photosynthetic bacteria generally possess one photosystem. The predominant chlorophylls in purple bacteria are bacteriochlorophylls *a* or *b,* whereas bacteriochlorophylls *c, d,* and *e* are additionally found in green bacteria (Gloe et al., 1975; Olson, 1980). No oxygen is produced in bacterial photosynthesis, and oxidisable substrates other than water (notably reduced sulphur compounds or organic compounds) serve as electron donors. Interestingly, some cyanobacteria are capable of anoxygenic photosynthesis by a process similar to that operative in photosynthetic bacteria (Padan, 1979). The capacity to switch from oxygenic to anoxygenic photosynthesis is advantageous to the survival of cyanobacteria in habitats alternating between photoaerobic and photoanaerobic conditions. A detailed appraisal of oxygenic and anoxygenic photosynthesis in prokaryotes is given by Shilo and Fattom (Chapter 6, this volume).

Interrelationships between photosynthetic and respiratory electron transfer systems have been proposed for both photosynthetic bacteria and cyanobacteria. In the Rhodospirillaceae, cytochrome c_2 apparently functions in photosynthetic and respiratory electron transport (Baccarini-Melandri and Zannoni, 1978; Connelly et al., 1973). Analogous dual roles have been suggested for cytochrome *c*-553, plastocyanin (Lockau, 1981), and plastoquinone (Eisbrenner and Bothe, 1979) in cyanobacteria.

Effects of Light and Dark Regimens on the Synthesis and Function of the Photosynthetic Apparatus

Light clearly has a dual role in photosynthesis. Not only does it serve as an energy source, but it also regulates development of the photosynthetic apparatus. In both photosynthetic bacteria and cyanobacteria, light can modify the composition of the photopigments and provoke changes in photosynthetic membrane synthesis. Such modifications are presumably stringently controlled in order to maximize the light-trapping ability at minimum biosynthetic expense.

Both the wavelength of light and light intensity can alter markedly the composite ratio of the phycobiliproteins, phycocyanin and phycoerythrin, which serve as photosynthetic accessory pigments in some cyanobacteria (Gantt, 1980; Tandeau de Marsac, 1977). Such adaptive responses are referred to as "complementary chromatic adaptation" and "intensity adaptation," respectively (see Bogorad, 1975; Glazer, 1977). At high light intensities, there is a reduction in the cellular content of both chlorophyll *a* and phycobiliproteins. Lamellar content of cyanobacteria is apparently controlled in concert with pigment content (Allen, 1968). In purple bacteria, an increase in light intensity results in a decreased synthesis of the pigments and proteins of the photosynthetic apparatus. Conversely, a decrease in light intensity is accommodated by an increase in photopigment content, the ratio of light-harvesting to reaction-centre bacteriochlorophyll increasing to balance light capture with the reduction in available light (Drews, 1978; Kaplan, 1978).

In facultative phototrophic bacteria, development of the photosynthetic apparatus is dramatically affected by molecular oxygen. Bacteriochlorophyll synthesis is repressed by oxygen in the dark or light. There is apparently a threshold level for oxygen which regulates pigment and protein synthesis (Kaplan, 1978). The rate of photopigment synthesis has been shown to increase as the partial pressure of oxygen decreases below the threshold (Cohen-Bazire and Sistrom, 1966). In the Rhodospirillaceae, sensitivity of the regulatory mechanism(s) to oxygen is strain specific. In *R. sphaeroides,* oxygen also represses the synthesis of intracytoplasmic membranes, whereas in *R. capsulata* tubular membranes can develop in response to oxygen (Lampe *et al.*, 1972). The mechanisms governing differentiation into tubular and vesicular intracytoplasmic membranes, on the one hand, and cytoplasmic membranes, on the other hand, have yet to be fully elucidated. Photosynthetic membrane development is considered in a number of reviews (Drews, 1978; Drews and Oelze, 1981; Kaplan, 1978; Kaplan and Arntzen, 1982; Ohad and Drews, 1982).

Modifications to electron transport systems have also been observed to accompany the growth of facultative photoheterotrophs under dark aerobic conditions. In *R. sphaeroides* an *a*-type cytochrome develops aerobically and presumably functions as a respiratory oxidase (Saunders and Jones, 1974a; Whale and Jones,

1970). *a*-Type cytochromes have also been reported in *Chloroflexus* and in species of cyanobacteria (Peschek, 1981) where they may serve as respiratory oxidases. High potential membrane-bound *b*-type cytochromes are apparently induced in response to aeration in *R. capsulata* (Zannoni *et al.*, 1974, 1976a,b). These *b*-type cytochromes have been proposed as putative terminal oxidases (Zannoni *et al.*, 1976a,b). In *R. palustris,* cytochrome *o* apparently serves as a terminal oxidase in respiration (King and Drews, 1975).

Whereas oxygenation affects the formation of photopigments and photosynthetic membranes in rhodopseudomonads, there is evidence for the retention of residual photosynthetic activity by aerobically grown cells (as judged by light-induced oxidation of reaction-centre bacteriochlorophyll and cytochrome *c* and light-induced carotenoid shifts) (Saunders and Jones, 1974b; Manwaring, 1981). Similarly, in cyanobacteria, dark heterotrophic growth does not lead to complete loss of photosynthetic functions (Evans, 1979; Evans *et al.*, 1978; Rippka, 1972). In *C. fritschii,* some PS I activity is maintained under dark heterotrophic conditions (Evans *et al.*, 1978). On transfer from dark to light conditions, restoration of complete PS I activity apparently occurs only after a lag of several hours' duration. It has been suggested (Evans, 1979) that the light triggers a specific rearrangement of reaction-centre and light-harvesting machineries of dark-grown cells, resulting ultimately in the recovery of full PS I efficiency.

Dark heterotrophic growth has differential effects on the synthesis of photosynthetic membranes and photopigments in different cyanobacterial strains (Tandeau de Marsac, 1977; Stanier and Cohen-Bazire, 1977). In *C. fritschii* there is less phycocyanin in dark-grown cells, and the lamellar arrangement is altered in comparison with light-grown cells (Evans *et al.*, 1976). However, in *Nostoc* strains, there is apparently little difference in thylakoid structure or phycobiliprotein content in photoautotrophic and dark heterotrophic cells (Hoare *et al.*, 1971). In view of the vast biosynthetic expense incurred in the maintenance of the photosynthetic apparatus, it is perhaps paradoxical that there appears to be no suppression of its development in the dark in *Nostoc*. Yet, controls exist to regulate phycobiliprotein composition in response to various stimuli, for instance light quality.

Nitrogen Fixation

Both cyanobacteria and photosynthetic bacteria include species with a capacity to fix nitrogen (for reviews, see Gallon, 1981; Johansson *et al.*, 1983; Stewart, 1980; Yoch, 1978). The process involves the multicomponent enzyme nitrogenase, which catalyzes the conversion of N_2 to ammonium in a reaction requiring ATP and a suitable reductant. Derepression of the genes for nitrogen fixation (*nif*) is triggered by removal of a combined N_2 source from the environment. In cyanobacteria, it is difficult to reconcile nitrogen fixation, which requires strictly anaerobic conditions, with the capacity for oxygen-evolving photosynthesis. The answer for certain cyanobacteria apparently lies in a spatial separation of the two processes, whereby heterocysts segregate the nitrogen-fixing machinery from neighboring vegetative cells and provide an oxygen-protected environment for the nitrogenase. However, the vegetative cells of some heterocystous cyanobacteria do have nitrogenase under anaerobic conditions (Haselkorn, 1978). Nonheterocystous cyanobacteria can also fix nitrogen. This normally occurs under anaerobic or microaerophilic conditions (Rippka and Waterbury, 1977; Rogerson, 1980), but strains of *Gloeothece* sp. (see Rippka *et al.*, 1979) and possibly other cyanobacteria (e.g., Pearson *et al.*, 1979; Singh, 1973) can maintain a functional nitrogenase under aerobic conditions. Presumably, therefore, alternative mechanisms exist to prevent oxygen inactivation of the nitrogenase. One possibility is a temporal separation of the processes of oxygen-evolving photosynthesis and nitrogen fixation (Gallon *et al.*, 1975; Millineaux *et al.*, 1981). However, it is more likely that a combination of mechanisms is operative (Kallas *et al.*, 1983; Tozum and Gallon, 1979).

In photosynthetic bacteria, nitrogen fixation occurs under anaerobic conditions (Yoch, 1978), although *R. capsulata* can apparently use atmospheric nitrogen as the sole nitrogen source under semiaerobic conditions (Meyer *et al.*, 1978). This implies a degree of oxygen tolerance for the nitrogenase of *R. capsulata*.

Genome Composition

There is an extremely wide variation in DNA base composition among photosynthetic prokaryotes. Cyanobacteria are a particularly heterogeneous group, with DNA base ratios ranging from 35 to 72 mole % (G + C) (Edelman *et al.*, 1967; Herdman *et al.*, 1979; Stanier *et al.*, 1971). For photosynthetic bacteria, the mole % G + C is 45 to 72 (Mandel *et al.*, 1971), whereas for *Prochloron*, a member of the Prochlorophyta (Lewin, 1977), it is 40.8% (Herdman, 1981).

Plasmid Biology

Various reports indicate the common occurrence of extrachromosomal (plasmid) DNA in photosynthetic prokaryotes. Both filamentous and unicellular cyanobacteria harbour extrachromosomal DNA (Friedberg and Seijffers, 1979; Lau and Doolittle, 1979; Lau *et al.*, 1980a,b; Reaston *et al.*, 1980; Roberts and Koths, 1976; Simon, 1978; van den Hondel *et al.*, 1979). In the majority of cases, multiple classes of plasmid DNA have been detected. To date, no correlation is apparent between the presence of plasmid DNA and the morphological complexity of cyanobacteria (Friedberg and Seijffers, 1979; Simon, 1978). Certain strains of *Synechococcus* carry homologous plasmids, as judged by molecular size and restriction cleavage patterns (Lau and Doolittle, 1979; van den Hondel *et al.*, 1979). Plasmids from other unicellular cyanobacteria share distinct regions of sequence homology (Lau *et al.*, 1980a). Such regions of homology may be analogous to transposable genetic elements that are known to exist in a number of bacterial plasmids (Campbell *et al.*, 1979). It is likely that the homologies observed among plasmids from different strains and different species of cyanobacteria have resulted from intra- or interspecies gene transfer.

Plasmid DNA has also been isolated from a number of species of photosynthetic bacteria, including *R. sphaeroides* (Fornari *et al.*, 1984; Gibson and Niederman, 1970; Pemberton and Tucker, 1977; Saunders *et al.*, 1976; Suyama and Gibson, 1966; Tucker and Pemberton, 1978), *R. capsulata* (Hu and Marrs, 1979; V. A. Saunders and S. J. Scahill, unpublished data), and *Rh. rubrum* (Kuhl and Yoch, 1981). Molecular weights of the plasmids range from 5 to 100 $\times 10^6$.

Some headway has been made in assigning functions to the plasmids of photosynthetic prokaryotes. Genes conferring resistance to certain antibiotics appear to be plasmid borne in strains of photosynthetic bacteria. By using DNA hybridization techniques, S. J. Scahill and R. P. Ambler (unpublished data) have located a penicillin-resistance determinant on a 5-megadalton (Md) plasmid from *R. capsulata* strain SP108. Penicillin resistance in this strain is attributable to the production and activity of an inducible β-lactamase, with a marked preference for benzyl penicillin as substrate (V. A. Saunders and S. J. Scahill, unpublished data). In addition, there is some evidence suggesting an association between the presence of plasmid DNA and streptomycin resistance in *Rh. rubrum* (Kuhl and Yoch, 1981). Other possible candidates for plasmid-determined functions include resistance to heavy metals, production of toxins, gas vacuolation, photoproduction of hydrogen, catabolic pathways, and conjugative ability. Such possibilities are currently being explored. It has also been speculated that plasmids may have a role in specifying the photosynthetic apparatus (Gibson and Niederman, 1970; Saunders *et al.*, 1976), but this remains equivocal. Kuhl and Yoch (1981) isolated photosynthetically incompetent (Pho⁻) strains of *Rh. rubrum* after treatment of cells with ethidium bromide. Such Pho⁻ strains had lost a 34-Md plasmid that was present in the photosynthetically competent parental strain. On this basis, Kuhl and Yoch (1981) proposed that genes for photosynthesis are plasmid associated in *Rh. rubrum*. This proposal would be more convincing if the putative photosynthetic plasmid could be introduced into the Pho⁻ strain with concomitant restoration of photosynthetic ability. Possibly the 34-Md plasmid has integrated into the chromosome and inactivated chromosomal genes concerned with photosynthesis in such Pho⁻ strains. In *R. sphaeroides* there appears to be some correlation between the occurrence of the Pho⁻ phenotype (usually involving the absence of light-harvesting bacteriochlorophyll–protein complexes) and rearranged plasmid DNA sequences (Fornari et al., 1984). The location of genes for photosynthesis in R. capsulata has been examined directly by using DNA hybridization probes containing the photopigment genes of R. capsulata (see "Conjugation" under "Gene Transfer Systems and Genetic Manipulation"). A much greater degree of homology was found between probe and chromosomal DNA as opposed to plasmid DNA from R. capsulata, favouring a chromosomal location for photopigment genes in this organism (Marrs, 1982).

Genes for nitrogen fixation are known to be plasmid encoded in a number of bacterial species (Nuti et al., 1979). The possibility that such genetic determinants are plasmid borne in nitrogen-fixing photosynthetic prokaryotes has been investigated. The total genomic DNA of Anabaena 7120 has been probed for homology with a radioactively labelled recombinant plasmid containing Klebsiella nif genes (Mazur et al., 1980). Although there was a considerable degree of homology between nif genes of Klebsiella and Anabaena, there was no evidence for plasmid involvement in nitrogen fixation in this cyanobacterium. Elucidation of the molecular organisation and expression of plasmids of photosynthetic prokaryotes should assist identification of the routes of gene flow between these organisms and the role of such plasmids in natural ecosystems.

Gene Transfer Systems and Genetic Manipulation

Transduction

To date, transductional analyses of photosynthetic prokaryotes remain strictly limited. A number of bacteriophages (and cyanophages) are known to infect specific photosynthetic prokaryotes (Marrs, 1978; Sherman and Brown, 1978; Wolk, 1973), and in certain cases to mediate the transfer of genetic material (Marrs, 1978; Saunders, 1978; Pemberton and Tucker, 1977). However, with the exception of the gene transfer agent (GTA) of R. capsulata (Marrs, 1974), these bacteriophages are not particularly proficient in promoting generalized transduction. The GTA, which is morphologically similar to a small bacterial virus, mediates genetic exchange exclusively between strains of R. capsulata in a process described as "capsduction" (Marrs, 1978, 1982). GTA-mediated gene transfer superficially resembles generalized transduction, but there are some notable differences. For instance, there is no transfer of the capacity to produce the GTA to recipients and no amplification of GTA-specific DNA during GTA production (Yen et al., 1979). The DNA of GTA particles is a linear duplex of about 3×10^6 daltons (Solioz and Marrs, 1977; Yen et al., 1979). GTA particles can apparently mediate the transfer of any region of the genome of R. capsulata and recombination frequencies of 10^{-3} per recipient cell have been achieved

(Yen et al., 1979). A genetic map of the photopigment region of *R. capsulata* (Fig. 1a) has been constructed using the GTA (Yen and Marrs, 1976). The GTA has proved to be a convenient tool for manipulating genes that specify both the photopigment system (Drews et al., 1976; Yen and Marrs, 1976) and the machinery for nitrogen fixation (Wall et al., 1975) in *R. capsulata*. This genetic vector may also play a natural role in promoting the traffic of genes and adaptation within populations of *R. capsulata*.

Transformation

The provision of efficient transformation systems is of vital importance for the development of gene-cloning systems in photosynthetic prokaryotes. Genetic transformation has been documented for a number of unicellular cyanobacteria, including *Anacystis nidulans* (Herdman, 1973; Herdman et al., 1970; Mitronova et al., 1973; Orkwiszewski and Kaney, 1974; Shestakov and Khyen, 1970), *Aphanocapsa* 6714 (Astier and Espardellier, 1976), *Gloeocapsa alpicola* (Devilly and Houghton, 1977), and *Agmenellum quadruplicatum* (Stevens and Porter, 1980) and for the filamentous cyanobacterium, *Nostoc muscorum* (Trehan and Sinha, 1981).

Progress in the development of transformation systems for photosynthetic bacteria has been relatively slow. A transformation system has been described for *R. sphaeroides* strain RS630 utilising DNA from the temperate bacteriophage RØ6P as the transformation probe (Tucker and Pemberton, 1980). This transformation process depends strictly upon simultaneous infection of the recipient by a closely related phage, RØ9. Optimum frequencies of transformation are attained if the recipient strain is already lysogenic for RØ9 when superinfected with the bacteriophage at multiplicities of infection between 1 and 10 RØ9 particles per recipient (Tucker and Pemberton, 1980). The development of such a phage-mediated transformation system opens up the possibility of designing a cloning system for *R. sphaeroides* utilising RØ6P or derivatives thereof as cloning vectors. More recently, a generalised transformation system has been developed for *R. sphaeroides*. $CaCl_2$-mediated uptake of plasmid DNA has been achieved, albeit at low frequency (Fornari and Kaplan, 1982; D. J. S. Virk and V. A. Saunders, unpublished data). However, transformation frequencies have been enhanced by treating cells with high concentrations (500 mM) of Tris prior to exposure to the transforming DNA (Fornari and Kaplan, 1982). A number of broad host range plasmids, including RSF1010 and pRK290, have been introduced into *R. sphaeroides* by using this methodology. Such a transformation system permits the cloning of specific genes directly in *R. sphaeroides*. The use of liposomes as vectors for introducing DNA into cells has received attention (Dimitriadis, 1979). Liposomes have been used by S. Kaplan and co-workers (personal communication) to transfer DNA to *R. sphaeroides* in a process that is

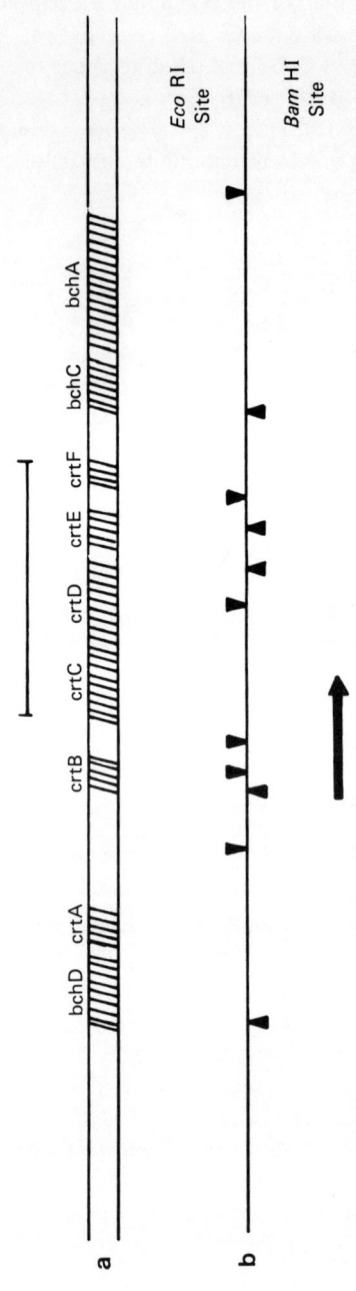

Fig. 1. Genetic (a) and restriction (b) maps of the photopigment region of *Rhodopseudomonas capsulata*. The shaded areas of the genetic map each denote a cluster of mutations that result in a particular phenotype. The restriction map shows the location of the restriction recognition sites for the *Eco* RI (▼) and *Bam* HI (▲) enzymes. The scaling factor is 2.9 Md per map unit. The thick arrow indicates the direction of transcription for the *crt* B region of the photopigment genes. Bar, 1.0 map unit. Modified from Marrs, 1983; Taylor *et al.*, 1983; Yen and Marrs, 1976.

insensitive to added DNAse. The transfer of DNA by encasement in liposomes may circumvent some of the potential barriers to transformation, including the lack of competence for DNA uptake and the activity of extracellular nucleases which may inactivate the transforming DNA. Such methodology may therefore be applicable in those cases in which transformation systems are not yet available.

Conjugation

Thus far, there has been no conclusive report of conjugation mediated by indigenous plasmids in either photosynthetic bacteria or cyanobacteria. However, conjugative transfer of promiscuous plasmids to photosynthetic prokaryotes from unrelated organisms has been documented. Such plasmids are providing a potent tool for the genetic manipulation of photosynthetic organisms. The P incompatibility (Inc) group plasmid R68.45 has been transferred to the cyanobacterium *Synechococcus* PCC6301 from *Escherichia coli* (Delaney and Reichelt, 1983). R68.45 does not appear to be present as an autonomous plasmid in the recipients. Instead, R68.45 integrates into the cyanobacterial genome. Integration probably involves the insertion sequence IS21(IS8) which is found on the plasmid. Such an integrated plasmid may prove useful in mobilizing the *Synechococcus* genome. Transfer of R plasmids [belonging principally to the Inc P group], presumably by conjugation, to various members of the Rhodospirillaceae has been achieved (see Table 1). Olsen and Shipley (1973) described the transfer of the R plasmid R1822 to *R. sphaeroides* and *Rh. rubrum*. Subsequently, Sistrom (1977) transferred R68.45 to strains of *R. sphaeroides* and *R. gelatinosa*. This R plasmid has also been transferred to *R. capsulata* (Marrs *et al.*, 1977; Yu *et al.*, 1981) and *Rh. vannielii* (L. E. Potts and C. S. Dow, personal communication). Whereas R68.45 is proficient at mobilising the chromosome of certain strains of *R. sphaeroides* (Sistrom, 1977), relatively low frequencies of chromosome transfer have been reported using R68.45 in *R. capsulata* (Yu *et al.*, 1981). Possibly variants of this plasmid will prove more effective in promoting chromosome transfer in the Rhodospirillaceae. Miller and Kaplan (1978) demonstrated the transfer of RP4 to *R. sphaeroides*. Interestingly, no transfer occurs if matings are performed under phototrophic conditions. Transfer of the closely related plasmid RP1 to *R. capsulata* has been achieved (Yu *et al.*, 1981). Tucker and Pemberton (1979a) screened a series of R plasmids (belonging to Inc P, Inc W, Inc C, Inc I, and Inc FII groups) for their ability to transfer to and promote chromosome transfer in *R. sphaeroides*. Only plasmids of Inc P and Inc W groups were freely transmissible to *R. sphaeroides*. However, these plasmids were relatively ineffective in mobilising chromosomal genes. Certain Inc P plasmids have been shown to promote the transfer of nonconjugative plasmids to the Rhodospirillaceae. For example, plasmids pML2 (Hershfield *et al.*, 1974), NTP16 (Anderson et al., 1977), and pTB70 (Windass et al., 1980) have been mobilised to strains of R. sphaeroides (D. J. S. Virk and V. A. Saunders, unpublished data).

A derivative of RP1 designated pBLM2, with an enhanced ability to mobilise the chromosome of R. capsulata, has been isolated by Marrs (1981). pBLM2-mediated chromosomal transfer appears to originate at more than one site on the chromosome of R. capsulata. Genes affecting tryptophan synthetase, rifampicin resistance, photosynthetic reaction-centre bacteriochlorophyll and carotenoid syntheses and cytochromes can all be mobilised by pBLM2. Recombination frequencies greater than 10^{-4} per donor cell have been achieved. Tight linkage was exhibited among the genes affecting photosynthesis, whereas no linkage was observed between these genes and those for cytochrome c_2 synthesis (Marrs, 1981).

pBLM2 has also been used to generate a series of R-prime derivatives carrying the photopigment region of R. capsulata (Marrs, 1981). Such R-prime plasmids retain the promiscuity of the parental plasmid and are thus transferable to a variety of gram-negative bacteria including *Escherichia coli*, *Pseudomonas fluorescens*, and R. sphaeroides. However, genes for bacteriochlorophyll synthesis do not appear to function in any of these hosts. Nevertheless, these R-prime plasmids provide a good source of DNA for cloning genes for photosynthesis. Dissection of such plasmids with restriction enzymes has enabled the isolation and subsequent cloning of specific segments of the chromosomes of R. capsulata in *E. coli* using pDPT42 and pDPT44 (both derived from pBR322) as cloning vectors (Clark et al., 1981; Marrs, 1983; Taylor et al., 1983). Identification of particular cloned regions has relied on mobilisation of the recombinant plasmids from the repository strains of *E. coli* to appropriate mutants of R. capsulata and subsequent screening of the transconjugants for specific complementation and recombination events. A physical map of the photopigment region has been defined on the basis of these data. (Fig. 1b) (Marrs, 1983; Taylor et al., 1983).

It has proved possible to transfer the hybrid plasmid comprising RP4 and phage Mu *cts* to strains of R. sphaeroides (Tucker and Pemberton, 1979b) and R. capsulata (Yu et al., 1981). Expression of the Mu genome occurs in these cases. Certain P-group plasmids carrying Mu have been used as "suicide plasmids" to introduce transposons into rhodopseudomonads. For example, a derivative of the Inc P group plasmid, pPH1J1, carrying Mu *cts* and Tn 5, designated pJB41J1 (Beringer et al., 1978), has been transferred to strains of R. sphaeroides (D. J. S. Virk and V. A. Saunders, unpublished data). The ability of this plasmid to become established in R. sphaeroides strain WS8 is drastically reduced relative to the parental plasmid, pPH1J1. Despite this inability to be maintained in WS8, transposable elements are carried into the strain. Acquisition of such elements appears to be associated with the induction of mutations in WS8. In matings of *E. coli* (pJB4J1) and R. sphaeroides WS8, pigment mutants are detected amongst the recipient population at a frequency of about 2% (D. J. S. Virk and V. A. Saunders, unpublished data).

Table 1. Examples of R plasmid transfers to the Rhodospirillaceae

Plasmid	Incompatibility group	Recipient	References
R1822	P	*Rhodopseudomonas sphaeroides*	Olsen and Shipley (1973)
RP4	P	*Rhodospirillum rubrum*	Miller and Kaplan (1978); Tucker and Pemberton (1979a)
		R. sphaeroides	
RP4::Mu *cts* 61	P	*R. capsulata*	Yu et al. (1981)
RP4::Mu *cts* 62	P	*R. sphaeroides*	Tucker and Pemberton (1979b)
RP1	P	*R. capsulata*	Yu et al. (1981)
pBLM2 (derivative of RP1)	P	*R. capsulata*	Marrs (1981)
RP1::Tn501	P	*R. sphaeroides*	Pemberton and Bowen (1981)
R68.45	P	*R. sphaeroides*	Tucker and Pemberton (1979a); Sistrom (1977)
		R. gelatinosa	Sistrom (1977)
		R. capsulata	Yu et al. (1981); Marrs et al. (1977)
		Rhodomicrobium vannielii	L. E. Potts and C. S. Dow (personal communication)
R751	P	*R. sphaeroides*	Tucker and Pemberton (1979a)
		R. rubrum	Quivey et al. (1981)
R751::Tn5	P	*R. rubrum*	Quivey et al. (1981)
R702	P	*R. sphaeroides*	Tucker and Pemberton (1979a)
RK2	P	*R. rubrum*	Quivey et al. (1981)
pPH1J1	P	*R. sphaeroides*	D. J. S. Virk and V. A. Saunders (unpublished data)
pJB4J1 (= pPH1J1::Mu *cts*::Tn5)	P	*R. sphaeroides*	D. J. S. Virk and V. A. Saunders (unpublished data)
R388	W	*R. sphaeroides*	Tucker and Pemberton (1979a)
S-a	W	*R. sphaeroides*	Tucker and Pemberton (1979a)

Other hybrid plasmids carrying transposons are transmissible to the Rhodospirillaceae (Table 1). The introduction and expression of transposable genetic elements in photosynthetic prokaryotes now provides scope for a variety of *in vivo* genetic manipulations. Indeed, Pemberton and Bowen (1981) have used plasmid RP1::Tn501 to promote high-frequency chromosome transfer in *R. sphaeroides*. Recombination frequencies vary between 10^{-3} and 10^{-7} per donor cell, depending on the chromosomal marker. It appears that RP1::Tn501 mediates polarized transfer of the chromosome of *R. sphaeroides* from one or possibly two origins. Furthermore, it seems that all regions of the genome of *R.*

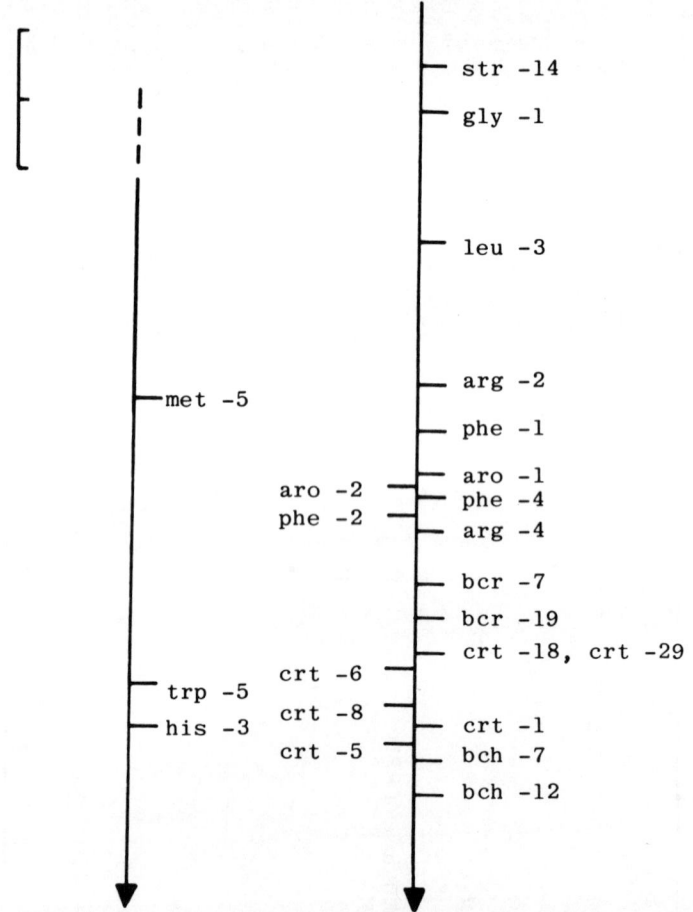

Fig. 2. Genetic map of two regions of the genome of *Rhodopseudomonas sphaeroides*. The map is based on linkage data obtained from RP1::Tn501-mediated chromosome transfer. Bar, 0.2 units. From Pemberton and Bowen (1981) with permission.

sphaeroides are accessible to mapping using RP1::Tn501 (Pemberton and Bowen, 1981). The mechanism of such RP1::Tn501-promoted chromosome mobilisation has yet to be established. One possibility is that the presence of the transposon on both the plasmid and the chromosome provides the necessary homology for integration of the conjugative plasmid to form a high frequency of recombination (Hfr) donor. However, this explanation seems insufficient to account for the enhanced chromosome-mobilising ability of RP1::Tn501 relative to RP1, since other transposons, for example Tn5, do not seem to confer such mobilizing ability on RP1 (Pemberton and Bowen, 1981). Irrespective of the precise role of Tn501 in the mobilisation of the chromosome of *R. sphaeroides* by RP1, the availability of Hfr donors has enabled the formulation of a map of the genome of this photosynthetic bacterium (Fig. 2). Interestingly, the genetic maps of both *R. sphaeroides* and *R. capsulata* exhibit a tight clustering of genes specifying bacteriochlorophyll and carotenoid syntheses. However, on the basis of interspecific DNA:DNA hybridizations (using cloned photosynthesis genes from *R. capsulata* as hybridization probes) the DNA sequences of these gene clusters do not appear to be closely related (Beatty and Cohen, 1983).

The use of transposable genetic elements that confer a chromosome-mobilising ability on conjugative plasmids could conceivably be extended in order to map the genomes of other photosynthetic organisms. Thus, localisation of genes encoding the various metabolic capabilities of photosynthetic bacteria, including photochemical nitrogen fixation, hydrogen production, photosynthesis, and respiration, should soon be possible.

Gene-Cloning Systems

Methods for gene cloning provide new and powerful approaches to the dissection of gene structure and function in prokaryotes and eukaryotes (for a review, refer to Timmis, 1981). Such methodology is of particular value in genetic studies of organisms that are not easily amenable to conventional genetic analyses.

Fundamental prerequisites for gene cloning are a cloning vector (normally a plasmid or bacteriophage genome) into which the specific DNA fragments can be spliced and an appropriate host strain to serve as a repository for the propagation of the DNA molecules. Desirable features for the cloning vector are that it should be small; easily isolated and purified in large quantities; possess at least one but preferably two or more selectable phenotypic traits; contain unique cleavage sites for a number of commonly used restriction enzymes; and permit efficient transcription of the cloned genes.

At the present time, the development of vectors utilising plasmids rather than phage genomes is favoured for cloning genes of photosynthetic prokaryotes (Table 2). A number of cyanobacterial plasmids have been tailored for cloning

Table 2. *Examples of potential cloning vectors for manipulation and analysis of genes of photosynthetic prokaryotes*

Vector	Replicon	Unique restriction recognition site	Host	References
1. Cyanobacteria				
pCH1 - pCH5	pUH24	Xho I	Anacystis nidulans	van den Hondel et al. (1980); van den Hondel et al. (1981)
pUC1	pUH24	Bam HI, Xho I	A. nidulans	van den Hondel et al. (1980); van den Hondel et al. (1981)
pUC3	pUH24	Bgl II, Xho I	A. nidulans	Kuhlemeier et al. (1983)
	pDU1	Bgl I	Nostoc PCC 7524	Reaston et al. (1982)
pUC104	pUC1/pACYC184	Eco RI, Sal I, Xho I	A. nidulans and Escherichia coli	Kuhlemeier et al. (1981)
pLS103	pUH24/pBR322	Hind III, Sal I	A. nidulans and E. coli	Sherman and van de Putte (1982)
Cosmid for packaging in λ heads				
pPUC29	pUC1/pACYC184	Eco RI, Xho I	A. nidulans and E. coli	van den Hondel et al. (1981)
2. Photosynthetic bacteria				
pSS101	5-Md plasmid of Rhodopseudomonas capsulata SP108	Bam HI, Eco RI, Hind III	R. capsulata	S. J. Scahill (personal communication)
	5-Md plasmid of R. capsulata/pBR325		E. coli	S. J. Scahill (personal communication)
pDPT42	pBR322	Bam HI	R. capsulata and E. coli	Clark et al. (1981); Marrs (1982)
pDPT44	pBR322	Eco RI	R. capsulata and E. coli	Clark et al. (1981); Marrs (1982)
pTB70	R300B(RSF1010)	Bam HI, Eco RI, Sal I	R. sphaeroides	D. J. S. Virk and V. A. Saunders (unpublished data)
pML2	Col E1		R. sphaeroides	D. J. S. Virk and V. A. Saunders (unpublished data)

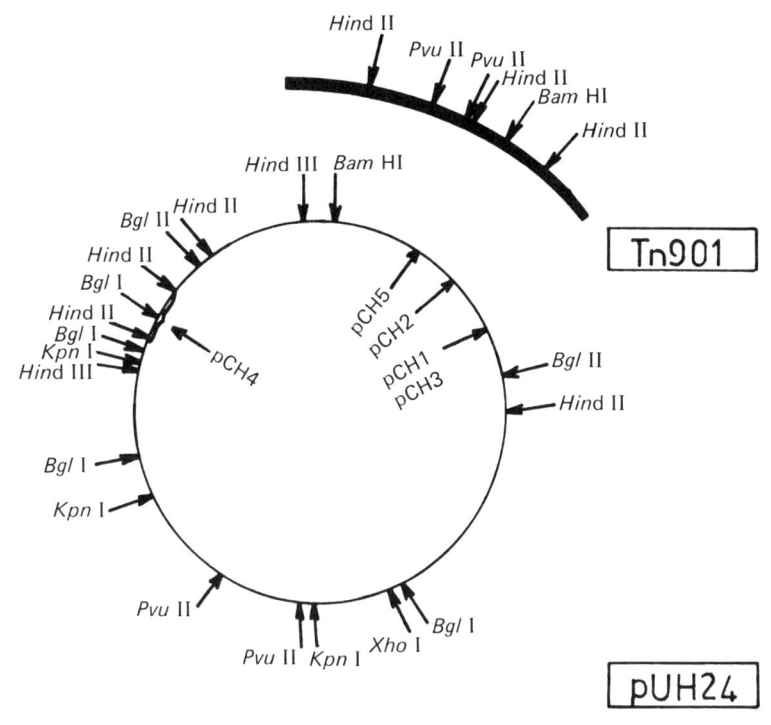

Fig. 3. Restriction maps of pUH24 and Tn901 and the sites of transposon insertion. Location of the restriction recognition sites of *Bam* HI, *Bgl* I, *Bgl* II, *Hin*d II, *Hin*d III, *Kpn* I, *Pvu* II, and *Xho* I in pUH24 is indicated by arrows outside the circle. Note: Only 4 of the 11 *Hin*d II sites are shown. The sites of Tn901 insertion in pUH24 are indicated by arrows inside the circle. From van den Hondel *et al.* (1980) with permission.

purposes. Vectors relying on insertional inactivation (Timmis *et al.*, 1974, 1978) as a means of detecting those clones containing the hybrid DNA molecules have been designed. The antibiotic resistance transposon Tn901 has been incorporated *in vivo* into a cryptic plasmid, pUH24, from *A. nidulans* R-2 generating a series of ampicillin-resistance plasmids, for example, pUCI (van den Hondel *et al.*, 1980, 1981) (Fig. 3 for a restriction map). Subsequent modification *in vitro* of the resulting hybrid plasmids provides plasmids (for example, pUC2) carrying two resistance genes (for ampicillin and streptomycin) and single cleavage sites for the restriction enzymes *Bam* HI and *Xho* I (van den Hondel *et al.*, 1981). pUC3 and pUC4 have also been obtained by deleting the ampicillin-resistance gene from pUC2. By coupling cyanobacterial plasmids to the *E. coli* vector, pACYC184, (Chang and Cohen, 1978) hybrid plasmids have been obtained [e.g., pUC104, a hybrid between pUC1 and pACYC184 (Fig. 4)] that can be propagated in either *A. nidulans* R-2 or *E. coli* (Kuhlemeier *et al.*, 1981, 1983).

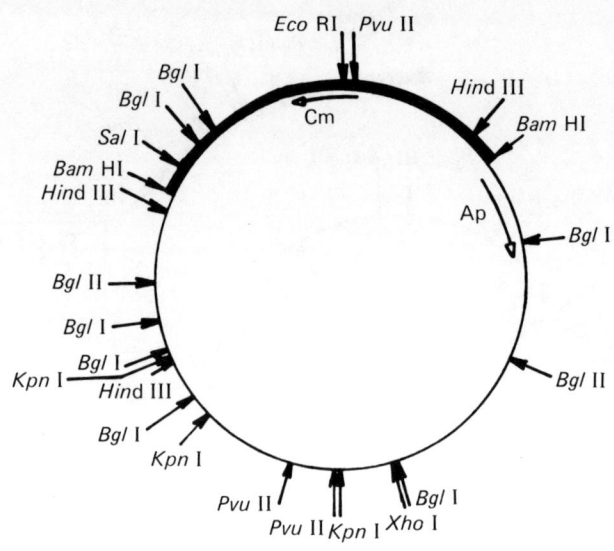

Fig. 4. Restriction map of pUC104. The thin line represents the pUC1 part (van den Hondel et al., 1980); the thick line represents the pACYC184 part (Chang and Cohen, 1978). Location of restriction recognition sites of *Bam* HI, *Bgl* I, *Bgl* II, *Eco* RI, *Hin*d III, *Kpn* I, *Pvu* II, *Sal* I, and *Xho* I in pUC104 is indicated by arrows outside the circle. From Kuhlemeier et al. (1981) with permission.

In order to overcome the instability of these recombinant plasmids in *A. nidulans* R2, a derivative strain, cured of the indigenous plasmid pUH24, is used as host. Further bifunctional vectors that can replicate in both *E. coli* and cyanobacteria are being constructed (Gendel et al., 1983; Sherman and van de Putte, 1982; van den Hondel et al., 1981). Cointegrate formation between the *E. coli* plasmid pBR322 (Bolivar et al., 1977) carrying a chloramphenicol transposon and plasmid pUH24 from *A. nidulans* R-2 has enabled the generation of a series of hybrid plasmids (Sherman and van de Putte, 1982). One such plasmid, pLS103, has unique restriction sites for *Sal* I and *Hin*d III and can be amplified in *E. coli* (but not in *A. nidulans*) with chloramphenicol. These vectors for *A. nidulans* are all based on the replicon pUH24. However, the simultaneous cloning of two different DNA fragments on separate vectors demands the availability of two compatible vectors. One approach is to construct additional vectors utilizing the origin of replication of another cryptic plasmid (pUH25) of *A. nidulans* R2. As a preliminary to the construction of a shuttle vector from pUH25, Laudenbach and colleagues (1983) have cloned the origin of replication of this plasmid. Plasmids isolated from a number of cyanobacterial strains [e.g., pDU1 from *Nostoc* PCC7524 (Reaston et al., 1980, 1982)] may prove useful in the formation of other cloning vectors. In addition to providing facilities for self-cloning, hybrid vectors permit exploitation of the specific advantages afforded by the *E. coli* host system.

Cosmids (Collins and Brüning, 1978; Collins and Hohn, 1978) have also been developed by grafting the cohesive ends of coliphage lambda (λ) onto hybrid cyanobacterial-enterobacterial plasmids (e.g., pUC104) *in vitro*. One such cosmid, pPUC29, is a 9.2-Md vector that carries the resistance markers ampicillin and chloramphenicol, and single cleavage sites for the restriction enzymes *Eco* RI and *Xho* I (van den Hondel *et al.*, 1981). In general, cosmids provide the means to clone relatively large fragments of DNA and consequently facilitate the construction of gene banks for specific organisms (Hohn and Hinnen, 1980). The utilisation of cosmids in conjunction with the *in vitro* phage-packaging procedure (Hohn and Murray, 1977) is enabling the generation of a gene bank for *A. nidulans* R-2 (C. A. M. J. J. van den Hondel, personal communication).

Among the plasmids of photosynthetic bacteria, the 5-Md plasmid of *R. capsulata* strain SP108 shows potential as a cloning vector. The plasmid has single restriction cleavage sites for *Eco* RI, *Bam* HI, and *Hind* III. Furthermore, hybrids have been constructed between this plasmid and pBR325. Such hybrids are stably maintained in *E. coli*, but to date successful transformation of *R. capsulata* by these hybrids has not been achieved (S. J. Scahill, unpublished data). Plasmid RØ6P isolated from *R. sphaeroides* can serve as transforming DNA for the transformation of this organism (see "Transformation", in "Gene Transfer Systems and Genetic Manipulation") and may, with appropriate modification, provide a useful cloning vector for *R. sphaeroides*. In addition, the replication region of the 28.0-Md plasmid from *R. sphaeroides* 2.4.1., which is present in more than one copy per cell, may be exploited in vector design. However, the majority of plasmids so far isolated from the photosynthetic bacteria do not appear to be particularly suitable for use as cloning vectors in their present form. They are relatively large, present as only a few copies per chromosome equivalent, and lack traceable phenotypic traits. Extensive modification would probably be necessary before the potential of such plasmids as cloning vectors could be tested.

As an alternative to the plasmids native to photosynthetic bacteria, certain plasmids commonly used for gene cloning in other bacteria, for example, those derived from the Col El replicon, such as pBR322 (Bolivar *et al.*, 1977), or those derived from the RSF1010 and R300B replicons, such as pKT210 (Bagdasarian *et al.*, 1979) and pTB70 (Barth, 1979; Windass *et al.*, 1980), may be applicable. Indeed, some of these plasmids can be transferred to and stably maintained in a number of photosynthetic bacteria. For instance, pTB70, NTP16, and the Col El derivative, pML2, can be mobilised to strains of *R. sphaeroides*. In addition, pDPT42 and pDPT44 can be transferred to strains of *R. capsulata* (see "Conjugation" in "Gene Transfer Systems and Genetic Manipulation"). These plasmids have been used to clone photosynthetic genes of *R. capsulata* in *E. coli* (Clark *et al.*, 1981; Marrs, 1983; Taylor *et al.*, 1983). The promiscuous plasmids RP4/RP1, which can be maintained in a number of photosynthetic bacteria (Table 1), might also be fashioned to provide appropriate cloning vectors. The wide host range of such plasmids would enable gene cloning in a number of genetic backgrounds.

A gene-cloning system may utilise as a host the same bacterial strain as that from which the cloned genes derive (= self-cloning). Alternatively, a different strain may be employed as a host. In both cases, an efficient mechanism for introduction of the cloning vector into the host is essential. This can occur either directly by transformation of the host concerned or by a two-stage process involving transformation of an intermediary host and subsequent transfer of hybrid plasmids to the host of choice. Efficient transformation systems are available for certain unicellular cyanobacteria (Herdman, 1973; Stevens and Porter, 1980), and for *R. sphaeroides* (Fornari and Kaplan, 1982).

In self-cloning experiments, the general recombination functions of the host may hamper isolation of cloned DNA in the extrachromosomal state. Recombination-defective (Rec$^-$) strains are thus favoured where recovery of the cloned DNA is a priority. In other cases in which heterologous DNA is cloned, the restriction system of the host may present an effective barrier to introduction of the foreign DNA. Restriction-defective (Res$^-$) mutants of the host are therefore desirable.

Towards Elucidation of the Control of Gene Expression

It is implicit from the foregoing discussions that environmental stimuli have profound effects on the development of photosynthetic prokaryotes and that such effects are manifested via a network of molecular controls. The availability of techniques for *in vitro* and *in vivo* genetic manipulation enables the adoption of a range of strategies to dissect these controls. Such strategies include:

1. Development of efficient gene-cloning systems to provide nucleic acids of absolute sequence purity for investigating gene expression and function. Cloned genes can provide useful probes, for example, for analysing patterns of transcription associated with specific differentiation events.

2. DNA sequence analysis of cloned genes enabling detailed structural analysis of the genes, including location of regulatory regions.

3. Isolation of a spectrum of regulatory (and other) mutants by *in vivo* and *in vitro* mutagenesis techniques. The use of transposable genetic elements as insertional mutagens may facilitate the isolation of mutants that have proved difficult to obtain by conventional means. Transposition mutagenesis has the advantage of permitting rapid mapping of mutations (see Sherratt, 1981).

4. Development of novel gene transfer systems enabling, for example, the performance of cis-trans complementation analyses.

This section assesses the progress made in elucidating processes controlling gene expression in photosynthetic prokaryotes by using such approaches.

Heterocyst Differentiation and Nitrogen Fixation

An approach used in elucidating the control of heterocyst and nitrogenase development in cyanobacteria has been to isolate and characterise mutants defective in heterocyst formation (het^-) and/or nitrogen fixation (nif^-) (Currier *et al.*, 1977; Padhy and Singh, 1978; Singh *et al.*, 1977; Singh and Singh, 1981; Wilcox *et al.*, 1975b). A major obstacle in such studies has been the dearth of operational gene transfer systems for detailed genetic analyses of the mutants. However, transfer of *nif* and *het* genes has been reported in *Nostoc muscorum* (Padhy and Singh, 1978; Singh and Singh, 1981; Stewart and Singh, 1975). On the basis of genetic recombinational analysis of *het* and *nif* genes, Singh and Singh (1981) propose that *nif* and *het* genes are not organised in a single operon (since independent inheritance of *het* or *nif* genes can occur), but may be subject to common regulation by positively acting regulatory element(s).

Various lines of inquiry indicate that glutamine synthetase or a regulator of its activity may be required for control of transcription of *nif* genes in a number of organisms (Dixon *et al.*, 1981; Johansson *et al.*, 1983; Stewart, 1980). In order to determine directly the transcriptional control of heterocyst differentiation and nitrogen fixation in cyanobacteria, Haselkorn's group is constructing recombinant DNA probes that incorporate specific cyanobacterial genes to measure transcription. The *nif* genes and the gene for glutamine synthetase of *Anabaena* 7120 have been identified by heterologous hybridisation with *Klebsiella nif* genes (Mazur *et al.*, 1980) and the *glnA* gene of *E. coli* (Fisher *et al.*, 1981), respectively, and cloned in *E. coli*. Hybridization probes containing nitrogenase structural genes from *Anabaena* have been used to localize *nif* genes in *Gloeothece* and *Fremyella* (Kallas *et al.*, 1982). It appears that *nif* genes of *Gloeothece* are clustered as they are in *Klebsiella* and *Rhizobium*. Whilst in *Fremyella* and *Anabaena* such genes are apparently dispersed on the genome. By using a similar approach, putative gene sequences for the nitrogenase of *R. sphaeroides* and of other photosynthetic bacteria have been identified and cloned (Fornari and Kaplan, 1983; Scolnik and Haselkorn, 1982). The availability of these and other DNA hybridization probes will assist the detection of RNA transcripts specific to differentiating cells and factors regulating the transcription programme. Furthermore, the cloning of particular genes and their mutagenesis *in vitro* opens up the possibility of using the process of homogenotisation [a component of reversed genetics (Weissmann, 1978)] for the generation of specific mutants of photosynthetic prokaryotes.

Controls on the Development of the Photosynthetic Apparatus

The Photosynthetic Apparatus of Photosynthetic Bacteria. Mapping studies of the genomes of *R. capsulata* (Marrs, 1981; Yen and Marrs, 1976) and *R. sphaeroides* (Pemberton and Bowen, 1981) reveal a tight clustering of genes for photopigment synthesis (Figs. 1 and 2). This has implications for the existence of common regulatory elements in the control of carotenoid and bacteriochlorophyll syntheses. The direction of transcription of genes within the photopigment region of *R. capsulata* has been deduced (Fig. 1b) from expression profiles of appropriate cloned genes in complementation analyses (Marrs, 1983; Taylor *et al.*, 1983). In order to examine the transcriptional regulation of genes for bacteriochlorophyll synthesis (*bch*) in *R. capsulata* Biel and Marrs (1983) have obtained strains carrying Mu d1 (Ap^r *lac*) phage (Casadaban and Cohen, 1979) insertions in various *bch* genes. In such strains the *lacZ* (β-galactosidase) gene of the phage is under the control of *bch* gene promoters. The effect of different light and oxygen regimes on the activity of *bch* gene promoters has been assessed by measuring the amount of β-galactosidase produced under the different conditions. Many of the genes governing bacteriochlorophyll synthesis, including *bchB, bchC,* and *bchH*, appear to be regulated in response to oxygen. However, to date only one operon, *bchCA*, has been identified. Although light influences photopigment composition, strains with *bch* :: Mu d fusions do not apparently regulate β-galactosidase production in response to light intensity. Light does not, therefore, appear to regulate transcription of *bch* genes directly. Cloned photosynthesis genes have been used to determine variation in species of messenger RNA in *R. capsulata* in response to oxygen tension (Clark *et al.*, 1984). Transcripts corresponding to genes (*rxcA* locus) for the L and M reaction-centre polypeptides and for the two light-harvesting complex 1 (LH1) polypeptides are greatly enhanced at low oxygen tension. Whereas transcripts from genes encoding bacteriochlorophyll biosynthetic enzymes and the H reaction-centre polypeptide (*rxcB* locus) are increased to a lesser extent. This would be consistent with the relative amounts of these gene products that are required for photosynthetic membrane synthesis: Biosynthetic enzymes catalytically produce bacteriochlorophyll, whilst pigment-binding antenna and reaction-centre proteins bind bacteriochlorophyll stoichiometrically. However, it is not clear whether these various transcripts are equally active in translation. It has been suggested (Clark *et al.*, 1984) that a regulatory element that responds to oxygen is located adjacent to the *rxcA* locus on the genome of *R. capsulata*.

Phycobilisomes of Cyanobacteria. It has been known for some time that chromatic adaptation in cyanobacteria requires *de novo* synthesis of phycobiliproteins. This suggests a transcriptional mode of regulation of the structural genes for phycobiliproteins controlled either directly or indirectly by the wavelength of light (Bogorad, 1975; Gendel *et al.*, 1979). Besides the biliproteins, other polypeptides of phycobilisomes appear to be altered during chromatic adaptation. It is possible, therefore, that a number of genes specifying components of the phycobilisomes are coordinately regulated in response to changes in the wavelength of light (Gendel *et al.*, 1979). Bogorad's group (1983) have isolated messenger RNAs (mRNAs) for phycobiliproteins. Such mRNAs will be useful for the identification and subsequent cloning of genes for these biliproteins, and in turn should assist clarification of the precise molecular controls governing chromatic adaptation.

Summary and Prospective

In essence, photosynthetic prokaryotes possess a wealth of metabolic and morphological capabilities that justify intensive genetic analysis. Although gene transfer systems for these organisms are underdeveloped at present, new and powerful genetic tools are becoming available by coupling *in vitro* recombinant DNA technology to the traditional genetic systems. Genetic organization in photosynthetic prokaryotes has already been probed using such methodology and maps of genes for photopigment synthesis (Marrs, 1981, 1983; Taylor *et al.*, 1983; Yen and Marrs, 1976), for the machinery for nitrogen fixation (Mazur *et al.*, 1980), and for ribosomal RNA (Tomioka *et al.*, 1981) are being constructed. DNA sequencing analysis is enabling the detection of regulatory regions and the determination of codon usage. In addition, the biological role of plasmids and their applicability to molecular genetic research is being assessed. It is expected that other biological functions will soon be ascribed to the plasmids of these photosynthetic organisms, which in turn will assist the development of novel cloning systems. The design of host–vector systems for gene-cloning purposes and the construction of gene banks for photosynthetic prokaryotes have become priorities. The availability of cloned genes in conjunction with *in vivo* and *in vitro* transcription–translation systems will permit a thorough functional analysis of the photosynthetic genome. Such trends in the genetics of this group highlight a need for more operational gene transfer systems and for a broader spectrum of mutants to complement and supplement the genetic tools currently available. Provision of a battery of techniques for *in vivo* and *in vitro* genetic manipulation will greatly facilitate dissection of the genetic controls over metabolism and differentiation. This should provide insights into genetic factors influencing the distribution and the survival potential of photosynthetic prokaryotes in their various ecological niches.

Acknowledgments

I am grateful to the Science Research Council for research grant No. GR/A 85797.

References

Adams, D. G. and Carr, N. G. (1981). The developmental biology of heterocyst and akinete formation in cyanobacteria. *CRC Crit. Rev. Microbiol.* **9,** 45–100.

Allen, M. M. (1968). Photosynthetic membrane system in *Anacystis nidulans. J. Bacteriol.* **96,** 836–841.

Anderson, E. S., Threlfall, E. J., Carr, J., McConnell, M. and Smith. H. R. (1977). Clonal distribution of resistance plasmid-carrying *Salmonella typhimurium* mainly in the Middle East. *J. Hyg.* **72,** 471–487.

Astier, C. and Espardellier, F. (1976). Mise en évidence d'un système de transfert génétique chez une cyanophycée du genre Aphanocapsa. *C. R. Hebd. Seances Acad. Sci. Ser. D* **282,** 795–797.

Baccarini-Melandri, A. and Zannoni, D. (1978). Photosynthetic and respiratory electron flow in the dual functional membrane of facultative photosynthetic bacteria. *J. Bioenerg. Biomemb.* **10,** 109–138.

Bagdasarian, M., Bagdasarian, M. M., Coleman, S. and Timmis, K. N. (1979). New vector plasmids for cloning in *Pseudomonas. In* "Plasmids of Medical, Environmental and Commercial Importance" (Eds. K. N. Timmis and A. Pühler), pp. 411–422. Elsevier/North-Holland Biomedical Press, Amsterdam.

Barth, P. T. (1979). RP4 and R300B as wide host-range plasmid cloning vehicles. *In* "Plasmids of Medical, Environmental and Commercial Importance" (Eds. K. N. Timmis and A. Pühler), pp. 399–410. Elsevier/North-Holland Biomedical Press, Amsterdam.

Beatty, J. T. and Cohen, S. N. (1983). Hybridization of cloned *Rhodopseudomonas capsulata* photosynthesis genes with DNA from other photosynthetic bacteria. *J. Bacteriol.* **154,** 1440–1445.

Beringer, J. E., Beynon, J. L., Buchanan-Wollaston, A. V. and Johnston, A. W. B., (1978). Transfer of the drug-resistance transposon Tn5 to *Rhizobium. Nature (London)* **276,** 633–634.

Biel, A. J. and Marrs, B. L. (1983). Transcriptional regulation of several genes for bacteriochlorophyll biosynthesis in *Rhodopseudomonas capsulata* in response to oxygen. *J. Bacteriol.* **156,** 686–694.

Bisalputra, T., Brown, D. L. and Weier, T. E. (1969). Possible respiratory sites in a bluegreen alga *Nostoc sphaericum* as demonstrated by potassium tellurite and tetranitroblue tetrazolium reduction. *J. Ultrastruct. Res.* **27,** 182–197.

Bogorad, L. (1975). Phycobiliproteins and complementary chromatic adaptation. *Annu. Rev. Plant Physiol.* **26,** 369–401.

Bogorad, L., Gendel, S. M., Haury, J. H. and Koller, K-P. (1983). Photomorphogenesis and complementary chromatic adaptation in *Fremyella diplosiphon. In* "Photosynthetic Prokaryotes. Cell Differentiation and Function" (Eds. G. C. Papageorgiou and L. Packer), pp. 119–126. Elsevier/North-Holland, Amsterdam.

Bolivar, F., Rodriguez, R. L., Greene, P. J., Betlach, M. C., Heyneker, H. L., Boyer, H. B., Crosa, J. H. and Falkow, S. (1977). Construction and characterization of new cloning vehicles. II. A multipurpose cloning system. *Gene* **2,** 95–113.

Bradley, S. and Carr, N. G. (1976). Heterocyst and nitrogenase development in *Anabaena cylindrica*. *J. Gen. Microbiol.* **96,** 175–184.
Brock, T. D. (1973). Evolutionary and ecological aspects of the cyanophytes. *In* "The Biology of Blue-Green Algae" (Eds. N. G. Carr and B. A. Whitton), pp. 487–500. Blackwell, Oxford.
Brown, A. H. and Webster, G. C. (1953). The influence of light on the rate of respiration of the blue-green alga *Anabaena*. *Am. J. Bot.* **40,** 753–758.
Campbell, A., Starlinger, P., Berg, D. E., Botstein, D., Lederberg, E. M., Novick, R. P. and Szybalski, W. (1979). Nomenclature of transposable elements in prokaryotes. *Plasmid* **2,** 466–473.
Carr, N. G. (1983). Biochemical aspects of heterocyst differentiation and function. *In* "Photosynthetic Prokaryotes. Cell Differentiation and Function" (Eds. G. C. Papageorgiou and L. Packer), pp. 265–280. Elsevier/North-Holland, Amsterdam.
Carr, N. G. and Bradley, S. (1973). Aspects of development in blue-green algae. *Symp. Soc. Gen. Microbiol.* **23,** 161–188.
Carr, N. G. and Whitton, B. A. (1982). "The Biology of Cyanobacteria." Blackwell, Oxford.
Casadaban, M. J. and Cohen, S. N. (1979). Lactose genes fused to exogenous promoters in one step using the Mu-*lac* bacteriophage: *in vivo* probe for transcriptional control sequences. *Proc. Natl. Acad. Sci. USA* **76,** 4530–4533.
Chang, A. C. Y. and Cohen, S. N. (1978). Construction and characterization of amplifiable DNA cloning vehicles derived from the p15A cryptic miniplasmid. *J. Bacteriol.* **134,** 1141–1156.
Clark, G., Taylor, D. P., Cohen, S. N. and Marrs, B. L. (1981). Restriction endonuclease map of the photopigment region of the *Rhodopseudomonas capsulata* chromosome. *Abstr. 81st Annu. Meet. Am. Soc. Microbiol.* H50, p. 122.
Clark, N. G., Davidson, E. and Marrs, B. L. (1984). Variation of levels of mRNA coding for antenna and reaction centre polypeptides in *Rhodopseudomonas capsulata* in response to changes in oxygen concentration. *J. Bacteriol.* **157,** 945–948.
Cohen-Bazire, G. and Sistrom, W. R. (1966). The procaryotic photosynthetic apparatus. *In* "The Chlorophylls" (Eds. L. P. Vernon and G. R. Seely), pp. 313–341. Academic Press, New York.
Collins, J. and Brüning, H. J. (1978). Plasmids useable as gene cloning vectors in an *in vitro* packaging by coliphage λ: "cosmids." *Gene* **4,** 85–107.
Collins, J. and Hohn, B. (1978). Cosmids: a type of plasmid gene-cloning vector that is packageable *in vitro* in bacteriophage λ heads. *Proc. Natl. Acad. Sci. U.S.A.* **75,** 4242–4246.
Connelly, J. L., Jones. O. T. G., Saunders, V. A. and Yates, D. W. (1973). Kinetic and thermodynamic properties of membrane-bound cytochromes of aerobically and photosynthetically grown *Rhodopseudomonas spheroides*. *Biochim. Biophys. Acta* **292,** 644–653.
Currier, T. C., Haury, J. F. and Wolk, C. P. (1977). Isolation and preliminary characterization of auxotrophs of a filamentous cyanobacterium. *J. Bacteriol.* **129,** 1556–1562.
Delaney, S. F. and Reichelt, B. Y. (1983). Integration of R68.45 into the genome of a cynaobacterium results in genome mobilization. *Heredity* **51,** 525–526.
Devilly, C. I. and Houghton, J. A. (1977). A study of genetic transformation in *Gloeocapsa alpicola*. *J. Gen. Microbiol.* **98,** 277–280.
Dimitriadis, G. J. (1979). Entrapment of plasmid DNA in liposomes. *Nucleic Acids Res.* **6,** 2697–2705.

Dixon, R., Kennedy, C. and Merrick, M. (1981). Genetic control of nitrogen fixation. *Symp. Soc. Gen. Microbiol.* **31,** 161–185.
Drews, G. (1978). Structure and development of the membrane system of photosynthetic bacteria. *Curr. Top. Bioenerg.* **8,** 161–207.
Drews, G. and Oelze. J. (1981). Organization and differentiation of membranes of phototrophic bacteria. *Adv. Microb. Physiol.* **22,** 1–92.
Drews, G., Dierstein, R. and Schumacher, A. (1976). Genetic transfer of the capacity to form bacteriochlorophyll-protein complexes in *Rhodopseudomonas capsulata*. *FEBS Lett.* **68,** 132–136.
Edelman, M., Swinton, D., Schiff, J. A., Epstein, H. T. and Zelden, B. (1967). Deoxyribonucleic acid of the blue-green algae (cyanophyta) *Bacteriol. Rev.* **31,** 315–331.
Eisbrenner, G. and Bothe, H. (1979). Modes of electron transfer from molecular hydrogen in *Anabaena cylindrica*. *Arch. Microbiol.* **123,** 37–45.
Evans, E. H. (1979). Photosystem I in dark- and light-grown cells of the cyanobacterium, *Chlorogloea fritschii*. *Abstr. 3rd Int. Symp. Photosynth. Prokaryotes, 1979* B20.
Evans, E. H., Foulds, I. and Carr, N. G. (1976). Environmental conditions and morphological variation in the blue-green alga *Chlorogloea fritschii*. *J. Gen. Microbiol.* **92,** 147–155.
Evans, E. H., Carr, N. G. and Evans, M. C. W. (1978). Changes in photosynthetic activity in the cyanobacterium *Chlorogloea fritschii* on transition from dark to light growth. *Biochim. Biophys. Acta* **501,** 165–173.
Fisher, R., Tuli, R. and Haselkorn, R. (1981). A cloned cyanobacterial gene for glutamine synthetase functions in *Escherichia coli*, but the enzyme is not adenylylated. *Proc. Natl. Acad. Sci. U.S.A.* **78,** 3393–3397.
Fogg, G. E. (1949). Growth and heterocyst production in *Anabaena cylindrica* Lemm. II. In relation to carbon and nitrogen metabolism. *Ann. Bot. (London)* [N.S.] **13,** 241–259.
Fornari, C. S. and Kaplan, S. (1982). Genetic transformation of *Rhodopseudomonas sphaeroides* by plasmid DNA. *J. Bacteriol.* **152,** 89–97.
Fornari, C. S., and Kaplan, S. (1983). Identification of nitrogenase and carboxylase genes in the photosynthetic bacteria and cloning of a carboxylase gene from *Rhodopseudomonas sphaeroides Gene* **25,** 291–299.
Fornari, C. S., Watkins, M. and Kaplan, S. (1984). Plasmid distribution and analyses in *Rhodopseudomonas sphaeroides*. *Plasmid* **11,** 39–47.
Foulds, I. J. and Carr, N. G. (1981). Unequal cell division preceding heterocyst development in *Chlorogloeopsis fritschii*. *FEMS Microbiol. Lett.* **10,** 223–226.
Friedberg, D. and Seijffers, J. (1979). Plasmids in two cyanobacterial strains. *FEBS Lett.* **107,** 165–168.
Gallon, J. R. (1981). Nitrogen fixation by photoautotrophs. In "Nitrogen Fixation" (Eds. W. D. P. Stewart and J. R. Gallon), pp. 197–238. Academic Press, London.
Gallon, J. R., Kurz, W. G. W. and Larue, T. A. (1975). The physiology of nitrogen fixation by a *Gloeocapsa* sp. In "Nitrogen Fixation by Free-living Micro-organisms" (Ed. W. D. P. Stewart), pp. 159–173. Cambridge Univ. Press, London and New York.
Gantt, E. (1980). Structure and function of phycobilisomes: light-harvesting pigment complexes in red and blue-green algae. *Int. Rev. Cytol.* **66,** 45–80.
Gendel, S., Ohad, I. and Bogorad, L. (1979). Control of phycoerythrin synthesis during chromatic adaptation. *Plant Physiol.* **64,** 786–790.
Gendel, S., Straus, N., Pulleyblank, D. and Williams, J. (1983). A novel shuttle cloning vector for the cyanobacterium *Anacystis nidulans*. *FEMS Microbiol. Lett.* **19,** 291–294.

Gibson, K. D. and Niederman, R. A. (1970). Characterization of two circular stallelite species of DNA in *Rhodopseudomonas spheroides*. *Arch. Biochem. Biophys.* **141**, 694–704.

Giddings, T. H., Jr. and Staehelin, L. A. (1978). Plasma membrane architecture of *Anabaena cylindrica*: occurrence of microplasmodesmata and changes associated with heterocyst development and the cell cycle. *Cytobiology* **16**, 235–249.

Giddings, T. H., Jr. and Staehelin, L. A. (1981). Observation of microplasmodesmata in both heterocyst-forming and non-heterocyst-forming filamentous cyanobacteria by freeze-fracture electron microscopy. *Arch. Microbiol.* **129**, 295–298.

Glazer, A. N. (1977). Structure and molecular organization of the photosynthetic accessory pigments of cyanobacteria and red algae. *Mol. Cell. Biochem.* **18**, 125–140.

Gloe, A., Pfennig, N., Brockmann, H., Jr. and Trowitzsch, W. (1975). A new bacteriochlorophyll from brown-coloured Chlorobiaceae. *Arch. Microbiol.* **102**, 103–109.

Golecki, J. R. and Oelze, J. (1980). Differences in the architecture of cytoplasmic and intracytoplasmic membranes of three chemotrophically and phototrophically grown species of the *Rhodospirillaceae*. *J. Bacteriol.* **144**, 781–788.

Haselkorn, R. (1978). Heterocysts. *Annu. Rev. Plant Physiol.* **29**, 319–344.

Herdman, M. (1973). Transformation in the blue-green alga *Anacystis nidulans* and the associated phenomena of mutation. In "Bacterial Transformation" (Ed. L. J. Archer), pp. 369–386. Academic Press, London.

Herdman, M. (1981). Deoxyribonucleic acid base composition and genome size of *Prochloron*. *Arch. Microbiol.* **129**, 314–316.

Herdman, M., Faulkner, B. M. and Carr, N. G. (1970). Synchronous growth and genome replication in the blue-green alga *Anacystis nidulans*. *Arch. Mikrobiol.* **73**, 238–249.

Herdman, M., Janvier, M., Waterbury, J. B., Rippka, R. and Stanier, R. Y. (1979). Deoxyribonucleic acid base composition of cyanobacteria. *J. Gen. Microbiol.* **111**, 63–71.

Hershfield, V., Boyer, H. W., Yanofsky, C., Lovett, M. A. and Helinski, D. R. (1974). Plasmid Col E1 as a molecular vehicle for cloning and amplification of DNA. *Proc. Natl. Acad. Sci. U.S.A.* **71**, 3455–3459.

Hoare, D. S., Ingram, L. O., Thurston, E. L. and Walkup, R. (1971). Dark heterotrophic growth of an endophytic blue-green alga. *Arch. Mikrobiol.* **78**, 310–321.

Hohn, B. and Hinnen, A. (1980). Cloning with cosmids in *E. coli* and yeast. *Genet. Eng.* **2**, 169–183.

Hohn, B. and Murray, K. (1977). Packaging recombinant DNA molecules into bacteriophage particles *in vitro*. *Proc. Natl. Acad. Sci. U.S.A.* **74**, 3259–3263.

Hu, N. T. and Marrs, B. L. (1979). Characterization of the plasmid DNAs of *Rhodopseudomonas capsulata*. *Arch. Microbiol.* **121**, 61–69.

Johansson, B. C., Nordlund, S. and Baltscheffsky, H. (1983). Nitrogen fixation and ammonia assimilation. In "The Phototrophic Bacteria" (Ed. J. G. Ormerod), pp. 120–145. Blackwell, Oxford.

Kallas, T., Rebière, M-C., Rippka, R. and Tandeau de Marsac, N. (1982). Different arrangments of the nitrogenase structural genes (*nif K, D, H*) in the unicellular cyanoacterium *Gloeothece* (PCC6909) and in the heterocystous cyanobacterium *Fremyella diplosiphon* (PCC 7601). *In Abstr. 4th Int. Symp. Photosynthet. Prokaryotes*, 1982 D30. Bombannes, France.

Kallas, T., Rippka, R., Coursin, T., Rebière, M-C., Tandeau de Marsac, N. and Cohen-Bazire, G. (1983). Aerobic nitrogen fixation by nonheterocystous cyanobacteria. In "Photosynthetic Prokaryotes. Cell Differentiation and Function" (Eds. G. C. Papageorgiou and L. Packer), pp. 281–302. Elsevier/North-Holland, Amsterdam.

Kampf, C. and Pfennig, N. (1980). Capacity of Chromatiaceae for chemotrophic growth. Specific respiration rates of *Thiocystis violacea* and *Chromatium vinosum*. *Arch. Microbiol*. **127**, 125-135.

Kaplan, S. (1978). Control and kinetics of photosynthetic membrane development. In "The Photosynthetic Bacteria" (Eds. R. K. Clayton and W. R. Sistrom), pp. 809-839. Plenum, New York.

Kaplan, S. and Arntzen, C. (1982). Photosynthetic membrane structure and function. In "Photosynthesis, Volume I: Energy Conversion by Plants and Bacteria" (Ed. Govindjee), pp. 65-152. Academic Press, New York.

Keating, K. I. (1977). Allelopathic influence on blue-green bloom sequence in a eutrophic lake. *Science* **196**, 885-886.

Keating, K. I. (1978). Blue-green algal inhibition of diatom growth: transition from mesotrophic to eutrophic community structure. *Science* **199**, 971-973.

Khoja, T. and Whitton, B. A. (1971). Heterotrophic growth of blue-green algae. *Arch. Mikrobiol*. **79**, 280-282.

King, M. T. and Drews, G. (1975). The respiratory electron transport system of heterotrophically-grown *Rhodopseudomonas palustris*. *Arch. Microbiol*. **102**, 219-231.

Kondratieva, E. N., Zhukov, V. G., Ivanovsky, R. N., Petushkova, Yu, P. and Monosov, E. Z. (1976). The capacity of phototrophic sulfur bacterium *Thiocapsa roseopersicina* for chemosynthesis. *Arch. Microbiol*. **108**, 287-292.

Krogmann, D. W. (1973). Photosynthetic reactions and components of thylakoids. In "The Biology of Blue-Green Algae" (Eds. N. G. Carr and B. A. Whitton), pp. 80-98. Blackwell, Oxford.

Kuhl, S. A. and Yoch, D. C. (1981). Loss of photosynthetic growth of *Rhodospirillum rubrum* associated with loss of a plasmid. *Abstr., 81st Annu. Meet. Am. Soc. Microbiol*. H126, p. 134.

Kuhlemeier, C. J., Borrias, W. E., van den Hondel, C. A. M. J. J. and van Arkel, G. A. (1981). Vectors for cloning in cyanobacteria: construction and characterization of two recombinant plasmids capable of transformation in *Escherichia coli* K12 and *Anacystis nidulans* R2. *Mol. Gen. Genet*. **184**, 249-254.

Kuhlemeier, K. J., Thomas, A. M. M., Van Der Ende, A., Van Leen, R. W., Borrias, W. E., Van Den Hondel, C. A. M. J. J. and Van Arkel, G. A. (1983). A host-vector system for gene cloning in the cyanobacterium *Anacystis nidulans* R2. *Plasmid* **10**, 156-163.

Lampe, H. H., Oelze, J. and Drews, G. (1972). Die Fraktionierung des Membransystems von *Rhodopseudomonas capsulata* und seine Morphogenese. *Arch. Mikrobiol*. **83**, 78-94.

Lau, R. H. and Doolittle, W. F. (1979). Covalently closed circular DNAs in closely related unicellular bacteria. *J. Bacteriol*. **137**, 648-652.

Lau, R. H., Sapienza, C. and Doolittle, W. F. (1980a). Cyanobacterial plasmids: their widespread occurrence and the existence of regions of homology between plasmids in the same and different species. *Mol. Gen. Genet*. **178**, 203-211.

Lau, R. H., Sapienza, C. and Doolittle, W. F. (1980b). Plasmids in cyanobacteria. In "Plasmids and Transposons: Environmental Effects and Maintenance Mechanisms" (Eds. C. Stuttard and K. R. Rozee), pp. 263-273. Academic Press, New York.

Laudenbach, D. E., Straus, N. A., Gendel, S. and Williams, J. P. (1983). The large endogenous plasmid of *Anacystis nidulans:* Mapping, cloning, and localization of the origin of replication. *Mol. Gen. Genet*. **192**, 402-407.

Lehmann, M. and Wober, G. (1976). Accumulation, mobilization and turn-over of glycogen in the blue-green bacterium *Anacystis nidulans*. *Arch. Microbiol*. **111**, 93-97.

Lewin, R. A. (1977). *Prochloron*, type genus of the prochlorophyta. *Phycologia* **16**, 217.
Lockau, W. (1981). Evidence for a dual role of cytochrome c-553 and plastocyanin in photosynthesis and respiration of the cyanobacterium, *Anabaena variabilis*. *Arch. Microbiol.* **128**, 336–340.
Madigan, M. T. and Gest, H. (1979). Growth of the photosynthetic bacterium *Rhodopseudomonas capsulata* chemoautotrophically in darkness with H_2 as the energy source. *J. Bacteriol.* **137**, 524–530.
Madigan, M. T., Cox, J. C. and Gest, H. (1980). Physiology of dark fermentative growth of *Rhodopseudomonas capsulata*. *J. Bacteriol.* **142**, 908–915.
Mandel, M., Leadbetter, E. R., Pfennig, N. and Trüper, H. G. (1971). Deoxyribonucleic acid base compositions of phototrophic bacteria. *Int. J. Syst. Bacteriol.* **21**, 222–230.
Manwaring, J. (1981). Pigment Composition in Heterotrophically and Photoheterotrophically Grown *Rhodopseudomonas capsulata*. Ph.D. Thesis, CNAA, Preston Polytechnic, U.K.
Marrs, B. (1974). Genetic recombination in *Rhodopseudomonas capsulata*. *Proc. Natl. Acad. Sci. U.S.A.* **71**, 971–973.
Marrs, B. L. (1978). Genetics and bacteriophage. In "The Photosynthetic Bacteria" (Eds. R. K. Clayton and W. R. Sistrom), pp. 873–883. Plenum, New York.
Marrs, B. (1981). Mobilization of the genes for photosynthesis from *Rhodopseudomonas capsulata* by a promiscuous plasmid. *J. Bacteriol.* **146**, 1003–1012.
Marrs, B. L. (1983). Genetics and molecular biology. In "Phototrophic Bacteria: Anaerobic Life in the Light" (Ed. J. G. Ormerod), pp. 186–214. Blackwell, Oxford.
Marrs, B. L., Wall, J. D. and Gest, H. (1977). Emergence of the biochemical genetics and molecular biology of photosynthetic bacteria. *Trends Biochem. Sci.* **2**, 105–108.
Mazur, B. J., Rice, D. and Haselkorn, R. (1980). Identification of blue-green algal nitrogen fixation genes by using heterologous DNA hybridization probes. *Proc. Natl. Acad. Sci. U.S.A.* **77**, 186–190.
Meyer, J., Kelley, B. C. and Vignais, P. M. (1978). Aerobic nitrogen fixation by *Rhodopseudomonas capsulata*. *FEBS Lett.* **85**, 224–228.
Miller, L. and Kaplan, S. (1978). Plasmid transfer and expression in *Rhodopseudomonas sphaeroides*. *Arch. Biochem. Biophys.* **187**, 229–234.
Millineaux, P. M., Gallon, J. R. and Chaplin, A. E. (1981). Acetylene reduction (nitrogen fixation) by cyanobacteria grown under alternating light-dark cycles. *FEMS Microbiol. Lett.* **10**, 245–247.
Mitronova, T. N., Shestakov, S. V. and Zhevner, V. D. (1973). Properties of a radiosensitive filamentous mutant of the blue-green alga *Anacystis nidulans*. *Mikrobiologiya* **42**, 519–524.
Nichols, J. M. and Carr, N. G. (1978). Akinetes of cyanobacteria. In "Spores VII" (Eds. G. Chambliss and J. C. Vary), pp. 335–343. Am. Soc. Microbiol., Washington, D.C.
Nuti, M. P., Lepidi, A. A., Prakash, R. K., Schilperoort, R. A. and Cannon, F. C. (1979). Evidence for nitrogen fixation (*nif*) genes on indigenous *Rhizobium* plasmids. *Nature (London)* **282**, 533–535.
Oelze, J. (1983). Structure of phototrophic bacteria; development of the photosynthetic apparatus. In "The Phototrophic Bacteria" (Ed. J. G. Ormerod), pp. 8–34. Blackwell, Oxford.
Ohad, I and Drews, G. (1982). Biogenesis of the photosynthetic apparatus in prokaryotes and eukaryotes. In "Photosynthesis, Volume 2: Development, carbon metabolism, and plant productivity" (Ed. Govindjee), pp. 89–140. Academic Press, New York.
Olsen, R. H. and Shipley, P. (1973). Host range and properties of the *Pseudomonas aeruginosa* R factor R1822. *J. Bacteriol.* **113**, 772–780.

Olson, J. M. (1980). Chlorophyll organization in green photosynthetic bacteria. *Biochim. Biophys. Acta* **594**, 33–51.
Orkwiszewski, K. G. and Kaney, A. R. (1974). Genetic transformation of the blue-green bacterium *Anacystis nidulans*. *Arch. Microbiol.* **98**, 31–37.
Ormerod, J. G. and Sirevag, R. (1983). Essential aspects of carbon metabolism. *In* "The Phototrophic Bacteria" (Ed. J. G. Ormerod), pp. 100–119. Blackwell, Oxford.
Padan, E. (1979). Facultative anoxygenic photosynthesis in cyanobacteria. *Ann. Rev. Plant Physiol.* **30**, 27–40.
Padhy, R. N. and Singh, P. K. (1978). Genetical studies on the heterocyst and nitrogen fixation of the blue-green alga *Nostoc muscorum*. *Mol. Gen. Genet.* **162**, 203–211.
Pearson, H. W., Howsley, R., Kjeldsen, C. K. and Walsby, A. E. (1979). Aerobic nitrogenase activity associated with a non-heterocystous filamentous cyanobacterium. *FEMS Microbiol. Lett.* **5**, 163–167.
Pelroy, R. A. and Bassham, J. A. (1972). Photosynthetic and dark carbon metabolism in unicellular blue-green algae. *Arch. Microbiol.* **86**, 25–28.
Pemberton, J. M. and Bowen, A. R. St. G. (1981). High-frequency chromosome transfer in *Rhodopseudomonas sphaeroides* promoted by broad-host-range plasmid RP1 carrying mercury transposon Tn501. *J. Bacteriol.* **147**, 110–117.
Pemberton, J. M. and Tucker, W. T. (1977). Naturally occurring viral R plasmid with a circular supercoiled genome in the extracellular state. *Nature (London)* **266**, 50–51.
Peschek, G. A. (1981). Occurrence of cytochrome aa_3 in *Anacystis nidulans*. *Biochim. Biophys. Acta* **635**, 470–475.
Peschek, G. A., Schmetterer, G. and Sleytr, U. B. (1981). Possible respiratory sites in the plasma membrane of *Anacystis nidulans*: ultracytochemical evidence. *FEMS Microbiol. Lett.* **11**, 121–124.
Pfennig, N. (1967). Photosynthetic bacteria. *Annu. Rev. Microbiol.* **21**, 285–324.
Pfennig, N. (1977). Phototrophic green and purple bacteria: a comparative systematic survey. *Annu. Rev. Microbiol.* **31**, 275–290.
Pierson, B. K. and Castenholz, R. W. (1978). Photosynthetic apparatus and cell membranes of the green bacteria. *In* "The Photosynthetic Bacteria" (Eds. R. K. Clayton and W. R. Sistrom), pp. 179–197. Plenum, New York.
Quivey, R., Jr., Meyer, R. J. and Tabita, F. R. (1981). Plasmid transfer into *Rhodospirillum rubrum*. *Abstr., 81st Annu. Meet. Am. Soc. Microbiol.* H113, p. 132.
Reaston, J., van den Hondel, C. A. M. J. J., van der Ende, A., van Arkel, G. A., Stewart, W. D. P. and Herdman, M. (1980). Comparison of plasmids from the cyanobacterium *Nostoc* PCC 7524 with two mutant strains unable to form heterocysts. *FEMS Microbiol. Lett.* **9**, 185–188.
Reaston, J., van den Hondel, C. A. M. J. J., van Arkel, G. A. and Stewart, W. D. P. (1982). A physical map of plasmid pDU1 from the cyanobacterium *Nostoc*, PCC 7524. *Plasmid* **7**, 101–104.
Remsen, C. C. (1978). Comparative subcellular architecture of photosynthetic bacteria. *In* "The Photosynthetic Bacteria" (Eds. R. K. Clayton and W. R. Sistrom), pp. 31–60. Plenum, New York.
Rippka, R. (1972). Photoheterotrophy and chemoheterotrophy among unicellular blue-green algae. *Arch. Mikrobiol.* **87**, 93–98.
Rippka, R. and Waterbury, J. B. (1977). The synthesis of nitrogenase by nonheterocystous cyanobacteria. *FEMS Microbiol. Lett.* **2**, 83–86.
Rippka, R., Dervelles, J., Waterbury, J. B., Herdman, M. and Stanier, R. Y. (1979). Generic assignments, strain histories and properties of pure cultures of cyanobacteria. *J. Gen. Microbiol.* **111**, 1–61.

Roberts, T. M. and Koths, K. E. (1976). The blue-green alga *Agmenellum quadruplicatum* contains covalently closed DNA circles. *Cell* **9**, 551–557.
Rogerson, A. C. (1980). Nitrogen-fixing growth by nonheterocystous cyanobacterium *Plectonema boryanum*. *Nature (London)* **284**, 563–564.
Saunders, V. A. (1978). Genetics of *Rhodospirillaceae*. *Microbiol. Rev.* **42**, 357–384.
Saunders, V. A. and Jones, O. T. G. (1974a). Properties of the cytochrome *a*-like material developed in the photosynthetic bacterium *Rhodopseudomonas spheroides*. *Biochim. Biophys. Acta* **333**, 439–445.
Saunders, V. A. and Jones, O. T. G. (1974b). Adaptation in *Rhodopseudomonas spheroides*. *FEBS Lett.* **44**, 169–172.
Saunders, V. A., Saunders, J. R. and Bennett, P. M. (1976). Extrachromosomal DNA in wild-type and photosynthetically incompetent strains of *Rhodopseudomonas spheroides*. *J. Bacteriol.* **125**, 1180–1187.
Schultz, J. E. and Weaver, P. F. (1981). Fermentation and anaerobic respiration by *Rhodospirillum rubrum* and *Rhodopseudomonas capsulata*. *J. Bacteriol.* **149**, 181–190.
Scolnik, P. and Haselkorn. R. (1982). Cloning and physical characterization of *Rhodopseudomonas capsulata* genes for nitrogen fixation and for glutamine synthetase. *In Abstr. 4th Int. Symp. Photosynth. Prokaryotes, 1982*. D11. Bombannes, France.
Sherman, L. A. and Brown, R. M., Jr. (1978). Cyanophages and viruses of eukaryotic algae. *Compr. Virol.* **12**, 145–234.
Sherman, L. A. and van de Putte, P. (1982). Construction of a hybrid plasmid capable of replication in the bacterium *Escherichia coli* and the cyanobacterium *Anacystis nidulans*. *J. Bacteriol.* **150**, 410–413.
Sherratt, D. (1981). *In vivo* genetic manipulation in bacteria. *Symp. Soc. Gen. Microbiol.* **31**, 35–47.
Shestakov, S. V. and Khyen, N. T. (1970). Evidence for genetic transformation in blue-green alga *Anacystis nidulans*. *Mol. Gen. Genet.* **107**, 372–375.
Simon, R. D. (1977). Sporulation in the filamentous cyanobacterium *Anabaena cylindrica*. The course of spore formation. *Arch. Microbiol.* **111**, 283–288.
Simon, R. D. (1978). Survey of extrachromosomal DNA found in filamentous cyanobacteria. *J. Bacteriol.* **136**, 414–418.
Singh, H. N., Ladha, J. K. and Kumar, H. D. (1977). Genetic control of heterocyst formation in the blue-green algae *Nostoc muscorum* and *Nostoc linckia*. *Arch. Microbiol.* **114**, 155–159.
Singh, P. K. (1973). Nitrogen fixation by the unicellular blue-green alga *Aphanothece*. *Arch. Mikrobiol.* **92**, 59–62.
Singh, R. K. and Singh, H. N. (1981). Genetic analysis of the *het* and *nif* genes in the blue-green alga *Nostoc muscorum*. *Mol. Gen. Genet.* **184**, 531–535.
Sistrom, W. R. (1977). Transfer of chromosomal genes mediated by plasmid R68.45 in *Rhodopseudomonas sphaeroides*. *J. Bacteriol.* **131**, 526–532.
Smith, A. J. and Hoare, D. S. (1977). Specialist phototrophs, lithotrophs and methylotrophs: a unity among a diversity of prokaryotes. *Bacteriol. Rev.* **41**, 419–448.
Smith, G. and Ownby, J. D. (1981). Cyclic AMP interferes with pattern formation in the cyanobacterium *Anabaena variabilis*. *FEMS Microbiol. Lett.* **11**, 175–180.
Solioz, M. and Marrs, B. (1977). The gene transfer agent of *Rhodopseudomonas capsulata*. Purification and characterization of its nucleic acid. *Arch. Biochem. Biophys.* **181**, 300–307.
Stanier, R. Y. and Cohen-Bazire, G. (1977). Phototrophic prokaryotes: the cyanobacteria, *Annu. Rev. Microbiol.* **31**, 225–274.

Stanier, R. Y., Kunisawa, R., Mandel, M. and Cohen-Bazire, G. (1971). Purification and properties of unicellular blue-green algae (order Chroococcales). *Bacteriol. Rev.* **35**, 171–205.
Stevens, S. E. and Porter, R. D. (1980). Transformation in *Agmenellum quadruplicatum*. *Proc. Natl. Acad. Sci. U.S.A.* **77**, 6052–6056.
Stewart, W. D. P. (1977). A botanical ramble among the blue-green algae. *Br. Phycol. J.* **12**, 89–115.
Stewart, W. D. P. (1980). Some aspects of structure and function in N_2-fixing cyanobacteria. *Annu. Rev. Microbiol.* **34**, 497–536.
Stewart, W. D. P. and Singh, H. N. (1975). Transfer of nitrogen-fixing (*nif*) genes in the blue-green alga *Nostoc muscorum*. *Biochem. Biophys. Res. Commun.* **62**, 62–69.
Sutherland, J. M., Herdman, M. and Stewart, W. D. P. (1979). Akinetes of the cyanobacterium *Nostoc* PCC 7524: macromolecular composition structure and control of differentiation. *J. Gen. Microbiol.* **115**, 273–287.
Suyama, Y. and Gibson, J. (1966). Satellite DNA in photosynthetic bacteria. *Biochem. Biophys. Res. Commun.* **24**, 549–553.
Tandeau de Marsac, N. (1977). Occurrence and nature of chromatic adaptation in cyanobacteria. *J. Bacteriol.* **130**, 82–91.
Taylor, D. P., Cohen, S. N., Clark, G. W. and Marrs, B. L. (1983). Alignment of genetic and restriction maps of the photosynthesis region of the *Rhodopseudomonas capsulata* chromosome by a conjugation-mediated marker rescue technique. *J. Bacteriol.* **154**, 580–590.
Tel-Or, E. and Stewart, W. D. P. (1977). Photosynthetic components and activities of nitrogen-fixing isolated heterocysts of *Anabaena cylindrica*. *Proc. R. Soc. London, Ser. B* **198**, 61–86.
Timmis, K. N. (1981). Gene manipulation *in vitro*. *Symp. Soc. Gen. Microbiol.* **31**, 49–109.
Timmis, K. N., Cabello, F. and Cohen, S. N. (1974). Utilization of two distinct modes of replication by a hybrid plasmid constructed *in vitro* from separate replicons. *Proc. Nat. Acad. Sci. U.S.A.* **71**, 4556–4560.
Timmis, K. N., Cabello, F. and Cohen, S. N. (1978). Cloning and characterization of *Eco* RI and *Hin*d III restriction endonuclease-generated fragments of antibiotic resistance plasmids R6-5 and R6. *Mol. Gen. Genet.* **162**, 121–137.
Tomioka, N., Shinozaki, K. and Sugiura, M. (1981). Molecular cloning and characterization of ribosomal RNA genes from a blue-green alga, *Anacystis nidulans*. *Mol. Gen. Genet.* **184**, 359–363.
Tozum, S. R. D. and Gallon, J. R. (1979). The effects of methyl viologen on *Gloeocapsa* sp. LB795 and their relationship to the inhibition of acetylene reduction (nitrogen fixation) by oxygen. *J. Gen. Microbiol.* **111**, 313–326.
Trehan, K. and Sinha, U. (1981). Genetic transfer in a nitrogen-fixing filamentous cyanobacterium. *J. Gen. Microbiol.* **124**, 349–352.
Trüper, H. G. and Pfennig, N. (1978). Taxonomy of the Rhodospirillales. *In* "The Photosynthetic Bacteria" (Eds. R. K. Clayton and W. R. Sistrom), pp. 19–27. Plenum, New York.
Tucker, W. T. and Pemberton, J. M. (1978). Viral R plasmid RØ6P: properties of the penicillinase plasmid prophage and the supercoiled, circular encapsidated genome. *J. Bacteriol.* **135**, 207–214.
Tucker, W. T. and Pemberton, J. M. (1979a). Conjugation and chromosome transfer in *Rhodopseudomonas sphaeroides* mediated by W and P group plasmids. *FEMS Microbiol. Lett.* **5**, 173–176.

Tucker, W. T. and Pemberton, J. M. (1979b). The introduction of RP4::Mu cts 62 into *Rhodopseudomonas sphaeroides*. *FEMS Microbiol. Lett.* **5**, 215–217.

Tucker, W. T. and Pemberton, J. M. (1980). Transformation of *Rhodopseudomonas sphaeroides* with deoxyribonucleic acid isolated from bacteriophage RØ6P. *J. Bacteriol.* **143**, 43–49.

Uffen, R. L. and Wolfe, R. S. (1970). Anaerobic growth of purple non-sulfur bacteria under dark conditions. *J. Bacteriol.* **104**, 462–472.

van den Hondel, C. A. M. J. J., Keegstra, W., Borrias, W. E. and van Arkel, G. A. (1979). Homology of plasmids in strains of unicellular cyanobacteria. *Plasmid* **2**, 323–333.

van den Hondel, C. A. M. J. J., Verbeek, S., van der Ende, A., Weisbeek, P. J., Borrias, W. E. and van Arkel, G. A. (1980). Introduction of transposon Tn901 into a plasmid of *Anacystis nidulans*: preparation for cloning in cyanobacteria. *Proc. Natl. Acad. Sci. U.S.A.* **77**, 1570–1574.

van den Hondel, C. A. M. J. J., Kuhlemeier, C. J., Borrias, W. E., van Arkel, G. A., Tandeau de Marsac, N. and Castets, A. M. (1981). Cloning vectors in the cyanobacterial strain *Anacystis nidulans*. R-2. *Soc. Gen. Microbiol. Q.* **8**, 138–139.

Van Gemerden, H. and Beeftink, H. H. (1983). Ecology of phototrophic bacteria *In* "The Phototrophic Bacteria" (Ed. J. G. Ormerod), pp. 146–185. Blackwell, Oxford.

Van Niel, C. B. (1944). The culture, general physiology, morphology and classification of the non-sulphur and brown bacteria. *Bacteriol. Rev.* **8**, 1–118.

Van Niel, C. B. (1954). The chemoautotrophic and photosynthetic bacteria. *Annu. Rev. Microbiol.* **8**, 105–132.

Wakim, B., Golecki, J. R. and Oelze, J. (1978). The unusual mode of altering the cellular membrane content by *Rhodospirillum tenue*. *FEMS Microbiol. Lett.* **4**, 199–201.

Wall, J. D., Weaver, P. F. and Gest, H. (1975). Genetic transfer of nitrogenase-hydrogenase activity in *Rhodopseudomonas capsulata*. *Nature (London)* **258**, 630–631.

Weckesser, J., Drews, G. and Tauschel, H-D. (1969). Zur Feinstruktur und Taxonomie von *Rhodopseudomonas gelatinosa*. *Arch. Mikrobiol.* **65**, 346–358.

Weissmann, C. (1978). Reversed genetics. *Trends Biochem. Sci.* **3**, N109–N111.

Westphal, K., Bock, E., Cannon, G. and Shively, J. M. (1979). Deoxyribonucleic acid in *Nitrobacter* carboxysomes. *J. Bacteriol.* **140**, 285–288.

Whale, F. R. and Jones, O. T. G. (1970). The cytochrome system of heterotrophically-grown *Rhodopseudomonas spheroides*. *Biochim. Biophys. Acta* **223**, 146–157.

Whittenbury, R. and Dow, C. S. (1977). Morphogenesis and differentiation in *Rhodomicrobium vannielii* and other budding and prosthecate bacteria. *Bacteriol. Rev.* **41**, 754–808.

Wilcox, M., Mitchison, G. J. and Smith, R. J. (1973a). Pattern formation in the blue-green alga *Anabaena*. I. Basic mechanisms. *J. Cell Sci.* **12**, 707–723.

Wilcox, M., Mitchison, G. J. and Smith, R. J. (1973b). Pattern formation in the blue-green alga *Anabaena*. II. Controlled proheterocyst regression. *J. Cell Sci.* **13**, 637–649.

Wilcox, M., Mitchison, G. J. and Smith, R. J. (1975a). Spatial control of differentiation in the blue-green alga *Anabaena*. *In* "Microbiology—1975" (Ed. D. Schlessinger), pp. 453–463. Am. Soc. Microbiol., Washington, D.C.

Wilcox, M., Mitchison, G. J. and Smith, R. J. (1975b). Mutants of *Anabaena cylindrica* altered in heterocyst spacing. *Arch. Microbiol.* **103**, 219–223.

Windass, J. D., Worsey, M. J., Pioli, E. M., Pioli, D., Barth, P. T., Atherton, K. T., Dart, E. C., Byrom, D., Powell, K. and Senior, P. J. (1980). Improved conversion of methanol to single-cell protein by *Methylophilus methylotrophus*. *Nature (London)* **287,** 396–401.

Wolk, C. P. (1973). Physiology and cytological chemistry of blue-green algae. *Bacteriol. Rev.* **37,** 32–101.

Wolk, C. P. (1975). Differentiation and pattern formation in filamentous blue-green algae. *In* "Spores VI" (Eds. P. Gerhardt, R. N. Costilow and H. L. Sadoff), pp. 85–96. Am. Soc. Microbiol., Washington, D.C.

Yen, H.-C. and Marrs, B. (1976). Map of genes for carotenoid and bacteriochlorophyll biosynthesis in *Rhodopseudomonas capsulata*. *J. Bacteriol.* **126,** 619–629.

Yen, H.-C. and Marrs, B. (1977). Growth of *Rhodopseudomonas capsulata* under anaerobic dark conditions with dimethyl sulfoxide. *Arch. Biochem. Biophys.* **181,** 411–418.

Yen, H.-C., Hu, N. T. and Marrs, B. L. (1979). Characterization of the gene transfer agent made by an overproducer mutant of *Rhodopseudomonas capsulata*. *J. Mol. Biol.* **131,** 157–168.

Yoch, D. C. (1978). Nitrogen fixation and hydrogen metabolism by photosynthetic bacteria. *In* "The Photosynthetic Bacteria" (Eds. R. K. Clayton and W. R. Sistrom), pp. 657–676. Plenum, New York.

Yu, P.-L., Cullum, J. and Drews, G. (1981). Conjugational transfer systems of *Rhodopseudomonas capsulata* mediated by R plasmids. *Arch. Microbiol.* **128,** 390–393.

Zannoni, D., Baccarini-Melandri, A., Melandri, B. A., Evans, E. H., Prince, R. C. and Crofts, A. R. (1974). Energy transduction in photosynthetic bacteria. The nature of cytochrome c oxidase in the respiratory chain of *Rhodopseudomonas capsulata*. *FEBS Lett.* **48,** 152–155.

Zannoni, D., Melandri, B. A. and Baccarini-Melandri, A. (1976a). Energy transduction in photosynthetic bacteria X. Composition and function of the branched oxidase system in wild type and respiration deficient mutants of *Rhodopseudomonas capsulata*. *Biochim. Biophys. Acta* **423,** 413–430.

Zannoni, D., Melandri, B. A. and Baccarini-Melandri, A. (1976b). Energy transduction in photosynthetic bacteria. XI. Further resolution of cytochromes of b type and the nature of the CO-sensitive oxidase present in the respiratory chain of *Rhodopseudomonas capsulata*. *Biochim. Biophys. Acta* **449,** 386–400.

Zannoni, D., Jasper, P. and Marrs, B. (1978). Light-induced O_2 reduction as a probe of electron transport between respiratory and photosynthetic components in membranes of *Rhodopseudomonas capsulata*. *Arch. Biochem. Biophys.* **191,** 625–631.

Index

A

Acetate kinase, 78
Acetivibrio cellulolyticus, 49
Acetobacterium sp., 23
Acetobacterium woodii, 46
Acetogenic bacteria, 22, 46, 47
Achromobacter carboxydans, 198, 199
Acinetobacter calcoaceticus, 201
Adhesion of microbes to surfaces, 9, 176–178
Aerobic aquatic environments, 15–21
Aerotolerance, 59
Agmenellum quadruplicatum, 135, 251
Agrobacterium tumefacieus, 201
Alcaligenes eutrophus, 68, 82, 132–134, 137, 141, 143, 146, 147, 153, 155–157, 187–209
Alcaligenes FOR, 132
Alcaligenes hydrogenophilis, 189, 197
Alcaligenes latus, 198, 199
Alcaligenes paradoxus, 198
Alcaligenes ruhlandii, 197, 198
Alcohol dehydrogenase, 201
Anoxygenic photosynthesis, 180–183, 226–227
Amino acids, 10–11, 15–17, 65, 82
Ammonifying bacteria, 15, 16
Amphiaerobes, 70
Anabaena spp., 107
Anabaena CA 135, 138
Anabaena 7120, 81, 144, 250
Anabaena cylindrica, 136, 138, 141–143, 150, 156
Anabaena flos-aquae, 156, 158, 159
Anabaena variabilis, 138, 148, 153, 159, 160, 244
Anabaena circularis, 179
Anacystis nidulans, 97, 109, 135, 152, 153, 156, 158, 159, 251, 258–261
Anaerobes, 21, 24, 45–47, 49, 53, 62, 82, 86
Anaerobic environments, 21, 39, 86–88
Anaerobic respiration, 41, 49, 83–84
Aphanizomenon, 107
Aphanocapsa, 138, 154, 156, 251

Aphanothece halophytica, 139, 140, 142, 143, 144, 147, 148, 177, 178
Apparent growth yield, Y_{app}, 7–8
Aquaspirillum autotrophicum, 198
Arginase, 82
Arginine deimidase, 82
Arthrobacter spp. 13, 134, 137, 198
Arthrobacter crystallopoites, 18, 85
L-Asparaginase, 85
Asterionella formosa, 112
Autecology, 11
Avicennia marina, 182
Azotobacter beijerinckii, 67, 68
Azotobacter vinelandii, 6, 66

B

Bacillus brevis, 82
Bacillus cereus, 87
Bacillus licheniformis, 82
Bacillus subtilis, 82
Bacterial growth rates and productivity, 6–8
Bacterivores, 4–5
Beggiatoa sp. 49, 212, 214, 222–224, 231
Beggiatoa alba, 132
Benthic cyanobacteria, 175–186
Bifidobacterium, 65
Butanediol dehydrogenase, 201

C

C_2-photorespiratory carbon oxidation cycle, 129–131, 156–158
Calothrix desertica, 177, 179
Calvin cycle, 109, 129–133, 148, 151, 193–195, 215, 231–232
Carbon dioxide fixation, 39, 40, 115, 129, 148, 151–156, 181, 193–195, 217, 222–223, 226, 230–232
Carbon monoxide-oxidizing bacteria, 132, 188, 198–199
Carboxysomes, 135–136, 141, 150–151, 153–155, 160, 243
Catalase, 63

Cell membrane, 17, 18, 61
Cell surface to volume ratio, 19, 20
Cellulose degradation, 51
Chemiosmosis, 36–39, 50
Chemolitho(auto)trophy, 40, 49, 129, 150, 187–209
Chemohetero(organo)trophy, 40, 42, 77, 213, 216–220
Chemostat, use of, 6, 8, 13–15, 20, 23, 45, 67–73, 152–153, 189–190, 217–220, 224–225
Chlamydomonas reinhardtii, 144
Chlorella sp. 69, 97, 108, 109, 110, 135, 156
Chlorella fusca, 143
Chlorobium sp., 49
Chlorobium thiosulfatophilum, 131, 136, 137, 157
Chloroflexus sp., 131, 247
Chlorogloeopsis fritschii, 135, 139, 141, 150, 154, 244, 247
Chromatic adaptation in cyanobacteria, 110–111, 264–265
Chromatium sp., 65, 133, 134, 137, 214, 230
Chromatium vinosum, 142, 146, 147, 156, 157
Citrate lyase, 81
Citrate synthase, 67
Citrobacter freundii, 73–76
Clostridium spp. 86
Clostridium acetobutylicum, 63
Clostridium cochlearium, 21
Clostridium papyrosolvens, 51
Clostridium tetanomorphum, 21
Clostridium thermoaceticum, 193
Clostridium thermocellum, 45
CO_2-concentrating mechanism, 153, 158–160
Coccochloris peniocystis, 158, 159, 160
Collagenase, 85
Colourless sulphur bacteria, 211–240
  physiological types, 212–220
  fieldwork, 220–234
Comamonas compransoris, 199
Cometabolism, 17
Conjugation, 198, 200, 253–257
Corynebacterium sp., 21
Corynebacterium glutamicum, 61
Cosmids, 259, 261
Coupled group transfer reactions, 37
Cyanobacteria, 39, 100, 105, 107, 109, 110, 143, 149, 158–159, 175–186, 226–227, 243–244

Cyanophora paradoxa, 139, 143, 151
Cyclotella, 108

D

Denitrification, 199
Derxia gummosa, 199
Desulfobacter postgatei, 23, 48
Desulfobacter propionicus, 23–26
Desulfotomaculum acetoxidans, 51
Desulphovibrio sp., 49
Desulfovibrio baculatus, 23, 25
Desulfovibrio desulfuricans, 82
Desulfovibrio gigas, 193
Desulfovibrio salexigens, 51
Desulfovibrio vulgaris, 52
Detritivores, 4–5
Diatoms, 100, 105, 108, 112
Dihydroxyacid dehydratase, 65
Dunaliella, 110

E

Ectothiorhodospira halophila, 137
Electrochemical potential, $\Delta\bar{\mu}_{H+}$, 17–18
Embden–Meyerhof pathway, 77, 81
Energy flow in ecosystems, role of microbes, 2–6, 50–53
Energy-generating mechanisms in microbes, 39–45
Energy limitation, 8
Energy sources for microbial growth, 35–57
Enrichment culture technique, 12, 51
Entner–Doudoroff pathway, 67, 77
Enterobacter aerogenes, 81
Erwina aroidea, 85
Escherichia coli, 6, 60, 61, 65, 71, 72, 77, 78, 80–85, 200, 253, 254, 258, 259, 263
Euglena, 108–142

F

Fermentation, 41, 45, 51, 63, 70, 71, 201–203
Flavobacterium sp., 51
Formate dehydrogenase, 83
Formate hydrogen-lyase, 83, 84
Fremyella diplosiphon, 111, 179, 263
Fructose bisphosphatase, 194, 195
Fructose-1,6-bisphosphate aldolase, 77
Fumarate reductase, 64, 74, 83

G

$\Delta G^{0'}$ values, 25, 44, 45
Gallionella ferruginea, 132
Gas vesicles of cyanobacteria, 112–113
Gene cloning, 251, 257–263
Gene expression in photosynthetic prokaryotes, 262–263
Genetics
 of *Alcaligenes* enzymes, 187–209
 of photosynthetic prokaryotes, 241–276
Genome composition, 248
Gloeochaete, 151
Glaucocystis, 151
Glaucosphaera vacuolata, 139, 143, 151
Glenodinium, 109
Gloeobacter violaceus, 139, 144
Gloeocapsa alpicola, 251
Gloeothece, 248, 263
Glucose-6-phosphate dehydrogenase, 67, 81
Glutaminase, 84
Glutamine phosphoribosylpyrophosphate amidotransferase, 82
Glutamine synthetase, 263
Glyceraldehyde-3-phosphate dehydrogenase, 195
Glycerate pathway, 133
Glycerol-3-phosphate dehydrogenase, 84
Glycollate dehydrogenase, 157
Glycollate
 excretion, 156–158, 194, 233
 formation and metabolism, 130, 156, 157
Gonyaulax polyedra, 108
Green algae, 100, 105
Growth
 of microbes under natural conditions, 6–11
 of phototrophs, 113–119
Guukensia demissa, 5
Grazing, 4
Growth-yield coefficient, 7

H

Herbicide degradation, 200–201
Heterocysts, 136, 150, 151, 243–344, 263
Hexokinase, 67
Hexose monophosphate pathway, 77
Hydrocarbons, utilization by microbes, 51, 62, 177
Hydrogen cyanide synthase, 85

Hydrogen
 electrode, 43
 production, 181
 utilizing bacteria, 46, 48, 49, 68, 150, 188–209
Hydrogen peroxide, 62, 63, 81, 112
Hydrogenase, 69, 80, 83, 189–199
Hydrogenomonas eutropha (see Alcaligenes eutrophus)
Hydrophilic cell envelopes, of cyanobacteria, 177
Hydrophobicity, of benthic cyanobacteria, 176–178
Hydrothermal vents, 231
Hydroxylase, 61
Hydroxyl radicals, 62
Hyphomicrobium, 18, 231

I

Inorganic sulphur oxidation, 212–236
Isocitrate dehydrogenase, 67, 132

K

Klebsiella spp., 77, 250, 263
Klebsiella aerogenes, 70, 71, 72, 77
Klebsiella pneumoniae, 84, 85

L

Lactate excretion, 202
Lactate dehydrogenase, 64, 80, 82, 201–203
Lauderia borealis, 108
Light–dark cycles and cyanobacterial growth, 115–119
Light-harvesting pigments, in aquatic environments, 105
Light
 interception, 102–104
 interception and cell composition/ morphology, 106, 246–247
 measurement, 98–100
 quality, responses of phytoplankton, 110–111, 265
 scattering, 104
 superoptimal effects on phytoplankton, 111–113
 underwater climate, 98–102, 178
 utilization by microorganisms, 97–127
Lissoclinum patella, 139

M

Maintenance requirement, 19
Malate dehydrogenase, 132
Metabolic activity of bacterial, variations in, 8–11
Methane oxygenases, 62
Methanobacillus omelianskii, 45
Methanobacterium bryantii, 63
Methanobacterium thermoautotrophicum, 63, 193
Methanobrevibacter arboriphilus, 52
Methanogens, 22, 42, 45–49, 52, 53, 63
Methanosarcina barkeri, 46, 49
Methanosinus trichosporium, 62
Methylococcus capsulatus, 132, 138, 140
Methylotrophs, 40, 49, 129, 132, 231
Microbes as a food source, 4–6
Microbial
  biomass, 3, 6, 10
  growth, energy sources for, 35–57
  predators, 6
Microcoleus chthonoplastes, 181
Microcyclus aquaticus, 132, 198, 199
Microcystis aeruginosa, 139, 144, 156, 160
Mixotrophs, 21, 152, 214, 218, 223, 234, 235
Molluscs, 5, 232–233

N

NADH oxidase, 64, 74
Neuraminidase, 85
Nickel, as a hydrogenase component, 192–193
nif (nitrogen fixation) genes, 84, 85, 247, 250, 263
Nitrate reductase, 74, 83
Nitrite reductase, 83
Nitrobacter sp., 150
Nitrobacter agilis, 137, 141, 143, 150
Nitrobacter winogrodsky, 135, 150
Nitrogen fixation, 136, 150, 181, 182, 188, 243–244, 247–248, 263
Nitrogenase, 66, 81, 84, 243–244
Nitrosococcus mobilis, 150
Nitrosomonas sp., 150
Nitrosovibrio tenuis, 150
Nitrospina gracilis, 150
Nocardia opaca, 134, 137, 191, 192, 195–197
Nostoc, 135, 139, 144, 154–155, 247, 258, 260
Nostoc canina, 139
Nostoc commune, 139
Nostoc muscorum, 263
Nutrient limitation, 13, 17, 18, 71, 152, 153
Nutritional interrelationships between microbes, 45–47

O

Oligotrophs, 20
Ornithine carbamoyltransferase, 82
Ornithine transaminase, 82
Oscillatoria sp., 107, 111, 156
Oscillatoria agardhii, 114, 116–118
Oscillatoria limnetica, 139, 180–182
Oscillatoria redekei, 108, 115–119
Oscillatoria rubescens, 111
2-Oxoglutarate dehydrogenase, 74
Oxygen, 18, 40, 41, 44, 59–96, 264
  electrode, 60, 71
  efficiency of microbial growth, 78–79
  metabolic regulation, 59–96, 179–182, 194, 224–230, 248
  tension, 60, 64, 65, 68–70, 71–72, 77, 85, 86, 112, 179, 194, 215, 221

P

Paracoccus denitrificans, 77, 132, 134, 137, 138
Pasteur effect, 81
Pavlova lutheri, 114
Peptostreptococcus anaerobius, 63
Peridinium cinctum, 110
Phanerochaete chrysosporium, 62
Phormidium sp., 177
Phormidium molle, 158–159
Phosphoenolpyruvate carboxylase, 132, 217
Phosphofructokinase, 81
Phosphoglyceromutase, 196
Phosphoribulokinase, 130, 133, 152, 155–156, 193–195, 217, 232
  catalytic and regulatory properties, 134–135
Phosphotransacetylase, 78
Phosphotransferase (PTS) system, 37
Photoautotrophs, 129
Photobacterium, 60
Photooxidative damage, 111–113, 157, 176, 182, 183
Photorespiration, 112, 157, 158–160
Photosynthesis, 107–119, 148, 244–247

Photosynthesis
  bacterial cell organization of, 242–243
  pigment changes in, 107–110
  units of, 110
Photosystem I, 109–110, 180, 245, 247
Photosystem II, 109–110, 136, 180, 245
Phototrophy, 39, 77
Phytoplankton, 2, 4, 10, 15, 97–127, 157–158, 160, 176
Plasmids, 80, 195–201, 203–204, 248–249, 253, 256–258, 210–261
Plectonema boryanum, 138, 154, 156, 177, 179
Poganophoran worms, 231–234
Polyhedral bodies, *see* Carboxysomes
Poly-β-hydroxybutyric acid, 18, 67, 68
Polyglucose, *see* Poly-β-hyroxybutyric acid
Primary production, 2
Prochloron, 139, 140, 150, 248
L-1,2-Propanediol oxidoreductase, 81
Propionibacterium pentosaceum, 64
Propionibacterium shermanii, 65
Proteus vulgaris, 71
Proton motive force, 38
Pseudomonas sp., 20, 51, 61
Pseudomonas aeruginosa, 67, 85, 214, 216
Pseudomonas carboxydoflava, 198, 199
Pseudomonas carboxydohydrogena, 109
Pseudomonas carboxydovarans, 198, 199
Pseudomonas facilis, 134, 198
Pseudomonas fluorescens, 201, 254
Pseudomonas oxalaticus, 133, 149
Pseudomonas palleronii, 189, 198
Pseudomonas pseudoflava, 134, 198
Pseudomonas putida, 201
Pseudomonas saccharophila, 198
Pseudomonas thermophila, 150
Pyruvate dehydrogenase, 67, 78
Pyruvate formate lyase, 77

R

Redox reactions, 36–38, 41, 43, 44, 80, 82, 84
Reductive tricarboxylic acid cycle, 131
Renobacter vacuolatum, 198, 199
Respiration, 41, 66, 69–71, 182, 244–245
Restriction enzymes, maps, sites, 252, 258–260
Rhizobium japonicum, 84, 138, 141, 142, 263
Rhodococcus erythropolis, 196

Rhodomicrobium vannielii, 134, 140, 143, 150, 156, 243, 255
Rhodopseudomonas acidophila, 132, 156
Rhodopseudomonas capsulata, 77, 80–81, 133–137, 242, 246–252, 253–255, 257, 258, 264
Rhodopseudomonas gelatinosa, 242, 253
Rhodopseudomonas palustris, 132, 134, 143, 156, 242, 247
Rhodopseudomonas sphaeroides, 77, 134, 136, 137, 143, 147, 201, 242, 246, 249, 251, 253–258, 261
Rhodopseudomonas viridis, 242
Rhodospirillum molischianum, 134
Rhodospirillum rubrum, 131, 134–137, 141, 143–149, 155, 156, 249, 253, 255
Rhodospirillum tenue, 134, 242
Ribulose-1,5-bisphosphate carboxylase/oxygenase, 69, 112, 130–133, 193–195, 217, 232, 233
  activation and catalysis, 144–148
  compartmentation, 150–151
  DNA sequences of large subunit gene, 145
  genes, 144
  purification and molecular properties, 136–144
  regulation, 148–151
  variations with nutritional modes, 152–156
Ribulose monophosphate cycle, 132
Riftia pachyptila, 214, 231, 233
Ruminococcus albus, 45

S

Saccharomyces carlsbergensis, 61
Saccharomyces cerevisiae, 61, 70, 81
Salmonella cranienberg, 71
Scenedesmus braunii, 143
Scytonema sp., 182
Selenomonas ruminantium, 45, 63, 64
Sewage fungas, 51
Siboglinum ekmani, 234
Siboglinum fiordicum, 234
Singlet oxygen, 63
Solemya panamensis, 214, 232, 233
Specialist (obligate) autotrophy, 151–152, 213
Specific absorption coefficient of phytoplankton, 103–104
Specific growth rate, 7, 13–15, 19, 20
Specific substrate uptake rate, 21
Spinacia oleracea, 143

INDEX

Spirillum sp., 18–20
Spirulina platensis, 177
Staphylococcus aureus, 80
Staphylococcus epidermidis, 18
Substrate affinity, 214, 230, 236
Sulfolobus acidocalcarius, 214, 230, 236
Sulphate reduction, 22, 23, 26, 42, 46–49, 51–53
Sulphate-reducing bacteria, 22, 23, 26, 42–49, 51–53
Sulphide, 49, 51–53, 176, 180–181, 220–230
Sulphur cycle, 53, 212–236
Superoxide
  anions, 62, 63, 82
  dismutase, 63, 78, 112, 182
Synechococcus, 139–148, 249, 253
Syntrophobacter wolinii, 46
Syntrophomonas wolfei, 46
Syntrophy, 45–46, 49, 50

T

Tetrahymena, 6
Tetrahydropteroyltriglutamate transmethylase, 78
Thiobacillus sp., 49, 86, 214, 228, 231
Thiobacillus A2, 138, 150, 214, 215, 224, 225, 235
Thiobacillus denitrificans, 138, 214, 235
Thiobacillus intermedius, 138, 150, 154, 220
Thiobacillus neapolitanus, 134, 135, 138, 143, 146, 150, 153, 156, 214, 224, 225, 233, 235
Thiobacillus novellus, 132, 138, 150, 214, 215
Thiobacillus pelophila, 20

Thiobacillus perometabolis, 214, 216
Thiobacillus thioparus, 20, 134, 150
Thiobacillus thiooxidans, 150
Thiocapsa roseopersicina, 137, 140, 143, 214
Thiomicrospira sp., 228, 231
Thiomicrospira pelophila, 214, 235
Thioredoxin, 81
Thiosphaera pantotropha, 235
Thiothrix sp., 49
Thiovulum sp., 224
Transduction, 250–251
Transformation, 251–253, 262
Transposons, 195–196, 256–257, 259
Tricarboxylic acid cycle, 38, 41, 42, 48, 66, 67, 70, 71, 74, 78, 79
Tryptophan synthetase, 254

U

Urease, 84

V

Veillonella alcalescens, 23, 64
Versatile (facultative) autotrophy, 152–155, 176, 179–182, 193–195, 212–220
Vibrio alginolyticus, 85
Vibrio succinogenes, 45

X,Y,Z

Xanthobacter autotrophicus, 69, 134, 198
Yield coefficient, 21
Zooplankton, 2, 5